A2-Level
Chemistry

The Revision Guide

Editor:
Sharon Keeley

Contributors:
Mike Bossart, Robert Clarke, Vikki Cunningham, Ian H. Davis, John Duffy, Chris Enos,
Emma Grimwood, Richard Harwood, Sarah Hilton, Tim Major, Lucy Muncaster, Sam Norman,
Stephen Phillips, Adrian Schmit, Jane Simoni, Ami Snelling, Paul Warren, Sharon Watson.

Proofreaders:
Jeremy Cooper, Lynette Harper, John Moseley, Glenn Rogers

Published by Coordination Group Publications Ltd.

This book covers the core material for:

AQA, OCR, OCR Salters, Edexcel and **Edexcel Nuffield.**

It also covers the following OCR optional modules:

- Biochemistry
- Methods of Analysis and Detection
- Transition Elements

There are notes on the pages to tell you if there's a bit you can ignore for your syllabus.

ISBN-10: 1 84146 366 3
ISBN-13: 978 1 84146 366 7

Groovy website: www.cgpbooks.co.uk
Jolly bits of clipart from CorelDRAW®
Printed by Elanders Hindson Ltd, Newcastle upon Tyne.

Contents

Section One — Energetics

Enthalpy Change 2
Lattice and Bond Enthalpies 4
Enthalpies and Dissolving 6
Entropy and Free Energy Change 8

Section Two — Kinetics

Reaction Rates 10
Activation Energy and Catalysts 12
Rate Equations 14
Orders of Reactions and Half-Life 16
Rates and Reaction Mechanisms 18

Section Three — Equilibria

Equilibria 20
Gas Equilibria 22
More on Equilibrium Constants 24
Acids and Bases 28
pH Calculations 30
Titrations and pH Curves 32
Indicators, pH Curves and Calculations 34
Buffers 36
Metal-Aqua Ions 38

Section Four — Elements of the Periodic Table

Period 3 Elements and Oxides 40
Period 3 Chlorides and Group 4 42
The Nitrogen Cycle and Water 44

Section Five — Electrochemistry and Transition Elements

Oxidation and Reduction 46
Electrode Potentials 48
The Electrochemical Series 50
Applying Electrochemistry 52
Transition Metals — The Basics 54
Transition Metals — Vanadium and Cobalt 56
Transition Metals — Chromium and Copper 58
Transition Metals — Titrations and Calculations 60
Complex Ions — The Basics 62
Complex Ions — Ligand Exchange 64
Transition Metal Ion Colour 66
Uses of Transition Metals 68

Section Six — Organic Chemistry

Isomerism 70
Reaction Mechanisms 72
Aldehydes and Ketones 74
Aldehydes, Ketones and Grignard Reagents 76
Carboxylic Acids 78
Esters 80
Acyl Chlorides 82
Aromatic Compounds 84
More Reactions of Aromatic Compounds 86
Phenols 88
Amines 90
Amines and Amides 92
Dyes 94
Amino Acids and Proteins 96
DNA 98
Addition Polymers 100
Condensation Polymers 102
Medicines 104

Section Seven — Analysis and Synthesis

Identifying Functional Groups 106
UV–Visible Spectra 108
Atomic Emission Spectroscopy 110
Chromatography and Electrophoresis 112
Mass Spectrometry 114
Infrared Spectrometry 116
NMR Spectroscopy — The Basics 118
More NMR Spectroscopy 120
Combined Spectral Analysis 122
Organic Synthesis 124
Practical Techniques 126
The Chemical Industry 128

Section Eight — Biochemistry

Carbohydrates 130
Lipids and Membrane Structures 132
Proteins 134
Enzymes 136
DNA and RNA Structure 138
DNA Replication and Protein Synthesis 140

Answers 142

Index 154

The Periodic Table 158

Enthalpy Change

There's no easing you in gently at A2 — the first section kicks off with a load of definitions to learn...

Lattice Enthalpy is a Measure of Ionic Bond Strength

In reactions, some bonds are **broken** and some bonds are **formed**. More often than not, this will cause a **change in energy**. As you might remember from AS, the chemistry term for this is **enthalpy change** (ΔH), and the units are **kJmol^{-1}**. The symbol $\ominus$ in $\Delta H^\ominus$ shows that the value refers to **standard conditions**.

If energy's given out, it's an **exothermic** reaction (ΔH is negative). If energy's taken in, it's **endothermic** (ΔH is positive).

Remember how **ionic compounds** form regular structures called **giant ionic lattices** — and how the positive and negative ions are held together by **electrostatic attraction**. Well, when **gaseous ions** combine to make a solid lattice, energy is given out — this is called the **lattice enthalpy**. It's quite handy as it tells you how **strong** the ionic bonding is.

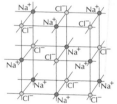

Part of the sodium chloride lattice

Here's the definition of **standard lattice enthalpy** that you need to know:

> The **standard lattice enthalpy**, $\Delta H^\ominus_{latt}$, is the enthalpy change when **1 mole** of a **solid ionic compound** is formed from its **gaseous ions** under standard conditions.

For example: $Na^+_{(g)} + Cl^-_{(g)} \rightarrow NaCl_{(s)}$ $\Delta H^\ominus = -780 \text{ kJmol}^{-1}$
$Mg^{2+}_{(g)} + O^{2-}_{(g)} \rightarrow MgO_{(s)}$ $\Delta H^\ominus = -3791 \text{ kJmol}^{-1}$

The **more negative** the lattice enthalpy, the **stronger** the bonding. So out of NaCl and MgO, **MgO** has stronger bonding.

You Can't Measure Lattice Enthalpy Directly

Lattice enthalpy **can't** be measured directly, but it **can** be calculated with a **Born-Haber cycle** — all will be revealed on page 4. But first you need to know these **definitions**:

> **Enthalpy change of formation**, ΔH_f, is the enthalpy change when **1 mole** of a **compound** is formed from its **elements** in their standard states under standard conditions, e.g. $2C_{(s)} + 3H_{2(g)} + \frac{1}{2}O_{2(g)} \rightarrow C_2H_5OH_{(l)}$

> The **bond dissociation enthalpy**, ΔH_{diss}, is the enthalpy change when **1 mole of bonds of the same type** are broken in **gaseous molecules**, e.g. $Cl_{2(g)} \rightarrow 2Cl_{(g)}$

> **Enthalpy change of atomisation of an element**, ΔH_{at}, is the enthalpy change when **1 mole of gaseous atoms** is formed from an element in its **standard state**, e.g. $\frac{1}{2}Cl_{2(g)} \rightarrow Cl_{(g)}$

> **Enthalpy change of atomisation of a compound**, ΔH_{at}, is the enthalpy change when **1 mole of gaseous atoms** is formed from a compound in its **standard state**, e.g. $NaCl_{(s)} \rightarrow Na_{(g)} + Cl_{(g)}$

> The **first ionisation enthalpy**, ΔH_{ie1}, is the enthalpy change when **1 mole of gaseous 1+ ions** is formed from **1 mole of gaseous atoms**, e.g. $Mg_{(g)} \rightarrow Mg^+_{(g)} + e^-$

> The **second ionisation enthalpy**, ΔH_{ie2}, is the enthalpy change when **1 mole of gaseous 2+ ions** is formed from **1 mole of gaseous 1+ ions**, e.g. $Mg^+_{(g)} \rightarrow Mg^{2+}_{(g)} + e^-$

> **First electron affinity**, ΔH_{e1}, is the enthalpy change when **1 mole of gaseous 1– ions** is made from **1 mole of gaseous atoms**, e.g. $O_{(g)} + e^- \rightarrow O^-_{(g)}$

> **Second electron affinity**, ΔH_{e2}, is the enthalpy change when **1 mole of gaseous 2– ions** is made from **1 mole of gaseous 1– ions**, e.g. $O^-_{(g)} + e^- \rightarrow O^{2-}_{(g)}$

> The **enthalpy change of hydration**, ΔH_{hyd}, is the enthalpy change when **1 mole of aqueous ions** is formed from **gaseous ions**, e.g. $Na^+_{(g)} \rightarrow Na^+_{(aq)}$

> The **enthalpy change of solution**, $\Delta H_{solution}$, is the enthalpy change when **1 mole of solute** is dissolved in **sufficient solvent** that no further enthalpy change occurs on further dilution, e.g. $NaCl_{(s)} \rightarrow NaCl_{(aq)}$

Enthalpy Change

Ionic Charge and Size Affects Lattice Enthalpy — Only OCR, Salters and Edexcel

The **larger the charges** on the ions, the **more energy** is released when an ionic lattice forms. More energy released means that the lattice enthalpy will be **more negative**. So the lattice enthalpies for compounds with 2+ or 2– ions (e.g. Mg^{2+} or S^{2-}) are **more negative** than those with 1+ or 1– ions (e.g. Na^+ or Cl^-).

For example, the lattice enthalpy of NaCl is only –780 kJmol⁻¹, but the lattice enthalpy of $MgCl_2$ is –2526 kJmol⁻¹. MgS has an even higher lattice enthalpy (–3299 kJmol⁻¹) because both magnesium and sulphur ions have double charges.

Magnesium oxide has a **very exothermic** lattice enthalpy too, so the compound is really resistant to heat. This makes it great as a **lining in furnaces**.

The **smaller** the **ionic radii** of the ions involved, the **more exothermic** (more negative) the **lattice enthalpy**. Smaller ions attract **more strongly** because their **charge density** is higher.

Charge Density Affects Thermal Decomposition — Only OCR, Salters and Edexcel

Group 2 carbonates decompose when heated to form a **solid oxide** and CO_2 **gas**, e.g. $MgCO_{3(s)} \rightarrow MgO_{(s)} + CO_{2(g)}$.

Be^{2+} has the **smallest radius** of the Group 2 cations, so $BeCO_3$ has the **most exothermic** lattice enthalpy, and the **strongest bonding** of the Group 2 carbonates. So you might think (but you'd be wrong) that $BeCO_3$ will need the **most heat** to make it decompose.

The actual trend is... | As you go **down** the group, the carbonates become **harder** to decompose with heat.

This is all to do with how much the cation **polarises** the carbonate ion. Polarisation occurs when a positive cation **drags** a negative anion's **electron cloud** towards it.

All of the Group 2 cations have the **same charge** (2+) but their sizes **increase** down the group, so the charge density is greatest on the **smallest ion** (Be^{2+}) and least on the largest (Ra^{2+}).

A small cation **polarises** a large carbonate ion a lot. The more polarised it becomes, the less stable it is — and so the easier it will be to **decompose**.

Boris was shocked to discover polarisation was setting in.

Magnesium is a small cation, so it polarises the carbonate a lot — this makes $MgCO_3$ unstable.

Barium's a fairly big cation, so it doesn't polarise the carbonate much. So $BaCO_3$ is quite stable and it takes lots of heat to decompose it.

Practice Questions

Q1 Define the following: a) lattice enthalpy, b) bond dissociation enthalpy, c) enthalpy of solution.
Q2 Give two factors that affect lattice enthalpy.
Q3 What is the trend in thermal decomposition of the Group 2 carbonates as you go down the group?

Exam Questions

1 a) Define first ionisation enthalpy. [3 marks]
b) Write an equation to represent the second ionisation enthalpy of aluminium. [2 marks]

2 a) Explain why magnesium oxide has a highly exothermic lattice enthalpy. [4 marks]
b) What property of magnesium oxide is explained by this high lattice enthalpy? [1 mark]
c) What use is made of magnesium oxide that depends upon this property? [1 mark]

3 a) The nitrate anion (NO^{3-}) is a large ion. Given this information, place the following in order of increasing thermal stability:
$Ba(NO_3)_2$ $Ca(NO_3)_2$ $Sr(NO_3)_2$ [1 mark]
b) Explain your answer to part (a). [4 marks]

You've just got to have an affinity for this stuff...

The worst thing about this page is all the definitions you have to learn. It's not enough to just have a vague idea what each one means; you have to know all the ins and outs — like whether it applies to gases, or to elements in their standard states. If you've forgotten, standard conditions are 298 K (otherwise known as 25 °C) and 101.3 kPa (1 atmosphere) pressure.

Lattice and Bond Enthalpies

Now you know all your enthalpy change definitions, here's how to use them... Enjoy.

Born-Haber Cycles can be Used to Calculate Lattice Enthalpies

Hess's law says that the **total enthalpy change** of a reaction is always the **same**, no matter which route is taken.

A **lattice enthalpy** is basically the **enthalpy change** when a mole of a **solid ionic compound** is formed from **gaseous ions**. You can't calculate a lattice enthalpy **directly**, so you have to use a **Born-Haber cycle** to figure out what the enthalpy change would be if you took **another, less direct, route**.

Here's how to draw a Born-Haber cycle for calculating the lattice enthalpy of **NaCl**:

$$Na^+_{(g)} + e^- + Cl_{(g)}$$

3 The electron affinity goes up here...

First ionisation energy of sodium (+494 kJmol⁻¹) $\Delta H4$

$$Na_{(g)} + Cl_{(g)} \text{ (gaseous atoms)}$$

$\Delta H5$

First electron affinity of chlorine (−349 kJmol⁻¹)

2 Then put the enthalpies of atomisation and ionisation above this.

Atomisation energy of sodium (+107 kJmol⁻¹) $\Delta H3$

$$Na_{(s)} + Cl_{(g)}$$

$$Na^+_{(g)} + Cl^-_{(g)}$$

Atomisation energy of chlorine (+122 kJmol⁻¹) $\Delta H2$

$$Na_{(s)} + \tfrac{1}{2}Cl_{2(g)} \text{ (standard states)}$$

$\Delta H6$ Lattice enthalpy of sodium chloride

1 Start with the enthalpy of formation here.

Enthalpy of formation of sodium chloride (−411 kJmol⁻¹) $\Delta H1$

$$NaCl_{(s)} \text{ (ionic lattice)}$$

4 ...and lattice enthalpy goes down here.

There are **two routes** you can follow to get from the elements in their **standard states** to the **ionic lattice**. The green arrow shows the **direct route** and the purple arrows show the **indirect route**. The enthalpy change for each is the **same**.

From Hess's law:
$$\Delta H6 = -\Delta H5 - \Delta H4 - \Delta H3 - \Delta H2 + \Delta H1$$
$$= -(-349) - (+494) - (+107) - (+122) + (-411) = -785 \text{ kJmol}^{-1}$$

You need a minus sign if you go the wrong way along an arrow.

Calculations involving Group 2 Elements are a Bit Different

Born-Haber cycles for compounds containing **Group 2 elements** have a few **changes** from the one above. Make sure you understand what's going on so you can handle whatever compound they throw at you.

Here's the Born-Haber cycle for calculating the lattice enthalpy of **magnesium chloride** ($MgCl_2$).

$$Mg^{2+}_{(g)} + 2e^- + 2Cl_{(g)}$$

1 Group 2 elements form 2+ ions — so you've got to include the second ionisation energy.

Second ionisation energy of magnesium

$$Mg^+_{(g)} + e^- + 2Cl_{(g)}$$

First electron affinity of chlorine × 2

First ionisation energy of magnesium

$$Mg_{(g)} + 2Cl_{(g)}$$

$$Mg^{2+}_{(g)} + 2Cl^-_{(g)}$$

3 ...and you need to double the first electron affinity of chlorine too.

Atomisation energy of magnesium

$$Mg_{(s)} + 2Cl_{(g)}$$

Atomisation energy of chlorine × 2

$$Mg_{(s)} + Cl_{2(g)}$$

Lattice enthalpy of magnesium chloride

2 There's 2 moles of chlorine ions in each mole of $MgCl_2$ — so you need to double the atomisation energy of chlorine...

Enthalpy of formation of magnesium chloride

$$MgCl_{2(s)}$$

Theoretical Lattice Enthalpies are Often Different from Experimental Values

Only Edexcel and Nuffield

If you assume that an ionic compound is made up of **nice spherical positive and negative ions**, you can work out what its lattice enthalpy **should be**. But, chances are you'll be way off the mark a lot of the time.

This is because the positive and negative ions are **not** usually exactly spherical — positive ions **polarise** neighbouring negative ions to different extents.

The **more polarisation** there is, the **more covalent** the bonding will be.

This is why theoretical lattice enthalpies are likely to be **different** from the values from Born-Haber cycles, especially for compounds that have lots of **covalent character**.

unpolarised ions — purely ionic bonding

polarised ions — partial covalent bonding

Fajan's rules say that a compound favours **ionic** bonding if: • the positive ions are **large** and the negative ions are small
• both types of ion have **small charges** (+1 and −1)

Lattice and Bond Enthalpies

This page is just for AQA — except for the questions of course.

Use **Bond Enthalpies** to Calculate Enthalpy Changes for **Covalent Compounds**

Energy is **taken in** when bonds are **broken** and **released** when bonds are **formed**, so —

| enthalpy change for a reaction = sum of enthalpies of bonds broken – sum of enthalpies of bonds formed |

If you need **more** energy to **break** bonds than is released when bonds are made it's an **endothermic reaction** and the enthalpy change, ΔH, is **positive**. If **more** energy is **released** than is taken in it's **exothermic** and ΔH is **negative**.

Example: Calculate the enthalpy change for the following reaction: $N_{2(g)} + 3H_{2(g)} \rightarrow 2NH_{3(g)}$

Bond	Mean bond enthalpy (kJmol⁻¹)
N≡N	945
H—H	436
N—H	391

It's easier to see what's going on if you **sketch out** the molecules involved:

$$N≡N + 3H-H \rightarrow 2H-N-H$$
$$\overset{|}{H}$$

Now add up the mean bond enthalpies for the reactant **bonds broken**...

$(1 \times N≡N) + (3 \times H–H) = (1 \times 945) + (3 \times 436) = 2253$ kJmol⁻¹

... and for the **new bonds formed** in the products

$(6 \times N–H) = (6 \times 391) = 2346$ kJmol⁻¹

So the **enthalpy change for the reaction** = 2253 – 2346 = **–93 kJmol⁻¹**.

It's negative, so it's exothermic.

Bond Enthalpy Calculations are Only **Approximations**

The bond enthalpy given above for N–H bonds is **not** exactly right for **every** N–H bond. A given type of bond will **vary in strength** from compound to compound and can even vary **within** a compound. Mean bond enthalpies are the **averages** of these bond enthalpies. Only the bond enthalpies of **diatomic molecules**, such as H_2 and HCl, will always be the same.

So calculations done using **mean bond enthalpies** will never be perfectly accurate. You get much **more exact** results from **experimental data** obtained from the **specific compounds**.

Practice Questions

Q1 Sketch out Born-Haber cycles for: a) LiF, b) $CaCl_2$.

Q2 Explain why theoretical lattice enthalpies are often different from experimentally determined lattice enthalpies.

Q3 How can bond enthalpies be used to calculate enthalpy changes for covalent compounds?

Exam Questions

1 Using the data below:
   a) Construct a Born-Haber cycle for potassium bromide (KBr). [4 marks]
   b) Use your Born-Haber cycle to calculate the lattice enthalpy of potassium bromide. [3 marks]

$\Delta H_f^{\ominus}$ [potassium bromide] = –394 kJmol⁻¹ $\Delta H_{at}^{\ominus}$ [bromine] = +112 kJmol⁻¹ $\Delta H_{at}^{\ominus}$ [potassium] = +89 kJmol⁻¹

$\Delta H_{ie1}^{\ominus}$ [potassium] = +419 kJmol⁻¹ $\Delta H_{e1}^{\ominus}$ [bromine] = –325 kJmol⁻¹

2 a) Calculate the enthalpy change for $CH_{4(g)} + 2O_{2(g)} \rightarrow CO_{2(g)} + 2H_2O_{(g)}$ using the bond enthalpies in the table below.

Bond	C–H	O=O	C=O	O–H
Mean bond enthalpy (kJmol⁻¹)	412	496	743	463

[3 marks]

   b) Give a reason why the enthalpy change calculated by this method is not likely to be as accurate as one calculated using an enthalpy cycle. [2 marks]

Using Born-Haber cycles — it's just like riding a bike...

All this energy going in and out can get a bit confusing. Remember these simple rules: 1) It takes energy to break bonds, but energy is given out when bonds are made. 2) A negative ΔH means energy is given out (it's exothermic). 3) A positive ΔH means energy is taken in (it's endothermic). 4) Never return to a firework once lit.

Enthalpies and Dissolving

Once you know what's happening when you stir sugar into your tea, your cuppa'll be twice as enjoyable.
OCR people can skip these two pages.

Dissolving Involves Enthalpy and Entropy Changes

When a solid **ionic lattice** dissolves in water these **two** things happen:

1) The bonds between the ions **break** — this is **endothermic**. The enthalpy change is the **opposite** of the **lattice enthalpy**.

2) Bonds between the ions and the water are **made** — this is **exothermic**. The enthalpy change here is called the **enthalpy change of hydration** or **enthalpy change of solvation**.

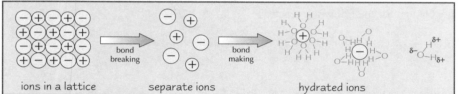

ions in a lattice   separate ions   hydrated ions

Oxygen is more electronegative than hydrogen, so it draws the bonding electrons toward itself, creating a dipole.

Most substances will **only** dissolve if the energy released is about **equal to**, or **greater than**, the amount of energy taken in.

The factors that affect **lattice enthalpy** (see page 3) also affect the hydration enthalpy. So the **smaller** and more **highly charged** the ion, the **more negative** the **hydration enthalpy**.

The enthalpy change isn't the only thing that decides if something will dissolve or not — **entropy change** is important too.

Dissolving normally causes an **increase** in entropy. But for **small highly charged ions** there may be a **decrease** because when water molecules surround the ions, it makes things **more orderly**. The entropy changes are usually **pretty small** but they can sometimes **make the difference** between something being soluble or insoluble.

See pages 8-9 for more on entropy.

Enthalpy Change of Solution can be Calculated   *Only AQA and Salters*

The **enthalpy change of solution** is the overall effect on the enthalpy when something dissolves.
You can work it out using an **enthalpy cycle** — but one drawn a bit differently from those on page 4.
You just need to know the **lattice enthalpy** of the compound and the enthalpies of **hydration of the ions**.

Here's how to draw the enthalpy cycle for working out the **enthalpy change of solution** for **sodium chloride**.

1 Put the ionic lattice and the dissolved ions on the top — connect them by the enthalpy change of solution. This is the direct route.

2 Connect the ionic lattice to the gaseous ions by the reverse of the lattice enthalpy.
The breakdown of the lattice has the opposite enthalpy change to the formation of the lattice.

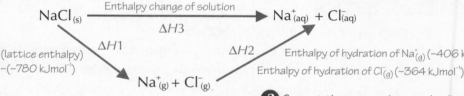

$NaCl_{(s)}$ → (Enthalpy change of solution, $\Delta H3$) → $Na^+_{(aq)} + Cl^-_{(aq)}$

$\Delta H1$ – (lattice enthalpy) –(–780 kJmol⁻¹)

$\Delta H2$ Enthalpy of hydration of $Na^+_{(g)}$ (–406 kJmol⁻¹)
Enthalpy of hydration of $Cl^-_{(g)}$ (–364 kJmol⁻¹)

$Na^+_{(g)} + Cl^-_{(g)}$

3 Connect the gaseous ions to the dissolved ions by the hydration enthalpies of **each** ion. This completes the indirect route.

From Hess's law: $\Delta H3 = \Delta H1 + \Delta H2 = +780 + (-406 + -364) = +10 \text{ kJmol}^{-1}$

The enthalpy change of solution is **slightly endothermic**, but this is compensated for by a small increase in **entropy**, so sodium chloride still dissolves in water.

And here's another. This one's for working out the **enthalpy change of solution** for **silver chloride**.

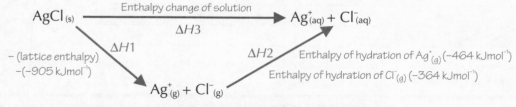

$AgCl_{(s)}$ → (Enthalpy change of solution, $\Delta H3$) → $Ag^+_{(aq)} + Cl^-_{(aq)}$

$\Delta H1$ – (lattice enthalpy) –(–905 kJmol⁻¹)

$\Delta H2$ Enthalpy of hydration of $Ag^+_{(g)}$ (–464 kJmol⁻¹)
Enthalpy of hydration of $Cl^-_{(g)}$ (–364 kJmol⁻¹)

$Ag^+_{(g)} + Cl^-_{(g)}$

From Hess's law: $\Delta H3 = \Delta H1 + \Delta H2 = +905 + (-464 + -364) = +77 \text{ kJmol}^{-1}$

This is much **more endothermic** than the enthalpy change of solution for sodium chloride. There is an **increase in entropy** again, but it's pretty small and not enough to make a difference — so silver chloride is **insoluble** in water.

Enthalpies and Dissolving

Group 2 Hydroxides and Sulphates have Different Solubility Trends

Edexcel only

THE SOLUBILITY OF HYDROXIDES INCREASES DOWN GROUP 2

For compounds with **small**, **singly charged negative ions**, such as OH⁻, the **most important** factor in deciding if they dissolve or not is the **lattice enthalpy**. The **more negative** the lattice enthalpy, the **lower** the solubility.

As you go down Group 2, the positive ions get **bigger**, lattice enthalpies **decrease** and the compounds get **more soluble**.

THE SOLUBILITY OF SULPHATES DECREASES DOWN GROUP 2

For compounds with **large doubly charged negative ions**, e.g. SO_4^{2-}, things are a bit different. The most important factor for these compounds is **hydration enthalpy**. The **more negative** the hydration enthalpy, the **higher** the solubility.

As you go down Group 2, the positive ions get **bigger**, hydration enthalpies **decrease** and the compounds get **less soluble**.

	hydroxide	sulphate
magnesium	least soluble	most soluble
calcium		
strontium		
barium	most soluble	least soluble

Ionic Compounds Only Dissolve in Polar Solvents *Salters only*

When something dissolves, bonds in the **solvent** and **solute** have to **break** and new bonds **between** the solvent and solute have to **form**. Stuff **won't** dissolve if the bonds to be **broken** are **stronger** than those that'll be **formed**.

There are **two** main types of solvent — **polar** (such as water) and **non-polar** (such as hexane).

Ionic substances won't dissolve in non-polar solvents
Non-polar molecules **don't** interact strongly enough with ions to **pull them away** from the lattice. The **electrostatic forces** between the ions are way **stronger** than any bonds that could form between the ions and the solvent molecules.

Covalent substances only dissolve in non-polar solvents
The bonds **between** covalent molecules tend to be **pretty weak**. They're **easily broken** by non-polar solvent molecules. Covalent substances **don't** tend to dissolve in **polar solvents** though. For example, **iodine** doesn't tend to dissolve in water — the **hydrogen bonds** between water molecules are **stronger** than the bonds that would form between water molecules and iodine molecules.

Practice Questions

Q1 What are the three main factors that determine whether an ionic substance will dissolve in water?

Q2 Prove that the enthalpy change of solution for $MgCl_{2(s)}$ is –122 kJmol⁻¹, given that:
$\Delta H^{\ominus}_{latt} [MgCl_{2(s)}] = -2526$ kJmol⁻¹, $\Delta H^{\ominus}_{hyd} [Mg^{2+}_{(g)}] = -1920$ kJmol⁻¹, $\Delta H^{\ominus}_{hyd} [Cl^-_{(g)}] = -364$ kJmol⁻¹.

Q3 What are the trends in solubility for Group 2 hydroxides and sulphates?

Exam Questions

1 Sodium chloride has a small endothermic enthalpy change of solution and is soluble. Calcium carbonate has a small exothermic enthalpy change of solution and is insoluble. Suggest an explanation for each observation. [2 marks]

2 a) Draw an enthalpy cycle for the enthalpy change of solution of $SrF_{2(s)}$. Label each enthalpy change. [5 marks]

b) Calculate the enthalpy change of solution for SrF_2 from the following data:
$\Delta H^{\ominus}_{latt} [SrF_{2(s)}] = -2492$ kJmol⁻¹, $\Delta H^{\ominus}_{hyd} [Sr^{2+}_{(g)}] = -1480$ kJmol⁻¹, $\Delta H^{\ominus}_{hyd} [F^-_{(g)}] = -506$ kJmol⁻¹ [2 marks]

c) What further data would you need to decide if $SrF_{2(s)}$ is likely to be soluble in water? [1 mark]

3 a) Predict whether or not potassium iodide is likely to be soluble in cyclohexane. [1 mark]

b) Explain your answer to part (a). [4 marks]

So the Wicked Witch of the West must have been ionic then...

Compared to the ones on page 4, these enthalpy cycles are an absolute breeze. You've got to make sure the definitions are firmly fixed in your mind though. Don't forget that a positive enthalpy change doesn't mean the stuff definitely won't dissolve — there might be an entropy change that'll make up for it. The delights of entropy are on the next two pages.

Entropy and Free Energy Change

Free energy — I could do with a bit of that. My gas bill is astronomical.

These two pages are just for AQA, Salters and Nuffield.

Entropy Tells you How Much Disorder there is

Entropy is a measure of the **number of ways** that **particles** can be **arranged** and the **number of ways** that the **energy** can be shared out between the particles.

Substances really **like** disorder, so the particles move to try to **increase the entropy**.

There are a few things that affect entropy:

PHYSICAL STATE affects Entropy

You have to go back to the good old **solid-liquid-gas** particle explanation thingy to understand this.

Solid particles just wobble about a fixed point — there's **hardly any** randomness, so they have the **lowest entropy**.

Gas particles whizz around wherever they like. They've got the most **random arrangements** of particles, so they have the **highest entropy**.

Dissolving a solid also increases its entropy — dissolved particles can **move freely** as they're no longer held in one place.

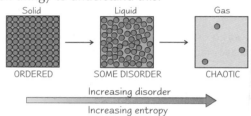

The AMOUNT OF ENERGY a Substance has affects Entropy too

Energy can be measured in **quanta** — these are fixed '**packages**' of energy.
The more energy quanta a substance has, the **more ways** they can be arranged and the greater the **entropy**.

MORE PARTICLES means More Entropy

It makes sense — the more particles you've got, the **more ways** they and their energy can be **arranged**.
— so in a reaction like $N_2O_{4(g)} \rightarrow 2NO_{2(g)}$, entropy increases because the **number of moles** increases.

Reactions Won't Happen Unless the Total Entropy Change is Positive

During a reaction, there's an entropy change between the **reactants and products** — the entropy change of the **system**.
The entropy of the **surroundings** changes too (because **energy** is transferred to or from the system).
The **TOTAL** entropy change is the sum of the entropy changes of the **system** and the **surroundings**.

The units of entropy are JK⁻¹mol⁻¹

$$\Delta S_{total} = \Delta S_{system} + \Delta S_{surroundings}$$

This equation isn't much use unless you know ΔS_{system} and $\Delta S_{surroundings}$. Luckily, there are formulas for them too:

This is just the difference between the entropies of the reactants and products.

$$\Delta S_{system} = S_{products} - S_{reactants}$$

and

$$\Delta S_{surroundings} = -\frac{\Delta H}{T}$$

ΔH = enthalpy change (in Jmol⁻¹)
T = temperature (in K)

> **Example:** Calculate the total entropy change for the reaction of ammonia and hydrogen chloride under standard conditions.
>
> $$NH_{3(g)} + HCl_{(g)} \rightarrow NH_4Cl_{(s)} \qquad \Delta H^{\ominus} = -315 \text{ kJmol}^{-1} \text{ (at 298K)}$$
>
> $S^{\ominus}[NH_{3(g)}] = 192.3 \text{ JK}^{-1}\text{mol}^{-1}$, $S^{\ominus}[HCl_{(g)}] = 186.8 \text{ JK}^{-1}\text{mol}^{-1}$, $S^{\ominus}[NH_4Cl_{(s)}] = 94.6 \text{ JK}^{-1}\text{mol}^{-1}$
>
> First find the entropy change of the **system**:
>
> $$\Delta S^{\ominus}_{system} = S^{\ominus}_{products} - S^{\ominus}_{reactants} = 94.6 - (192.3 + 186.8) = -284.5 \text{ JK}^{-1}\text{mol}^{-1}$$
>
> This shows a negative change in entropy. It's not surprising as 2 moles of gas have combined to form 1 mole of solid.
>
> Now find the entropy change of the **surroundings**:
>
> $$\Delta H^{\ominus} = -315 \text{ kJmol}^{-1} = -315 \times 10^3 \text{ Jmol}^{-1}$$
>
> Put $\Delta H^{\ominus}$ in the right units.
>
> $$\Delta S^{\ominus}_{surroundings} = -\frac{\Delta H^{\ominus}}{T} = \frac{-(-315 \times 10^3)}{298} = +1057 \text{ JK}^{-1}\text{mol}^{-1}$$
>
> Finally you can find the **total** entropy change:
>
> $$\Delta S^{\ominus}_{total} = \Delta S^{\ominus}_{system} + \Delta S^{\ominus}_{surroundings} = -284.5 + (+1057) = +772.5 \text{ JK}^{-1}\text{mol}^{-1}$$
>
> The total entropy has **increased**. The entropy increase in the surroundings was big enough to make up for the entropy decrease in the system.

Entropy and Free Energy Change

Entropy Increase May Explain Spontaneous Endothermic Reactions

A spontaneous (or feasible) change is one that'll **just happen** by itself — you don't need to give it energy.
But the weird thing is, some **endothermic** reactions are **spontaneous**. You'd normally have to supply **energy** to make an endothermic reaction happen, but if the **entropy** increases enough, the reaction will happen by itself.

- Water evaporates at room temperature. This change needs **energy** to break the bonds between the molecules — but because it's **changing state** (from a liquid to a gas), the entropy increases.
- The reaction of sodium hydrogencarbonate with hydrochloric acid is a **spontaneous endothermic reaction**. Again there's an **increase in entropy**.

$$NaHCO_{3(s)} + H^+_{(aq)} \rightarrow Na^+_{(aq)} + CO_{2(g)} + H_2O_{(l)}$$

| 1 mole solid | 1 mole aqueous ions | 1 mole aqueous ions | 1 mole gas | 1 mole liquid |

The product has more particles — and gases and liquids have more entropy than solids too.

For Spontaneous Reactions ΔG must be Negative or Zero   AQA only

Free energy change, **ΔG**, is a measure used to predict whether a reaction is **feasible**.
If **ΔG** is **negative or equal to zero**, then the reaction might happen by itself.
Free energy change takes into account the changes in **enthalpy** and **entropy**.
And of course, there's a formula for it:

$$\Delta G = \Delta H - T\Delta S_{total}$$

ΔH = enthalpy change (in $Jmol^{-1}$)
T = temperature (in K)

Even if ΔG shows that a reaction is theoretically feasible, it might have a really high activation energy and be so slow that you wouldn't notice it happening at all.

Example: Calculate the free energy change for the following reaction at 298 K.

$$MgCO_{3(g)} \rightarrow MgO_{(s)} + CO_{2(g)} \qquad \Delta H^{\ominus} = +117\,000\ Jmol^{-1},\ \Delta S^{\ominus}_{total} = +175\ Jmol^{-1}$$

$$\Delta G = \Delta H - T\Delta S_{total} = +117\,000 - [(298 \times (+175)] = +64\,850\ Jmol^{-1}$$

ΔG is positive — so the reaction isn't feasible at this temperature.

Practice Questions

Q1 What does the term 'entropy' mean?

Q2 In each of the following pairs choose the one with the greater entropy value.
   a) 1 mole of $NaCl_{(aq)}$ and 1 mole of $NaCl_{(s)}$ b) 1 mole of $Br_{2(l)}$ and 1 mole of $Br_{2(g)}$
   c) 1 mole of $Br_{2(g)}$ and 2 moles of $Br_{2(g)}$

Q3 Write down the formulas for the following:
   a) total entropy change   b) entropy change of the surroundings   c) free energy change

Exam Questions

1 a) Based on just the equation, predict whether the reaction below
      is likely to be spontaneous. Give a reason for your answer.   [2 marks]
      $$Mg_{(s)} + \tfrac{1}{2}O_{2(g)} \rightarrow MgO_{(s)}$$

   b) Use the data on the right to calculate the entropy change for
      the system above.   [3 marks]

   c) Does the result of the calculation indicate than the reaction
      will be spontaneous? Give a reason for your answer.   [2 marks]

Substance	Entropy — standard conditions ($JK^{-1}mol^{-1}$)
$Mg_{(s)}$	32.7
$\tfrac{1}{2}O_{2(g)}$	102.5
$MgO_{(s)}$	26.9

2 $S^{\ominus}[H_2O_{(l)}] = 70\ JK^{-1}mol^{-1}$, $S^{\ominus}[H_2O_{(s)}] = 48\ JK^{-1}mol^{-1}$, $\Delta H^{\ominus} = -6\ kJmol^{-1}$
   For the reaction $H_2O_{(l)} \rightarrow H_2O_{(s)}$:
   a) Calculate the total entropy change at   (i) 250 K   (ii) 300 K   [5 marks]
   b) Will this reaction be spontaneous at 250 K or 300 K? Explain your answer.   [2 marks]

Being neat and tidy is against the laws of nature...

*There's a scary amount of scary looking formulas on these pages. They aren't too hard to use, but watch out for your units.
Make sure the temperature's in kelvin — if you're given one in °C, you need to subtract 273 from it to get it in kelvin.
And check that all your enthalpy and entropy values involve joules, not kilojoules (so $Jmol^{-1}$, not $kJmol^{-1}$, etc.).*

Reaction Rates

The rate of a reaction is just how quickly it happens. Lots of things can make it go faster or slower.

Particles **Must** Collide to **React**

1) For the stuff on reaction rates to make any sense, you need to have a good grasp of **collision theory**. This basically says that particles in liquids and gases will **only** react if —

- they **collide**.
- they collide in the **right direction** — they need to be **facing** each other the right way.
- they collide with at least a certain **minimum** amount of kinetic **energy**.

2) The **minimum amount of kinetic energy** particles need to react is called the **activation energy**, E_a. The particles need this much energy to **break the bonds** of the reactants and start the reaction.

Increasing the *Concentration* or *Temperature* makes Reactions *Faster*

1) If you increase the **concentration** of reactants in a **solution** there'll be more particles, and these will on average be **closer together**. More particles, closer together means there'll be more **collisions** and **more chances** to react. Increasing the **pressure** of a **gas** has the same effect.

2) Increasing the **temperature** means that the particles will tend to have more **kinetic energy**, so:

- they'll fly about **faster** and will **collide more often**, speeding up the reaction.
- a **greater proportion** of them will have the **activation energy** to **react**. Have a gander at the blue box below.

If you plot a graph of the **numbers of molecules** in a substance with different **kinetic energies** you get a nice curve called a **Maxwell-Boltzmann distribution**. Increasing the temperature changes the **shape** of this curve:

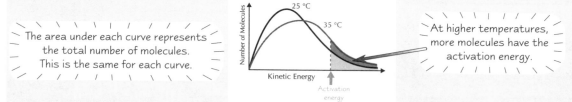

The area under each curve represents the total number of molecules. This is the same for each curve.

At higher temperatures, more molecules have the activation energy.

At higher temperatures the curve is **flatter**, since the particles have a **greater spread of energies**. The curve is also shifted to the **right** because the average energy has increased.

Take care if you have to draw a Maxwell-Boltzmann distribution. Your curve must start at **(0, 0)** because **no** molecules have **zero energy**. The other end of the curve **approaches** the x-axis but **never** touches it. Another thing to notice is that the activation energy **doesn't change** with temperature.

The **Reaction Rate** tells you How Fast **Reactants** are Converted to **Products**

The **reaction rate** is the **change in the amount** of reactants or products **per unit time** (normally per second). If the reactants are in **solution**, the rate'll be **change in concentration per second** and the units will be **moldm⁻³s⁻¹**.

If you draw a graph of the **amount of reactant or product against time** for a reaction, the rate at any point is given by the **gradient** at that point on the graph. If the graph's a curve, you'll have to draw a **tangent** to the curve and find the gradient of that.

> A tangent is a line that just touches a curve and has the same gradient as the curve does at that point.

A graph of the **concentration of a reactant against time** might look something like this:

The gradient of the blue tangent is the rate of the reaction after **30 seconds**.

$$\text{Gradient} = \frac{-0.8}{60} = -0.013 \text{ moldm}^{-3}\text{s}^{-1}$$

So, the rate after 30 seconds is **0.013 moldm⁻³s⁻¹**

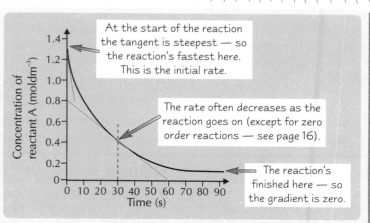

At the start of the reaction the tangent is steepest — so the reaction's fastest here. This is the initial rate.

The rate often decreases as the reaction goes on (except for zero order reactions — see page 16).

The reaction's finished here — so the gradient is zero.

Reaction Rates

There are **Loads** of Ways to **Follow the Rate of a Reaction**   Not AQA or OCR

Although there are quite a few ways to follow reactions, not every method works for every reaction. You've got to **pick a property** that **changes** as the reaction goes on.

1) **Gas volume**

If a **gas** is given off, you could **collect it** in a gas syringe and record how much you've got at **regular time intervals**. For example, this'd work for the reaction between an **acid** and a **carbonate** in which **carbon dioxide gas** is given off.

CO_2 gas
acid
carbonate

2) **Loss of mass**

If a **gas** is given off, the system will **lose mass**. You can measure this at regular intervals with a **balance**.

CO_2 gas
Balance

3) **Colour change**

You can sometimes track the colour change of a reaction using a gadget called a **colorimeter**. For example, in the reaction between propanone and iodine, the **brown** colour fades.

$$CH_3COCH_{3(aq)} + I_{2(aq)} \rightarrow CH_3COCH_2I_{(aq)} + H^+_{(aq)} + I^-_{(aq)}$$
colourless   brown
colourless

4) **Electrical conductivity**

If the **number of ions** changes, so will the **electrical conductivity**. This happens in the reaction between propanone and iodine above.

Practice Questions

Q1 Outline the main points of collision theory.

Q2 Define the term activation energy.

Q3 Suggest three ways of following the rate of a reaction. For each, give an example of a reaction that could be followed that way.

Exam Questions

1   The rate of the acid-catalysed reaction between bromine, Br_2, and methanoic acid, HCOOH, was investigated.

$$Br_{2(aq)} + HCOOH_{(aq)} \xrightarrow{H^+_{(aq)}} 2H^+_{(aq)} + 2Br^-_{(aq)} + CO_{2(g)}$$

a) Suggest one method that could be used to follow the reaction rate.   [2 marks]

b) Outline how the rate of reaction with respect to Br_2, at any particular time, could be determined.   [3 marks]

c) Explain the likely effect on the rate of the reaction of reducing the initial concentration of Br_2.   [3 marks]

2   The rate of decomposition of hydrogen peroxide was followed by monitoring the concentration of hydrogen peroxide.

$$2H_2O_{2(aq)} \rightarrow 2H_2O_{(l)} + O_{2(g)}$$

Time (minutes)	0	20	40	60	80	100
$[H_2O_2]$ (moldm^{-3})	2.00	1.00	0.50	0.25	0.125	0.0625

a) Suggest an alternative method that could have been used to follow the rate of this reaction.   [2 marks]

b) Using the data above, plot a graph and determine the rate of the reaction after 30 minutes.   [6 marks]

This stuff'll speed up your revision rate...

I'd like to promise you that Section 2's going to be an exhilarating roller-coaster of delight. But you'd never believe me. Some of this stuff should cause AS bells to start ringing, but you'll have to have a really good look at the bit on finding rates from concentration-time graphs — it'll make the rest of this section slightly less traumatic.

Activation Energy and Catalysts

Catalysts are definitely the best thing since ready-grated cheese — especially if you happen to own an ammonia factory.

OCR people can skip these two pages.

Catalysts Increase the Rate of Reactions

You can use **catalysts** to make chemical reactions happen **faster**. Learn this definition:

> A **catalyst** increases the **rate** of a reaction by providing an **alternative reaction pathway** with a **lower activation energy**. The catalyst is **chemically unchanged** at the end of the reaction.

1) Catalysts are **great**. They **don't** get used up in reactions, so you only need a **tiny bit** of catalyst to catalyse a **huge** amount of stuff. Many **do** take part in reactions, but they're **remade** at the end.

2) Many catalysts are **very fussy** about which reactions they catalyse — they have **high specificity**.

3) Catalysts are amazingly important in **industrial processes** — they save a fortune.
For example, **iron** is used as a catalyst in the **Haber process** (that's the one for making **ammonia**; see page 25).

Enthalpy Profiles and Boltzmann Distributions Show Why Catalysts Work

If you look at an **enthalpy profile** together with a **Maxwell-Boltzmann Distribution**, you can see **why** catalysts work.

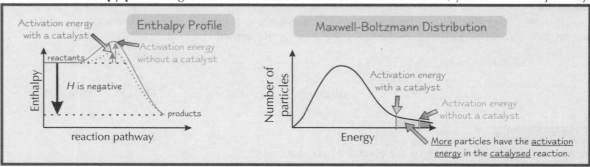

The catalyst provides an **alternative pathway** with **lower activation energy**, meaning there are **more particles** with **enough energy** to react when they collide. So, in a certain amount of time, **more particles will react**.

There are Homogeneous Catalysts and Heterogeneous Catalysts

NOT Edexcel

1) A **homogeneous catalyst** is in the **same state** as the **reactants**. So, if the reactants are **gases**, the catalyst must be a **gas** too. And if the reactants are **aqueous** (dissolved in water), the catalyst has to be **aqueous** too.
When **enzymes** catalyse reactions in your body **cells**, everything's **aqueous** — so it's **homogeneous catalysis**.

2) **Heterogeneous catalysts** are in a **different physical state** from the reactants.
Here are some examples of heterogeneous catalysts:

Transition metals are often used as catalysts — see page 68.

- **vanadium pentoxide** in the **contact process** for making **sulphuric acid**
- **nickel** in the **hydrogenation of vegetable oils**
- **platinum** and **rhodium** in **catalytic converters** in cars. Catalytic converters change nasty gases like **nitric oxide** and **carbon monoxide** into nitrogen and carbon dioxide.

$$2NO_{(g)} + 2CO_{(g)} \xrightarrow{Pt_{(s)} / Rh_{(s)}} N_{2(g)} + 2CO_{2(g)}$$

Solid heterogeneous catalysts can provide a **surface** for a reaction to take place on.
Here's how it works —

1) **Reactant molecules** arrive at the **surface** and **bond** with the solid catalyst. This is called **adsorption**.

2) The bonds between the **reactant's** atoms are **weakened** and they **break up**. This forms **radicals**.
These radicals then **get together** and make **new molecules**.

This example shows you how a catalytic converter changes nitric oxide and carbon monoxide to nitrogen and carbon dioxide.

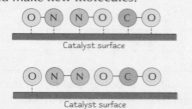

The adsorption must be strong enough to weaken reactant bonds, but not too strong or it won't let go of the atoms.

In catalytic converters, the catalyst is usually in the form of a **mesh** or a **fine powder** to increase the **surface area**. Alternatively it might be spread over an **inert support**, such as ceramic.

Heterogeneous catalysts can be poisoned. This happens if a **poison** clings to the catalyst's surface **more strongly** than the reactant does, **preventing** the catalyst from getting involved in the reaction it's meant to be **speeding up**. For instance, **sulphur** can poison the **iron catalyst** used in the **Haber process**.

Activation Energy and Catalysts

This page is just for Salters and those of you doing the OCR Biochemistry Option — except for the questions, of course.

Enzymes are Biological Catalysts

Enzymes are **protein-based** catalysts found in animal and plant **cells**. Like any other catalysts, they **increase reaction rates** by providing a reaction route with a **lower activation energy**.

Enzymes have heaps of uses. Here are a few examples:

- They're used in **biological washing powders** to digest stains. E.g. **proteases** digest proteins in egg or blood.
- The enzyme **rennin** from bacteria is used to make **cheese**.
- Enzymes in **yeast** are used with barley to make **beer**.
- The enzyme **pectinase** is used in **fruit juice** manufacture. It increases juice yield and clarity.

Enzymes will only work with **specific substrates** (reactants) — usually only one.
This is because, for the enzyme to work, the substrate has to **fit** into the enzyme's **active site**.
If the substrate's shape doesn't match the active site's shape, then the reaction **won't** be catalysed.

*The **lock and key** model is used to picture this — make sure you take a look. See page 136.*

Enzymes need the right conditions or they won't work

Changes in the environment of an enzyme can sometimes reduce or completely stop its activity.

Temperature has a BIG influence on enzyme activity

1) Like with any reaction, increasing the temperature of an enzyme-controlled reaction increases the rate. **BUT**, if the temperature goes **above** a certain level, some of the **bonds** that hold the enzyme in shape **break**.

2) This is bad news for the enzyme because the **active site changes shape** and the enzyme and substrate **no longer fit together**. Once this happens, the enzyme is **denatured** and it no longer works as a catalyst.

pH also affects enzyme activity

1) All enzymes have an **optimum pH value**. Above and below the optimum pH, the enzyme activity is messed up because groups such **-COOH** and **-NH$_2$** in the active sites of enzymes **lose or gain protons**.

2) This can prevent the substrate **temporarily bonding** with the active site. It can also **change the shape** of the active site so the substrate **no longer fits** at all.

Inhibitors prevent enzyme activity

There are two types of inhibitors — competitive and non-competitive. **Competitive inhibitors** have a **similar shape** to the substrate and bond to the **active sites**. This **blocks** the active site so the substrate can't fit in. **Non-competitive inhibitors** bond to the enzyme away from its active site, but this causes the active site to change shape so that the substrate won't fit any more. Many **heavy metal ions** can act as non-competitive inhibitors, e.g. Hg^{2+}.

Practice Questions

Q1 What do the terms 'catalyst' and 'activation energy' mean?

Q2 Explain the difference between heterogeneous and homogeneous catalysis.

Q3 Give three ways that the activity of an enzyme can be reduced.

Mr Smith found he got to Tesco's much more quickly if he used a carrot as a catalyst

Exam Question

1 Hydrogen peroxide, H_2O_2, is a toxic substance that can accumulate in cells. Its decomposition in cells is catalysed by the enzyme catalase.

$$2H_2O_{2(aq)} \xrightarrow{\text{catalase}} 2H_2O_{(l)} + O_{2(g)}$$

a) Explain how catalase increases the rate of decomposition of hydrogen peroxide. [3 marks]

b) Solid manganese(IV) oxide also acts as a catalyst for this reaction. What type of catalyst is manganese(IV) oxide? [1 mark]

c) Enzymes are said to show high specificity. Explain what you understand by this term. [2 marks]

Bagpuss, Sylvester, Tom — the top three on my catalyst...

I bet you thought you'd seen the last of old Max Boltzmann's curve, but here it is again. Make sure you know how catalysts work and remember that enzymes are just special types of catalysts, so they work in the same way. Enzymes are complete fusspots and like everything just so — if it gets a bit hot, they fall apart at the seams and the wrong pH spells disaster too.

Rate Equations

This is when it all gets a bit mathsy. You've just got to take a deep breath, dive in, and don't bash your head on the bottom.

The **Rate Equation** links **Reaction Rate** to **Reactant Concentrations**

Rate equations look ghastly, but all they're really telling you is how the **rate** is affected by the **concentrations of reactants**. For a general reaction: $A + B \rightarrow C + D$, the **rate equation** is:

The units of rate are $moldm^{-3}s^{-1}$.

$$\text{Rate} = k[A]^m[B]^n$$

Remember — square brackets mean the concentration of whatever's inside them.

1) **m** and **n** are the **orders of the reaction** with respect to reactant A and reactant B.
   m tells you how the **concentration of reactant A** affects the **rate** and **n** tells you the same for **reactant B**.

   If [A] doubles and the rate **stays the same**, the order with respect to A is **0**.

   If [A] doubles and the rate **also doubles**, the order with respect to A is **1**.

   If [A] doubles and the rate **quadruples**, the order with respect to A is **2**.

2) The **overall order of the reaction** is **m + n**.

3) You can only find **orders of reaction** from **experiments**. You **can't** work them out from chemical equations.

4) **k** is the **rate constant** — the bigger it is, the **faster** the reaction. The rate constant is **always the same** for a certain reaction at a **particular temperature** — but if you **increase** the temperature, the rate constant's going to rise too. The **units** vary, so you have to **work them out**. The example further down the page shows you how.

Example:
The chemical equation below shows the acid-catalysed reaction between propanone and iodine.

$$CH_3COCH_{3(aq)} + I_{2(aq)} \xrightarrow{H^+_{(aq)}} CH_3COCH_2I_{(aq)} + H^+_{(aq)} + I^-_{(aq)}$$

This reaction is **first order** with respect to propanone and $H^+_{(aq)}$ and **zero order** with respect to iodine. Write down: a) the rate equation, b) the overall order of the reaction.

Even though $H^+_{(aq)}$ is a catalyst, rather than a reactant, it can still be in the rate equation.

a) The **rate equation** is: rate $= k[CH_3COCH_{3(aq)}]^1[H^+_{(aq)}]^1[I_{2(aq)}]^0$

   But $[X]^1$ is usually written as **[X]**, and $[X]^0$ equals **1** so is usually **left out** of the rate equation.

   So you can **simplify** the rate equation to: **rate $= k[CH_3COCH_{3(aq)}][H^+_{(aq)}]$**

Think about the indices laws from maths.

   b) The overall order of the reaction is $1 + 1 + 0 = 2$

You can Calculate the **Rate Constant** from the **Orders** and **Rate of Reaction**

Once the rate and the orders of the reaction have been found by experiment, you can work out the **rate constant, *k***.

Example:
The reaction below was found to be second order with respect to NO and zero order with respect to CO and O_2.
The rate is 1.76×10^{-3} $moldm^{-3}s^{-1}$, when $[NO_{(g)}] = [CO_{(g)}] = [O_{2(g)}] = 2.00 \times 10^{-3}$ $moldm^{-3}$.

$$NO_{(g)} + CO_{(g)} + O_{2(g)} \rightarrow NO_{2(g)} + CO_{2(g)}$$

First write out the **rate equation**:

$$\text{Rate} = k[NO_{(g)}]^2[CO_{(g)}]^0[O_{2(g)}]^0 = k[NO_{(g)}]^2$$

Next insert the **concentration** and the **rate**. Rearrange the equation and calculate the value of *k*:

$$\text{Rate} = k[NO_{(g)}]^2, \text{ so, } 1.76 \times 10^{-3} = k \times (2.00 \times 10^{-3})^2 \Rightarrow k = \frac{1.76 \times 10^{-3}}{(2.00 \times 10^{-3})^2} = 440$$

Find the **units for k** by putting the other units in the rate equation:

$$\text{Rate} = k[NO_{(g)}]^2, \text{ so } moldm^{-3}s^{-1} = k \times (moldm^{-3})^2 \Rightarrow k = \frac{moldm^{-3}s^{-1}}{(moldm^{-3})^2} = \frac{s^{-1}}{moldm^{-3}} = dm^3mol^{-1}s^{-1}$$

So the answer is: $k = 440$ $dm^3mol^{-1}s^{-1}$

Rate Equations

A Large Activation Energy means a Small Rate Constant   *Edexcel only*

Reactions with **large activation energies** have **small rate constants**.
You need to **increase the temperature** a lot to increase the **rate constant** enough for a reaction to happen.

The **Arrhenius equation** links the **rate constant** with **activation energy** and **temperature**.
This is probably the **worst** equation there is in A2 Chemistry. But the good news is, you **don't** have to learn it —
you just have to understand what it's showing you. It is:

$$k = Ae^{-E_a/RT}$$

k = rate constant
E_a = activation energy (J)
T = temperature (K)
R = gas constant (8.31 JK^{-1}mol^{-1})
A = another constant

It's an exponential relationship. This 'e' is the e^x button on your calculator.

1) As the activation energy, E_a, gets **bigger**, k gets **smaller**.
   So a **large E_a** leads to a **slow rate**.
   You can **test** this out by trying **different numbers**
   for E_a in the equation...aaah, go on, be a devil.

2) The equation also shows that as the temperature **rises**,
   k **increases**. Try this one out too.

Use the Arrhenius Equation to Calculate the Activation Energy   *Nuffield only*

Putting the **Arrhenius equation** into **logarithmic form** makes it a bit easier to use.

$$\ln k = \ln A - \frac{E_A}{RT} = \text{(a constant)} - \frac{E_A}{RT}$$

There's a handy '$\ln$' button on your calculator for this.

If you plot the graph of **$\ln k$** (logarithm to the base e of k) against $\frac{1}{T}$ you get a straight line with a **gradient** $\frac{-E_a}{R}$.

And once you know the gradient, you're only a pogo stick bounce away from finding the **activation energy**.

Example:
The graph on the right shows $\ln k$ against $\frac{1}{T}$ for the decomposition of hydrogen iodide.

Calculate the activation energy for this reaction. R = 8.31 JK^{-1}mol^{-1}.

The gradient, $\dfrac{-E_a}{R} = \dfrac{-15}{0.0008} = -18\,750$

So, $E_a = -(-18\,750 \times 8.31) = 155\,812.5$ Jmol$^{-1} \approx 156$ kJmol^{-1}

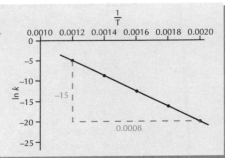

Practice Questions

Q1 Explain what the terms 'order of reaction' and 'rate constant' mean.

Q2 Predict the effect on k and the reaction rate of raising the temperature of a chemical reaction.

Q3 If a chemical reaction has a large E_a, what can you say about k and the reaction rate?

Exam Question

1   The following reaction is second order with respect to NO and first order with respect to H$_2$.
$$2NO_{(g)} + 2H_{2(g)} \rightarrow 2H_2O_{(g)} + N_{2(g)}$$

a) Write a rate equation for the reaction and deduce the overall order for the reaction.   [3 marks]

b) The rate of the reaction at 800 °C was determined to be 0.00267 moldm^{-3}s^{-1} when [H$_{2(g)}$] = 0.0020 moldm^{-3}
    and [NO (g)] = 0.0040 moldm^{-3}.

    i) Calculate a value for the rate constant at 800 °C, including units.   [3 marks]
    ii) Predict the effect on the rate constant of decreasing the temperature of the reaction to 600 °C.   [1 mark]

A2 Chemistry's a quagmire of constants — this is just the first of many...

*As far as rate equations go, forget the weird chemistryness of them and just treat them like straightforward algebra equations.
If you're a maths whizz, then all this log stuff probably makes total sense to you. But if not, then make sure you can bluff
your way through by remembering the main points and being able to press your calculator keys in the right order.*

Orders of Reactions and Half-Life

*This page is on about **chemical** half-life, not radioactive half-life — there's not a single Geiger counter in sight.*
They're the same sort of idea though — they just measure how quickly different things fall.

Orders can be Worked Out from the Shape of Concentration-Time Graphs...

You can figure out the **order of reaction** with respect to a reactant by looking at the shape of a **concentration-time graph** and seeing whether the **half-life**, $t_{1/2}$, is **constant**. The half-life is just the **time taken** for the **concentration** of a **reactant** to decrease to **half** of its original value.

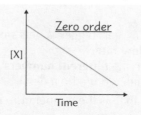

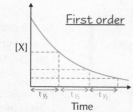

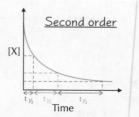

The rate **doesn't change** as concentration falls — the graph is a **straight line**.

The graph is an **exponential decay**. $t_{1/2}$ is **constant** — it always takes the **same amount of time** for the concentration to halve.

The half-life **isn't** constant — it **increases** as the reaction goes on.

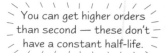
You can get higher orders than second — these don't have a constant half-life.

...and from Rate-Concentration Graphs too

You can draw a **rate-concentration graph** from a **concentration-time graph**. Here's how:

1) Find the **gradient** at various points on the graph. This gives you the **rate** at that particular **concentration**. With a **straight-line graph**, this is easy. But if it's a **curve**, you need to draw **tangents** and find their gradients.

2) Now plot a graph of each **rate** against the **concentration at that point** — and that's all there is to it.

The shapes of **rate-concentration graphs** show the **orders** really clearly — you don't even need to worry about half-lives.

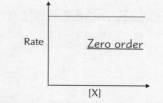

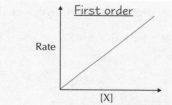

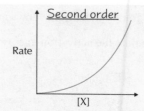

The graph is **horizontal** since changing concentration has **no effect** on rate.

Rate is **proportional** to concentration — so the graph's a **straight line through the origin**.

The graph is a **curve** since rate is proportional to $[X]^2$.

You need to be able to Figure Out the Rate Equation from a Graph

Have a good look at this example:

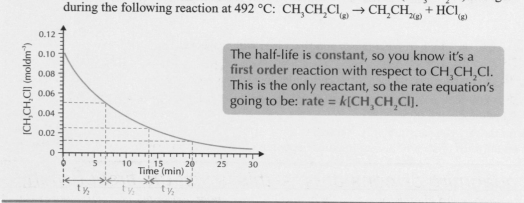

Example: The graph below shows how the concentration of chloroethane (CH_3CH_2Cl) changes during the following reaction at 492 °C: $CH_3CH_2Cl_{(g)} \rightarrow CH_2CH_{2(g)} + HCl_{(g)}$

The half-life is **constant**, so you know it's a **first order** reaction with respect to CH_3CH_2Cl. This is the only reactant, so the rate equation's going to be: **rate = $k[CH_3CH_2Cl]$**.

Harold the half-life found it difficult to find a hat that fitted well.

Orders of Reactions and Half-Life

The Initial Rates Method can be used to work out Rate Equations too

The **initial rate of a reaction** is the rate right at the **start** of the reaction. You can find this from a **concentration-time** graph by calculating the **gradient** of the **tangent** at **time = 0**.

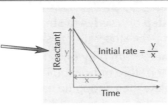

Here's a quick explanation of how to use the **initial rates method**:

1) Repeat the experiment several times using **different initial concentrations** of reactants. You should usually only change **one** of the concentrations at a time, keeping the rest constant.

2) Calculate the **initial rate** for each experiment using the method above.

3) Finally, see how the **initial concentrations** affect the **initial rates** and figure out the **order** for each reactant. The example below shows you how to do this. Once you know the **orders**, you can work out the rate equation.

Example:
The table on the right shows the results of a series of initial rate experiments for the reaction:

$$NO_{(g)} + CO_{(g)} + O_{2(g)} \rightarrow NO_{2(g)} + CO_{2(g)}$$

Write down the rate equation for the reaction.

Experiment number	$[NO_{(g)}]$ (moldm^{-3})	$[CO_{(g)}]$ (moldm^{-3})	$[O_{2(g)}]$ (moldm^{-3})	Initial rate (moldm^{-3}s^{-1})
1	2.0×10^{-2}	1.0×10^{-2}	1.0×10^{-2}	0.176
2	4.0×10^{-2}	1.0×10^{-2}	1.0×10^{-2}	0.704
3	2.0×10^{-2}	2.0×10^{-2}	1.0×10^{-2}	0.176
4	2.0×10^{-2}	1.0×10^{-2}	2.0×10^{-2}	0.176

1) Look at experiments 1 and 2 — when $[NO_{(g)}]$ doubles (but all the other concentrations stay constant), the rate **quadruples**. So the reaction is **second order** with respect to NO.

2) Look at experiments 1 and 3 — when $[CO_{(g)}]$ doubles (but all the other concentrations stay constant), the rate **stays the same**. So the reaction is **zero order** with respect to CO.

3) Look at experiments 1 and 4 — when $[O_{2(g)}]$ doubles (but all the other concentrations stay constant), the rate **stays the same**. So the reaction is **zero order** with respect to O_2.

4) Now that you know the order with respect to each reactant you can write the rate equation: **rate = $k[NO_{(g)}]^2$** .

k can be calculated by inserting the concentrations and the initial rate from one of the experiments into the rate equation.

Practice Questions

Q1 Sketch reactant concentration-time graphs for zero, first and second order reactions.

Q2 Outline how you could use rate-concentration graphs to determine the order of a reaction with respect to a reactant.

Q3 Define the term half-life. What order would a reactant with a constant half-life have?

Exam Question

1 The table shows the results of a series of initial rate experiments for the reaction between substances D and E.

Experiment	[D] (moldm^{-3})	[E] (moldm^{-3})	Initial rate $\times 10^{-3}$ (moldm^{-3}s^{-1})
1	0.2	0.2	1.30
2	0.4	0.2	5.19
3	0.2	0.4	2.61

a) Find the order of the reaction with respect to reactants D and E. Explain your reasoning. [4 marks]

b) Write the rate equation for the reaction. [1 mark]

c) Calculate a value for the rate constant, k, including units. [3 marks]

Book number 8 — Harry Potter and the Order of the Reaction...

Make sure you don't get your concentration-time graphs muddled up with your rate-concentration graphs — they're totally different. You need to be able to draw a rate-concentration graph from a concentration-time graph too. More drawing tangents here. It's a bit of a faff — you basically have to slide your ruler around until your tangent looks about right.

Rates and Reaction Mechanisms

And now ladies and gentlemen — the grand finale of Section Two... **It's just for OCR,** *Edexcel* **and** *Nuffield* **though.**

The **Rate-Determining Step** is the **Slowest Step** in a Multi-Step Reaction

Mechanisms can have **one step** or a **series of steps**. In a series of steps, each step can have a **different rate**. The **overall rate** is decided by the step with the **slowest** rate — the **rate-determining step**.

It's a bit like a busy supermarket with only one checkout open. It **doesn't matter** how many customers come into the store or how quickly they choose their items, the number of customers that can complete their shopping each day will **mainly** be decided by the number than can get through the checkout — this is the **rate-determining step**.

An example of a multi-step reaction is the free radical substitution of halogens in alkanes — see page 72.

Reactants in the **Rate Equation** Affect the **Rate**

The rate equation is handy for working out the **mechanisms** of a chemical reaction.

You need to be able to pick out which reactants from the chemical equation are involved in the **rate-determining step**. Here are the **rules** for doing this:

> If a reactant appears in the **rate equation**, it must be affecting the **rate**.
> So this reactant, or something derived from it, must be in the **rate-determining step**.
>
> If a reactant **doesn't** appear in the **rate equation**, then it **won't** be involved in the **rate-determining step** (and neither will anything derived from it).

Catalysts can appear in rate equations, so they can be in rate-determining steps too.

Some **important points** to remember about rate-determining steps and mechanisms are:
1) The rate-determining step **doesn't** have to be the first step in a mechanism.
2) The reaction mechanism **can't** usually be predicted from **just** the chemical equation.

You Can Predict the **Rate Equation** from the **Rate-Determining Step**...

> The **order of a reaction** with respect to a reactant shows the **number of molecules** of that reactant which are involved in the **rate-determining step**.

So, if a reaction's second order with respect to X, there'll be two molecules of X in the rate-determining step.

For example, the mechanism for the reaction between **chlorine free radicals** and **ozone**, O_3, consists of **two steps**:

$$Cl\bullet_{(g)} + O_{3(g)} \rightarrow ClO\bullet_{(g)} + O_{2(g)} \text{ — slow (rate-determining step)}$$
$$ClO\bullet_{(g)} + O\bullet_{(g)} \rightarrow Cl\bullet_{(g)} + O_{2(g)} \text{ — fast}$$

$Cl\bullet$ and O_3 must both be in the rate equation, so the rate equation is likely to be: **rate** $= k[Cl\bullet]^m[O_3]^n$.
There's only **one** $Cl\bullet$ molecule and **one** O_3 molecule in the rate-determining step, so the **orders**, m and n, are both **1**.
So the predicted rate equation is **rate** $= k[Cl\bullet][O_3]$.

...And You Can Predict the **Mechanism** from the **Rate Equation**

Knowing exactly which reactants are in the **rate-determining step** gives you an idea of the reaction **mechanism**.

For example, the nucleophile **OH⁻** can substitute for **Br** in 2-bromo-2-methylpropane. Here are two possible mechanisms:

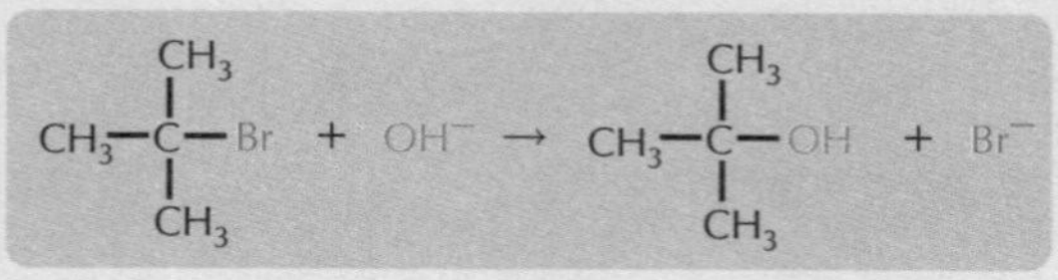

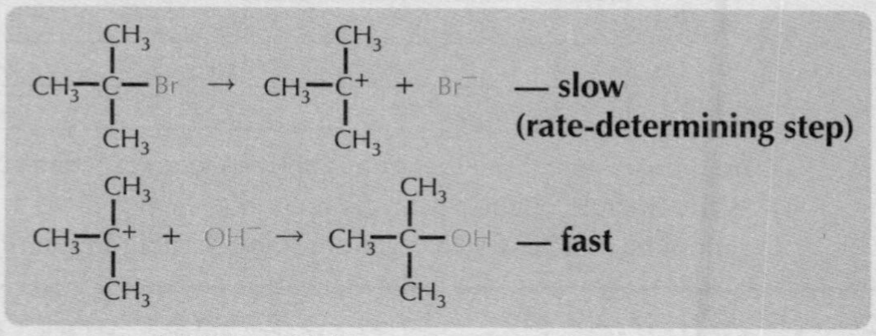

The actual **rate equation** was worked out by rate experiments: **rate** $= k[(CH_3)_3CBr]$
OH⁻ isn't in the **rate equation**, so it **can't** be involved in the rate-determining step.
The **second mechanism** is most likely to be correct because OH⁻ **isn't** in the rate-determining step.

Rates and Reaction Mechanisms

You have to Take Care when Suggesting a Mechanism

If you're suggesting a mechanism, **watch out** — things might not always be what they seem.

For example, when nitrogen(V) oxide, N_2O_5, decomposes, it forms nitrogen(IV) oxide and oxygen:

$$2N_2O_{5(g)} \rightarrow 4NO_{2(g)} + O_{2(g)}$$

From the chemical equation, it looks like **two** N_2O_5 molecules react with each other. So you might predict that the reaction is **second order** with respect to N_2O_5... but you'd be wrong.

Experimentally, it's been found that the reaction is **first order** with respect to N_2O_5 — the rate equation is: **rate = $k[N_2O_5]$**. This shows that there's only one molecule of N_2O_5 in the rate-determining step.

One **possible mechanism** that fits the rate equation is:

Only one molecule of N_2O_5 is in the rate-determining step, fitting in with the rate equation.

$$N_2O_{5(g)} \rightarrow NO_{2(g)} + NO_{3(g)} \text{ — slow (rate-determining step)}$$
$$NO_{3(g)} + N_2O_{5(g)} \rightarrow 3NO_{2(g)} + O_{2(g)} \text{ — fast}$$

The two steps add up to the overall chemical equation. You can cancel the $NO_{3(g)}$ as it appears on both sides.

Many Reactions go through a Transition State   *Edexcel only*

During a chemical reaction, bonds have to be **rearranged**. At some point in most reactions, an **energy maximum is reached**, where bonds are being broken and formed. This is called a **transition state**, since the reactants are in transition to products. Some reactions have **more than one** transition state.

The reaction between **bromomethane** and **hydroxide ions** is a **one-step reaction**. In the **transition state**, the C–Br bond is breaking and the C–OH bond is forming.

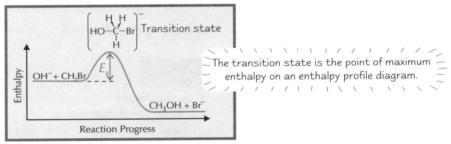

The transition state is the point of maximum enthalpy on an enthalpy profile diagram.

Practice Questions

Q1 Explain the following terms: a) rate-determining step
b) transition state

Q2 If a reaction is zero order with respect to a reactant, will that reactant be in the rate-determining step? Explain your answer.

Q3 Explain how you could use a reaction mechanism to predict the reaction's rate equation.

Exam Question

1   The following reaction is first order with respect to H_2 and first order with respect to ICl.
$$H_{2(g)} + 2ICl_{(g)} \rightarrow I_{2(g)} + 2HCl_{(g)}$$

a) Write the rate equation for this reaction.   [1 mark]

b) The mechanism for this reaction consists of two steps.
i) Identify the molecules that are in the rate-determining step. Justify your answer.   [3 marks]
ii) A chemist suggested the following mechanism for the reaction.

$$2ICl_{(g)} \rightarrow I_{2(g)} + Cl_{2(g)} \quad \text{slow}$$
$$H_{2(g)} + Cl_{2(g)} \rightarrow 2HCl_{(g)} \quad \text{fast}$$

Suggest, with reasons, whether this mechanism is likely to be correct.   [2 marks]

I found rate-determining step aerobics a bit on the slow side...

These pages show you how rate equations, orders of reaction and reaction mechanisms all tie together and how each actually means something in the grand scheme of A2 Chemistry. It's all very profound. So get it all learnt and answer the questions and then you'll have plenty of time to practise the quickstep for your Strictly Come Dancing routine.

Equilibria

There's a lot of toing and froing coming up. You won't know whether you're coming or going.

At **Equilibrium** the Amounts of Reactants and Products **Stay the Same**

1) Lots of changes are **reversible** — they can go **both ways**. To show a change is reversible, you stick in a ⇌.

2) As the **reactants** get used up, the **forward** reaction **slows down** — and as more **product** is formed, the **reverse** reaction **speeds up**. After a while, the forward reaction will be going at exactly the **same rate** as the backward reaction.

   The amounts of reactants and products **won't be changing** any more, so it'll seem like **nothing's happening**. It's a bit like you're **digging a hole** while someone else is **filling it in** at exactly the **same speed**. This is called a **dynamic equilibrium**.

3) Equilibria can be set up in **physical** systems, e.g.:

   When **liquid bromine** is shaken in a closed flask, some of it changes to orange **bromine gas**. After a while, **equilibrium** is reached — bromine liquid is **still** changing to bromine gas and bromine gas is still changing to bromine liquid, but they are changing at the **same rate**.

   $$Br_{2(l)} \rightleftharpoons Br_{2(g)}$$

   ...and **chemical** systems, e.g.:

   If **hydrogen gas** and **iodine gas** are mixed together in a closed flask, **hydrogen iodide** is formed.

   $$H_{2(g)} + I_{2(g)} \rightleftharpoons 2HI_{(g)}$$

   Imagine that **1.0 mole** of hydrogen gas is mixed with **1.0 mole** of iodine gas at a constant temperature of **600 K**. When this mixture reaches equilibrium, there will be **1.6 moles** of hydrogen iodide and **0.2 moles** of both hydrogen gas and iodine gas. No matter how long you leave them at this temperature, the **equilibrium** amounts **never change**. As with the physical system, it's all a matter of the forward and backward rates **being equal**.

4) A **dynamic equilibrium** can only happen in a **closed system** at a **constant temperature**.   *A closed system just means nothing can get in or out.*

K_c is the **Equilibrium Constant**

If you know the **molar concentration** of each substance at equilibrium, you can work out the **equilibrium constant**, K_c. Your value of K_c will only be true for that particular **temperature**.

Before you can calculate K_c, you have to write an **expression** for it. Here's how:

For the general reaction $aA + bB \rightleftharpoons dD + eE$, $K_c = \dfrac{[D]^d [E]^e}{[A]^a [B]^b}$

The products go on the top line. The square brackets, [ ], mean concentration in $moldm^{-3}$.

The lower-case letters a, b, d and e are the number of moles of each substance.

So for the reaction $H_{2(g)} + I_{2(g)} \rightleftharpoons 2HI_{(g)}$, $K_c = \dfrac{[HI]^2}{[H_2]^1 [I_2]^1}$. This simplifies to $K_c = \dfrac{[HI]^2}{[H_2][I_2]}$.

Calculate K_c by **Sticking Numbers** into the Expression

If you know the **equilibrium concentrations**, just bung them in your expression. Then with a bit of help from the old calculator, you can work out the **value** for K_c. The **units** are a bit trickier though — they **vary**, so you have to work them out after each calculation.

> **Example:** If the volume of the closed flask in the hydrogen iodide example above is 2.0 dm^3, what is the equilibrium constant for the reaction at 600 K? The equilibrium concentrations are:
>
> $[HI] = 0.8$ $moldm^{-3}$, $[H_2] = 0.1$ $moldm^{-3}$, and $[I_2] = 0.1$ $moldm^{-3}$.
>
> Just stick the concentrations into the **expression** for K_c: $K_c = \dfrac{[HI]^2}{[H_2][I_2]} = \dfrac{0.8^2}{0.1 \times 0.1} = 64$ ⟵ This is the value of K_c.
>
> To work out the **units** of K_c put the units in the expression instead of the numbers:
>
> $K_c = \dfrac{(moldm^{-3})^2}{(moldm^{-3})(moldm^{-3})} = 0$, so there are **no units** for K_c because the concentration units cancel.
>
> So K_c is just **64**.

Equilibria

You Might Need to **Work Out** the **Equilibrium Concentrations**

You might have to figure out some of the **equilibrium concentrations** before you can find K_c:

Example: 0.20 moles of phosphorus(V) chloride decomposes at 600 K in a vessel of 5.00 dm³. The equilibrium mixture is found to contain 0.08 moles of chlorine. Write the expression for K_c and calculate its value, including units.

$$PCl_{5(g)} \rightleftharpoons PCl_{3(g)} + Cl_{2(g)}$$

First find out how many moles of PCl_5 and PCl_3 there are at equilibrium:

The **equation** tells you that when **1 mole of PCl_5** decomposes, **1 mole of PCl_3** and **1 mole of Cl_2** are formed. So if 0.08 moles of chlorine are produced at equilibrium, then there will be **0.08 moles** of PCl_3 as well. 0.08 mol of PCl_5 must have decomposed, so there will be **0.12 moles** left (0.2 − 0.08).

Divide each number of moles by the volume of the flask to give the molar concentrations:

$[PCl_3] = [Cl_2] = 0.08 \div 5.00 = \textbf{0.016 moldm}^{-3}$ $\qquad [PCl_5] = 0.12 \div 5.00 = \textbf{0.024 moldm}^{-3}$

Put the concentrations in the expression for K_c and calculate it: $\quad K_c = \dfrac{[PCl_3][Cl_2]}{[PCl_5]} = \dfrac{[0.016][0.016]}{[0.024]} = \textbf{0.011}$

Now find the units of K_c: $\quad K_c = \dfrac{(moldm^{-3})(moldm^{-3})}{moldm^{-3}} = \textbf{moldm}^{-3}$ $\qquad$ So $K_c = \textbf{0.011 moldm}^{-3}$

K_c can be used to Find **Concentrations** in an **Equilibrium Mixture**

Example: When ethanoic acid was allowed to reach equilibrium with ethanol at 25 °C, it was found that the equilibrium mixture contained 2.0 moldm⁻³ ethanoic acid and 3.5 moldm⁻³ ethanol. The K_c of the equilibrium is 4.0 at 25 °C. What are the concentrations of the other components?

$$CH_3COOH_{(l)} + C_2H_5OH_{(l)} \rightleftharpoons CH_3COOC_2H_{5\,(l)} + H_2O_{(l)}$$

Put all the values you know in the K_c expression: $\quad K_c = \dfrac{[CH_3COOC_2H_5][H_2O]}{[CH_3COOH][C_2H_5OH]} \Rightarrow 4.0 = \dfrac{[CH_3COOC_2H_5][H_2O]}{2.0 \quad 3.5}$

Rearranging this gives: $\quad [CH_3COOC_2H_5][H_2O] = 4.0 \times 2.0 \times 3.5 = 28.0$

From the equation, you know that $[CH_3COOC_2H_5] = [H_2O]$, so: $\quad [CH_3COOC_2H_5] = [H_2O] = \sqrt{28} = 5.3$ moldm⁻³

The concentration of $CH_3COOC_2H_5$ and H_2O is 5.3 moldm⁻³

Practice Questions

Q1 Describe what a dynamic equilibrium is.

Q2 Write the expression for K_c for the following reaction: $2NO_{(g)} + O_{2(g)} \rightleftharpoons 2NO_{2(g)}$

Exam Questions

1 The equilibrium constant for the reaction $2HI_{(g)} \rightleftharpoons H_{2(g)} + I_{2(g)}$ is 0.0167 at 450 °C. How many moles of hydrogen iodide will be in equilibrium with 2.0 moles hydrogen and 0.3 moles iodine at 450 °C? [3 marks]

2 Nitrogen dioxide dissociates according to the equation $2NO_{2(g)} \rightleftharpoons 2NO_{(g)} + O_{2(g)}$. When 42.5 g of nitrogen dioxide were heated in a vessel of volume 22.8 dm³ at 500 °C, 14.1 g of oxygen were found in the equilibrium mixture.

a) Calculate i) the number of moles of nitrogen dioxide originally. [1 mark]
ii) the number of moles of each gas in the equilibrium mixture. [3 marks]

b) Write an expression for K_c for this reaction. Calculate the value for K_c at 450 °C and give its units. [5 marks]

A big K_c means heaps of product...

Most organic reactions and plenty of inorganic reactions are reversible. Sometimes the backwards reaction's about as speedy as a dead snail though, so some reactions might be thought of as only going one way. It's like if you're walking forwards, continental drift could be moving you backwards at the same time, just reeeeally slowly.

Gas Equilibria

Gases are different from solutions — they're more floaty and less soggy. So they've got a special equilibrium constant, K_p.

The **Total Pressure** is **Equal** to the **Sum** of the **Partial Pressures**

In a mixture of gases, each individual gas exerts its own pressure — this is called its **partial pressure**.

> The **total pressure** of a gas mixture is the **sum** of all the **partial pressures** of the individual gases.

You might have to put this fact to use in pressure calculations:

Example: When 3.0 moles of the gas PCl_5 is heated, it decomposes into PCl_3 and Cl_2. $PCl_{5(g)} \rightleftharpoons PCl_{3(g)} + Cl_{2(g)}$
In a sealed vessel at 500 K, the equilibrium mixture contains chlorine with a partial pressure of 264 kPa.
If the total pressure of the mixture is 714 kPa, what is the partial pressure of PCl_5?

> From the equation you know that PCl_3 and Cl_2 are produced in equal amounts, so the partial pressures of these two gases are the **same** at equilibrium — they're both 264 kPa.
>
> Total pressure = $p(PCl_5) + p(PCl_3) + p(Cl_2)$ ← *p is often used to mean partial pressure.*
>
> $714 = p(PCl_5) + 264 + 264$
>
> So the partial pressure of $PCl_5 = 714 - 264 - 264 =$ **186 kPa**

Partial Pressures can be Worked Out from **Mole Fractions**   *AQA and OCR only*

'**Mole fraction**' might sound a bit complicated, but it's just the **proportion** of a gas mixture that is a particular gas.
So if you've got four moles of gas in total, and two of them are gas A, the mole fraction of gas A is ½.
There are two formulas you've got to know:

> Mole fraction of a gas in a mixture = $\dfrac{\text{number of moles of gas}}{\text{total number of moles of all gases in the mixture}}$
>
> Partial pressure of a gas = mole fraction of gas × total pressure of the mixture

Example: When 3.0 mol of PCl_5 is heated in a sealed vessel as above, the equilibrium mixture contains 1.75 mol of chlorine.
If the total pressure of the mixture is 714 kPa, what is the partial pressure of PCl_5?

> PCl_3 and Cl_2 are produced in equal amounts, so there'll be **1.75 moles** of PCl_3 too.
> 1.75 moles of PCl_5 must have decomposed so **1.25 moles** of PCl_5 must be left at equilibrium (3.0 – 1.75).
> This means that the total number of moles of gas at equilibrium = 1.75 + 1.75 + 1.25 = **4.75**
>
> So the mole fraction of $PCl_5 = \dfrac{1.25}{4.75} =$ **0.263**
>
> The partial pressure of PCl_5 = mole fraction × total pressure = 0.263 × 714 = **188 kPa**

The **Equilibrium Constant** K_p is Calculated from **Partial Pressures**

The expression for K_p is just like the one for K_c — except you use partial pressures instead of concentrations.

For the equilibrium $aA_{(g)} + bB_{(g)} \rightleftharpoons dD_{(g)} + eE_{(g)}$: $K_p = \dfrac{p(D)^d\, p(E)^e}{p(A)^a\, p(B)^b}$ *There are no square brackets because they're partial pressures, not molar concentrations.*

So to **calculate K_p**, it's just a matter of sticking the partial pressures in the expression.
You have to work out the **units** each time though, just like for K_c.

Example: Calculate K_p for the decomposition of PCl_5 gas at 500 K (as shown above).
The partial pressures of each gas are: $p(PCl_5) = 186$ kPa, $p(PCl_3) = 264$ kPa, $p(Cl_2) = 264$ kPa

> $K_p = \dfrac{p(PCl_3)\,p(Cl_2)}{p(PCl_5)} = \dfrac{264 \times 264}{186} = 375$

The units for K_p are worked out by putting the units into the
expression instead of the numbers, and cancelling (like for K_c): $K_p = \dfrac{\text{kPa kPa}}{\text{kPa}} = \text{kPa}$ So, $K_p = 375$ kPa

Gas Equilibria

Here are a Few More **Example Calculations**...

You might be given the K_p and have to use it to calculate **equilibrium partial pressures**.

Example: An equilibrium exists between ethanoic acid monomers (CH_3COOH) and dimers ($(CH_3COOH)_2$).
At 160 °C the K_p for the reaction $(CH_3COOH)_{2(g)} \rightleftharpoons 2CH_3COOH_{(g)}$ is 180 kPa.
At this temperature the partial pressure of the dimer, $(CH_3COOH)_2$, is 28.5 kPa.
Calculate the partial pressure of the monomer in this equilibrium and state the total pressure exerted by the equilibrium mixture.

$$K_p = \frac{p(CH_3COOH)^2}{p((CH_3COOH)_2)}. \text{ This rearranges to give: } p(CH_3COOH)^2 = K_p \times p((CH_3COOH)_2) = 180 \times 28.5 = 5130$$

$$\Rightarrow p(CH_3COOH) = \sqrt{5130} = 71.6 \text{ kPa}$$

So the total pressure of the equilibrium mixture = 28.5 + 71.6 = **100.1 kPa** ← *Add the two partial pressures together to get the total pressure.*

Example: When dinitrogen tetroxide (N_2O_4) is heated in a sealed flask at 100 °C, an equilibrium is set up with nitrogen dioxide (NO_2) only. At this temperature K_p is 1800 kPa.
If the partial pressure of NO_2 is 220 kPa, calculate the partial pressure of N_2O_4 and state the total pressure exerted by the equilibrium mixture.

The equilibrium's just between N_2O_4 and NO_2, so the equation is $N_2O_{4(g)} \rightleftharpoons 2NO_{2(g)}$ ← *Remember — you've got to balance the equation.*

So, $K_p = \dfrac{p(NO_2)^2}{p(N_2O_4)}$. This rearranges to $p(N_2O_4) = \dfrac{p(NO_2)^2}{K_p} = \dfrac{220^2}{1800} = 26.9 \text{ kPa}$

The total pressure of the equilibrium mixture = 220 + 26.9 = **246.9 kPa**

Practice Questions

Q1 What is meant by partial pressure?

Q2 How do you work out the mole fraction of a gas?

Q3 Write the expression for K_p for the following equilibrium: $PCl_{5(g)} \rightleftharpoons PCl_{3(g)} + Cl_{2(g)}$

Exam Questions

1 At high temperatures, SO_2Cl_2 dissociates according to the equation $SO_2Cl_{2(g)} \rightleftharpoons SO_{2(g)} + Cl_{2(g)}$.
When 1.50 moles of SO_2Cl_2 dissociates at 700 K, the equilibrium mixture contains SO_2 with a partial pressure of 60.2 kPa.
The mixture has a total pressure of 141 kPa.

  a) Write an expression for K_p for this reaction. [1 mark]

  b) Calculate the partial pressure of SO_2Cl_2 and the partial pressure of Cl_2 in the equilibrium mixture. [4 marks]

  c) Calculate a value for K_p for this reaction and give its units. [3 marks]

2 When nitric oxide and oxygen were mixed in a 2:1 mole ratio, an equilibrium was set up at a
constant temperature in a sealed flask, according to the equation $2NO_{(g)} + O_{2(g)} \rightleftharpoons 2NO_{2(g)}$.
The partial pressure of the nitric oxide (NO) at equilibrium was 36 kPa and the total pressure in the flask was 99 kPa.

  a) Deduce the partial pressure of oxygen in the equilibrium mixture. [2 marks]

  b) Calculate the partial pressure of nitrogen dioxide in the equilibrium mixture. [2 marks]

  c) Write an expression for the equilibrium constant, K_p, for this reaction and calculate its value at
this temperature. State its units. [4 marks]

Baked beans unbalance your gas equilibrium...

Partial pressures are just like concentrations for gases. The more of a substance you've got in a solution, the higher the concentration, and the more of a gas you've got in a container, the higher the partial pressure. It's all to do with how many molecules you've got crashing into the sides. With gases though, you've got to keep the lid on tight or they'll escape.

More on Equilibrium Constants

Awkward things these equilibria — they never let you do what you want with them.

Le Chatelier's Principle *Predicts what will happen if* **Conditions are Changed**

If you **change** the **concentration**, **pressure** or **temperature** of a reversible reaction, you're going to **alter** the **position of equilibrium**. This just means you'll end up with **different amounts** of reactants and products at equilibrium.

If the position of equilibrium moves to the **left**, you'll get more **reactants**. $H_{2(g)} + I_{2(g)} \rightleftharpoons 2HI_{(g)}$

If the position of equilibrium moves to the **right**, you'll get more **products**. $H_{2(g)} + I_{2(g)} \rightleftharpoons \mathbf{2HI}_{(g)}$

Le Chatelier's principle tells you how the **position of equilibrium** will change if a **condition changes**:

If there's a change in **concentration**, **pressure** or **temperature**, the equilibrium will move to help **counteract** the change.

So, basically, if you **raise the temperature**, the position of equilibrium will shift to try to **cool things down**. And if you **raise the pressure or concentration**, the position of equilibrium will shift to try to **reduce it again**.

Temperature *Changes Alter* K_c *and* K_p — *Concentration and* **Pressure** *Changes Don't*

CONCENTRATION   *Not Edexcel*

The value of the **equilibrium constant**, K_c, is **fixed** at a given temperature. So if the concentration of one thing in the equilibrium mixture **changes** then the concentrations of the others must change to keep the value of K_c the same.

$$CH_3COOH_{(l)} + C_2H_5OH_{(l)} \rightleftharpoons CH_3COOC_2H_{5(l)} + H_2O_{(l)}$$

If you increase the concentration of CH_3COOH then the equilibrium will move to the right to get rid of the extra CH_3COOH — so more $CH_3COOC_2H_5$ and H_2O are produced. This keeps the equilibrium constant the same.

PRESSURE   (changing this only really affects **equilibria involving gases**) *Not Edexcel*

Increasing the pressure shifts the equilibrium to the side with the **fewest** gas molecules — this **reduces** the pressure. **Decreasing** the pressure shifts the equilibrium to the side with **most** gas molecules. This **raises** the pressure again. K_p stays the **same**, no matter what you do to the pressure.

The removal of his dummy was a change that Maxwell always opposed.

There's 3 moles on the left, but only 2 on the right. So an increase in pressure would shift the equilibrium to the right. $\Longrightarrow 2SO_{2(g)} + O_{2(g)} \rightleftharpoons 2SO_{3(g)}$

TEMPERATURE

1) If you **increase** the temperature, you **add heat**. The equilibrium shifts in the **endothermic** (positive ΔH) direction to absorb this heat.

2) **Decreasing** the temperature **removes heat**. The equilibrium shifts in the **exothermic** (negative ΔH) direction to try to replace the heat.

3) If the forward reaction's **endothermic**, the reverse reaction will be **exothermic**, and vice versa.

4) If the change means **more product** is formed, K_c and K_p will **rise**. If it means **less product** is formed, then K_c and K_p will **decrease**.

The reaction below is exothermic in the forward direction. If you increase the temperature, the equilibrium shifts to the left to absorb the extra heat. This means that less product's formed.

Exothermic $\Longrightarrow$
$2SO_{2(g)} + O_{2(g)} \rightleftharpoons 2SO_{3(g)}$ $\Delta H = -197 \text{ kJmol}^{-1}$
$\Longleftarrow$ Endothermic

$$K_p = \frac{p(SO_3)^2}{p(SO_2)^2 p(O_2)}$$

There's less product, so K_p decreases.

Catalysts have **NO EFFECT** on the position of equilibrium.
They **can't** increase yield — but they **do** mean equilibrium is approached **faster**.

A **Heterogeneous Equilibrium** *has Stuff in* **Different States**   *For Edexcel only*

Up until now, we've only worried about **homogeneous equilibria**, in which the reactants and products are all in the **same state**, e.g. all gases. In **heterogeneous equilibria**, **not** everything's in the same state.

If you're writing an expression for K_c or K_p for a **heterogeneous equilibrium**, you don't include **solids** or **liquids**.

E.g. for the **heterogeneous equilibrium** $NH_4HS_{(s)} \rightleftharpoons NH_{3(g)} + H_2S_{(g)}$, $K_p = p(NH_3)\,p(H_2S)$ There's no bottom line as the reactant is a solid.

More on Equilibrium Constants

The **Haber Process** Combines **Nitrogen** and **Hydrogen** to make **Ammonia**

$$N_{2(g)} + 3H_{2(g)} \rightleftharpoons 2NH_{3(g)} \quad \Delta H = -93 \text{ kJmol}^{-1}$$ ← The forward reaction's exothermic.

The Haber process bosses want to make **as much** ammonia as they can, as **quickly** as possible, so that they make bags of money. They've worked out the conditions that'll net them the most dough — they are:

Pressure: 200 atmospheres, **Temperature:** 450 °C, and **Catalyst:** Iron

The **Temperature** Chosen is a **Compromise**

1) Because it's an **exothermic reaction**, **lower** temperatures favour the forward reaction. This means **more** hydrogen and nitrogen is converted to ammonia — you get a better **yield**.

2) The trouble is, **lower temperatures** mean a **slower rate of reaction** — and you'd be **daft** to try to get a **really high yield** of ammonia if it's going to take you 10 years. So the 450 °C is a **compromise** between **maximum yield** and **a faster reaction**.

3) As the gases **leave** the reactor, the **temperature** is **reduced** so that the **ammonia** can be **liquefied** and **removed**. The **hydrogen** and **nitrogen** which didn't react are **recycled**. Thanks to this recycling, a very respectable **98%** of these gases ends up being converted to ammonia.

High Pressure would give a **Big Yield** — but it'd be **Expensive**

1) **Higher pressures** favour the **forward reaction**, hence the **200 atmospheres** operating pressure. This is because the equilibrium moves to the side with **fewer molecules**. There are **four molecules** of gas on the reactant side ($N_{2(g)} + 3H_{2(g)}$) to every **two molecules** on the product side ($2NH_{3(g)}$).

2) **Increasing** the **pressure** also **increases** the **rate** of reaction.

3) Cranking up the pressure as high as you can sounds like a great idea so far. **But** very **high pressures** are really **expensive** to produce. You also need **strong pipes** and **containers** to **withstand** the **high pressure**. So **200 atmospheres** is a **compromise**. In the end, it all comes down to **minimising costs**.

The iron catalyst doesn't affect how much product is made — it does makes the reaction reach equilibrium much more quickly though.

Carbon Dioxide Dissolves in the Oceans in **Equilibrium Reactions** | *Salters only*

Here's how **carbon dioxide** in the air gets dissolved in water:

$$CO_{2(g)} \rightleftharpoons CO_{2(aq)} \longrightarrow CO_{2(aq)} + H_2O_{(l)} \rightleftharpoons H^+_{(aq)} + HCO_3^-{}_{(aq)} \longrightarrow HCO_3^-{}_{(aq)} \rightleftharpoons H^+_{(aq)} + CO_3^{2-}{}_{(aq)}$$

Overall, the reaction is: $CO_{2(g)} + H_2O_{(l)} \rightleftharpoons 2H^+_{(aq)} + CO_3^{2-}{}_{(aq)}$

If the amount of CO_2 in the air **increased**, you'd expect the equilibrium to move to the **right** — this'd mean more H^+ ions in the oceans, so they'd be more **acidic**. Luckily, the oceans **don't** become more acidic because the carbonate ions **precipitate** out as **limestone** ($CaCO_3$). This pushes the equilibrium back to the **left**.

Practice Questions

Q1 State Le Chatelier's principle.

Q2 What happens to the position of an equilibrium and the value of K_c or K_p when the temperature is decreased?

Exam Question

1 The following equilibrium was established at temperature T_1: $2SO_{2(g)} + O_{2(g)} \rightleftharpoons 2SO_{3(g)}$ $\Delta H = -196$ kJmol^{-1}. K_p at T_1 was found to be 0.67 kPa^{-1}.

a) When equilibrium was established at a different temperature, T_2, the value of K_p was found to have increased. State which of T_1 or T_2 is the lower temperature and explain why. [3 marks]

b) The experiment was repeated exactly the same in all respects at T_1, except a flask of smaller volume was used. How would this change affect the yield of sulphur trioxide and the value of K_p? [2 marks]

If the sea became too acidic it'd pickle all the fish...

...and your swimsuit'd slowly dissolve if you went swimming. So it'd be a bit of a disaster really. The reason why the carbonate ions precipitate out is to do with the solubility product — which is another constant that you're going to meet on page 26. I bet you can't wait — hmmm... perhaps swimming in a sea of acid is looking more inviting by the second.

More on Equilibrium Constants

If you want to saturate a solution, you could just keep chucking stuff in and wait till it stops dissolving.
Or you could do some nifty calculations and get it spot on. **All the stuff on these two pages is just for** *Salters.*

Solubility Product *is Another* Equilibrium Constant

When a **sparingly** soluble ionic solid dissolves in water, an equilibrium is set up between the **ions in solution** and the **undissolved solid**. Silver chloride's a good example:

Sparingly soluble means a tiny bit soluble.

$$AgCl_{(s)} \rightleftharpoons Ag^+_{(aq)} + Cl^-_{(aq)} \implies K = \frac{\left[Ag^+_{(aq)}\right]\left[Cl^-_{(aq)}\right]}{\left[AgCl_{(s)}\right]}$$

Silver chloride dissolves by such a **tiny amount** that its concentration is considered to be **constant**, so **K × [AgCl$_{(s)}$] = another constant.**

If you multiply two constants together, you always get a third constant.

This new constant is known as the **solubility product**, **K$_{sp}$**. So **K$_{sp}$ = [Ag$^+_{(aq)}$][Cl$^-_{(aq)}$]**.

K$_{sp}$ gives the **product** of the **maximum concentrations** of ions possible in a **saturated solution**. The bigger the value of K$_{sp}$, the **more soluble** the solid is. As with K$_c$ and K$_p$, the temperature must be given for K$_{sp}$, as it **varies** with temperature.

Solubility products let you work out whether a solution's **saturated or not**, and how much solid you have to add for a **precipitate** to form. Here are the rules for deciding this:

1) If [Ag$^+$][Cl$^-$] is **less** than K$_{sp}$(AgCl), the solution is **not saturated** and more AgCl can dissolve.
2) If [Ag$^+$][Cl$^-$] is **equal to** K$_{sp}$(AgCl), the solution is **saturated** — no more AgCl will dissolve.
3) If you make [Ag$^+$][Cl$^-$] **greater** than K$_{sp}$(AgCl), then AgCl will precipitate.

Here are a Few Example Calculations...

For these examples, you need to know that the solubility product of silver chloride is 1.0×10^{-10} mol^{2}dm^{-6} at 18 °C.

$$AgCl_{(s)} \rightleftharpoons Ag^+_{(aq)} + Cl^-_{(aq)}$$

Example: How many grams of silver chloride will dissolve in 1 dm^3 of water at 18°C?

From the equation you know that when 1 mole of AgCl dissolves, it makes 1 mole of Ag$^+$ ions and 1 mole of Cl$^-$ ions, so [Ag$^+$] = [Cl$^-$]. You can use this fact to find the value of [Ag$^+$] or [Cl$^-$]:

$$K_{sp}(AgCl) = [Ag^+][Cl^-] = 1.0 \times 10^{-10} \text{ mol}^2\text{dm}^{-6}$$
$$\text{So } [Ag^+]^2 = 1.0 \times 10^{-10} \implies [Ag^+] = \sqrt{1.0 \times 10^{-10}} = 1.0 \times 10^{-5} \text{ moldm}^{-3}$$

If there are 1.00×10^{-5} moles of Ag$^+$ ions in 1 dm^3 of water, **1.0×10^{-5} moles** of AgCl must have dissolved. The question asked for the mass though:

mass = number of moles × M$_r$ M$_r$ of AgCl = 108 + 35.5 = 143.5.
So mass = $(1.0 \times 10^{-5}) \times 143.5 = 1.435 \times 10^{-3}$ g

So, at 18 °C, **1.4×10^{-5} g** of AgCl will dissolve in a dm^3 of water.

Example: Will silver chloride precipitate when equal volumes of 0.004 moldm^{-3} silver nitrate and 0.005 moldm^{-3} sodium chloride are mixed at 18 °C?

If you mix equal volumes the total volume doubles.

Initially, there's 0.004 moles of silver ions in 1 dm^3. Adding another 1 dm^3 of solution means there'll only be 0.002 moles of silver ions per dm^3 — so the concentration halves if the volume is doubled.
The concentration of the chloride ions halves too — it changes from 0.005 moldm^{-3} to 0.0025 moldm^{-3}.

So when the solutions are mixed, [Ag$^+$] = 0.002 moldm^{-3} and [Cl$^-$] = 0.0025 moldm^{-3}.
[Ag$^+$][Cl$^-$] = 0.002 × 0.0025 = 5.00×10^{-6} mol^{2}dm^{-6}. This value's greater than K$_{sp}$(AgCl) at 18 °C, **so AgCl will precipitate**.

Silver Chloride *is* Less Soluble *in* Hydrochloric Acid *than in* Water

How many moles of silver chloride will dissolve in each dm^3 of 0.01 moldm^{-3} hydrochloric acid at 18 °C?

You can assume that all the chloride ions are from the hydrochloric acid because the number of chloride ions from silver chloride is negligible compared to this. So [Cl$^-$] = 0.01 moldm^{-3}.

HCl is a strong acid, so every mole of HCl releases a mole of Cl$^-$.

$$K_{sp}(AgCl) = 1.00 \times 10^{-10} = [Ag^+][Cl^-] \implies 1.00 \times 10^{-10} = [Ag^+] \times 0.01 \implies [Ag^+] = \frac{1.00 \times 10^{-10}}{0.01} = 1.00 \times 10^{-8} \text{ moldm}^{-3}$$

If there's 1.00×10^{-8} moles of silver ions per dm^3, **1.00×10^{-8} moles** of AgCl must dissolve per dm^3 at 18 °C.

More on Equilibrium Constants

Solutes can Dissolve in **More Than One** Solvent

Imagine you've got a mixture of **two immiscible solvents** like water and tetrachloromethane (CCl_4), and you shake them up with substance X which dissolves in **both** solvents.

Immiscible liquids won't mix together.

After a while, the concentration of X in the water and the concentration of X in the CCl_4 will remain **constant**, no matter how much more shaking is done — **equilibrium** has been reached. $\implies X_{(aq)} \rightleftharpoons X_{(CCl_4)}$

The **ratio** of the solute concentrations in the two solvents is always the same, so long as the **temperature** stays the same. So, if you **divide** one concentration by the other, you get a **constant**. This is yet another **equilibrium constant**, K, which is known as the **partition coefficient**.

$$K = \frac{[X_{(CCl_4)}]}{[X_{(aq)}]}$$

The partition coefficient has no units because the concentration units cancel out.

Example: The partition coefficient of iodine between tetrachloromethane and water is 88.0 at 20 °C. At equilibrium the concentration of iodine in tetrachloromethane is 13.2 gdm^{-3}. The expression for the partition coefficient is given on the right. What is the concentration of iodine in water? $K = \frac{[I_{2(CCl_4)}]}{[I_{2(aq)}]}$

$$K = \frac{[I_{2(CCl_4)}]}{[I_{2(aq)}]} \implies 88.0 = \frac{13.2}{[I_{2(aq)}]} \implies [I_{2(aq)}] = \frac{13.2}{88.0} = \textbf{0.15 gdm}^{-3}$$

The Idea of **Partition** is Used in **Solvent Extraction** and **Pest Control**

Organic compounds are generally heaps **more soluble** in **organic solvents** than in water. To separate an organic compound from an **aqueous solution**, you add an organic solvent that is immiscible with water, and shake the mixture. Almost all the compound will dissolve in the organic solvent. **Two layers** will be formed — these can be separated in a **separating funnel**. Then you can distil off the organic solvent to leave the pure compound.

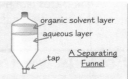
organic solvent layer
aqueous layer
tap — A Separating Funnel

Pests eat crops, which really annoys farmers. So chemists create **pesticides** that kill off the little blighters. They take compounds that they know are harmful to pests and try to make them work even better by altering their structure. For a pesticide to work well, the **partition coefficient** has to be large so that **enough** will dissolve in the fatty layers of the pest to kill it, even though the pesticide's spread as a very **dilute aqueous solution**.

Chemists don't just have to worry about how good a pesticide is at killing creepy crawlies — they have to consider how it affects the **environment** too. For example, they have to think about whether it's **biodegradable**, or if it'll leach into **water supplies**, or if it'll build up in the **food chain**... They don't want to end up harming things other than the pests.

Practice Questions

Q1 Write an expression for the solubility product of calcium sulphate ($CaSO_4$).

Q2 Explain how to calculate the partition coefficient for a substance that is dissolved between two solvents.

Q3 Why is the idea of partition important when developing pesticides?

Exam Questions

1   The solubility product of lead(II) sulphate at 20 °C is 2.5×10^{-8} mol^{2}dm^{-6}.
    a) Write an expression for the solubility product of lead(II) sulphate ($PbSO_4$). [1 mark]
    b) Calculate the solubility (in gdm^{-3}) of lead(II) sulphate in water at 20 °C. [4 marks]
    c) Calculate the solubility (in gdm^{-3}) of lead(II) sulphate in 0.15 moldm^{-3} H_2SO_4 at 20 °C. [3 marks]

2   The solubility product of calcium sulphate at 25 °C is 2.0×10^{-5} mol^{2}dm^{-6}.
    Will a precipitate of calcium sulphate form when 50 cm^3 of 0.01 moldm^{-3} calcium chloride ($CaCl_2$)
    and 50 cm^3 0.006 moldm^{-3} sodium sulphate (Na_2SO_4) are mixed at 25 °C? [3 marks]

3   The partition coefficient of iodine between tetrachloromethane and water is 85.5 at 25 °C.
    What mass of iodine would remain in the water if 1.00 g of iodine were shaken with
    50.0 cm^3 of water and 50.0 cm^3 of tetrachloromethane? $K = \frac{[I_{2(CCl_4)}]}{[I_{2(aq)}]}$
    Assume that saturation point is not reached for either solvent. [3 marks]

You must be reaching saturation point by now...

Silver chloride's less soluble in hydrochloric acid than it is in water because there are already loads of chloride ions in hydrochloric acid — these push the equilibrium $AgCl_{(s)} \rightleftharpoons Ag^+_{(aq)} + Cl^-_{(aq)}$ back over to the left. In the same way, magnesium sulphate's less soluble in sulphuric acid than it is in water. This is called the common ion effect. So there you go.

Acids and Bases

Remember this stuff? Well, it's all down to Brønsted and Lowry — they've got a lot to answer for.

An Acid **Releases** Protons — a Base **Accepts** Protons

Brønsted-Lowry acids are **proton donors** — they release **hydrogen ions** (H^+) when they're mixed with water. You never get H^+ ions by themselves in water though — they're always combined with H_2O to form **hydroxonium ions, H_3O^+**.

HA is just any old acid. ⟹ $HA_{(aq)} + H_2O_{(l)} \rightarrow H_3O^+_{(aq)} + A^-_{(aq)}$

Brønsted-Lowry bases do the opposite — they're **proton acceptors**. When they're in solution, they grab **hydrogen ions** from water molecules.

B is just a random base. ⟹ $B_{(aq)} + H_2O_{(l)} \rightarrow BH^+_{(aq)} + OH^-_{(aq)}$

Acids and Bases can be **Strong** or **Weak**

These are really all reversible reactions, but the equilibrium lies extremely far to the right.

1) **Strong acids ionise almost completely** in water — **nearly all** the H^+ ions will be released. **Hydrochloric acid** is a strong acid — $HCl_{(g)} + water \rightarrow H^+_{(aq)} + Cl^-_{(aq)}$.
   Strong bases (like sodium hydroxide) **ionise almost completely** in water too. E.g. $NaOH_{(s)} + water \rightarrow Na^+_{(aq)} + OH^-_{(aq)}$.

2) **Weak acids** (e.g. ethanoic or citric) ionise only very **slightly** in water — so only small numbers of H^+ ions are formed. An **equilibrium** is set up which lies well over to the **left**. E.g. $CH_3COOH_{(aq)} \rightleftharpoons CH_3COO^-_{(aq)} + H^+_{(aq)}$.
   Weak bases (such as ammonia) **only slightly ionise** in water too. E.g. $NH_{3(aq)} + H_2O_{(l)} \rightleftharpoons NH_4^+_{(aq)} + OH^-_{(aq)}$.
   Just like with weak acids, the equilibrium lies well over to the **left**.

Protons are **Transferred** when **Acids** and **Bases** React

Acids **can't** just throw away their protons — they can only get rid of them if there's a **base** to accept them. In this reaction the **acid**, HA, **transfers** a proton to the **base**, B: $HA_{(aq)} + B_{(aq)} \rightleftharpoons BH^+_{(aq)} + A^-_{(aq)}$

It's an **equilibrium**, so if you add more **HA** or **B**, the position of equilibrium moves to the **right**. But if you add more **BH⁺** or **A⁻**, the equilibrium will move to the **left**. This is all down to **Le Chatelier's principle** — see page 24.

When an acid is added to **water**, water acts as the **base** and accepts the proton:

$HA_{(aq)} + H_2O_{(l)} \rightleftharpoons H_3O^+_{(aq)} + A^-_{(aq)}$ ← *The equilibrium's far to the left for weak acids, and far to the right for strong acids.*

Acids and Bases form **Conjugate Pairs**   *Not AQA or Salters*

So when an acid's added to water, the equilibrium shown on the right is set up.
In the **forward reaction**, HA acts as an **acid** as it **donates** a proton. In the **reverse reaction**, A⁻ acts as a **base** and **accepts** a proton from the H_3O^+ ion to form HA.
HA and A⁻ are called a **conjugate pair** — HA is the **conjugate acid** of A⁻ and A⁻ is the **conjugate base** of the acid, HA. H_2O and H_3O^+ are a conjugate pair too.

conjugate pair
$HA + H_2O \rightleftharpoons H_3O^+ + A^-$
acid   base   acid   base
conjugate pair

The acid and base of a conjugate pair can be linked by an H^+, like this: $HA \rightleftharpoons H^+ + A^-$ or this: $H^+ + H_2O \rightleftharpoons H_3O^+$

Here's the equilibrium for aqueous HCl.
Cl⁻ is the conjugate base of $HCl_{(aq)}$.

conjugate pair
$HCl_{(aq)} + H_2O_{(l)} \rightleftharpoons H_3O^+_{(aq)} + Cl^-_{(aq)}$
acid   base   acid   base
conjugate pair

An equilibrium with **conjugate pairs** is also set up when a **base** dissolves in water.

The base B takes a proton from the water to form **BH⁺** — so B is the **conjugate base** of BH⁺, and BH⁺ is the **conjugate acid** of B. H_2O and OH⁻ also form a **conjugate pair**.

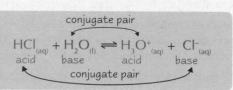

conjugate pair
$B + H_2O \rightleftharpoons BH^+ + OH^-$
base   acid   acid   base
conjugate pair

Acids and Bases

Water can Behave as an Acid AND a Base

Water can act as an **acid** by **donating** a proton — but it can also act as a **base** by accepting a **proton**.
So in water there'll always be both **hydroxonium ions** and **hydroxide ions** swimming around at the **same time**.

> The equilibrium below exists in water:
>
> $$H_2O_{(l)} + H_2O_{(l)} \rightleftharpoons H_3O^+_{(aq)} + OH^-_{(aq)}$$ or more simply $$H_2O_{(l)} \rightleftharpoons H^+_{(aq)} + OH^-_{(aq)}$$
>
> And, just like for any other equilibrium reaction, you can apply the equilibrium law and write an expression for the **equilibrium constant**: $$K_c = \frac{[H^+][OH^-]}{[H_2O]}$$

Water only dissociates a **tiny amount**, so the equilibrium lies well over to the **left**. There's so much water compared to the amounts of H^+ and OH^- ions that the concentration of water is considered to have a **constant** value.

So if you multiply K_c (a constant) by $[H_2O]$ (another constant), you get a **constant**. This new constant is called the **ionic product of water** and it is given the symbol K_w.

$$K_w = K_c \times [H_2O] = [H^+][OH^-] \Rightarrow \boxed{K_w = [H^+][OH^-]}$$

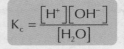 The units of K_w are always mol^2dm^{-6}.

K_w always has the **same value** for an aqueous solution at a **given temperature**.
It's important that you know its value at standard temperature, **25 °C**:   At 25 °C (298 K), $K_w = 1.0 \times 10^{-14}\ mol^2dm^{-6}$

You **Edexcel** folks need to know that K_w can also be given as pK_w, where $pK_w = -\log_{10}K_w$.
The advantage of pK_w values is that they're a decent size so they're easy to work with —
for example, the pK_w of water at 25 °C is **14** ($-\log_{10}(1.0 \times 10^{-14})$).

A Neutral Solution has Equal H⁺ and OH⁻ Concentrations

> A neutral solution is one in which $[H^+] = [OH^-]$.

If $[H^+]$ is greater than $[OH^-]$ the solution is **acidic**, and if $[OH^-]$ is greater than $[H^+]$ the solution is **alkaline**.

Neutralisation is an Exothermic Reaction   *Edexcel only*

When an acid and a base neutralise each other, energy is given out — this is the **enthalpy change of neutralisation**.
If a **strong** acid is neutralised by a **strong** base, the enthalpy change of neutralisation is about 57 kJmol⁻¹ (at 298 K).
But if either the acid or base is **weak**, then the enthalpy change is **lower** — the weaker the acid or base, the lower the enthalpy change. If a **weak** acid neutralises a **weak** base, then the enthalpy change will be **very low**.

Practice Questions

Q1 What's the difference between a strong acid and a weak acid?

Q2 What is the conjugate base of $HCl_{(aq)}$?

Q3 What does K_w mean and what is its value at 298 K?

Exam Questions

1   a)   How did Brønsted and Lowry define: (i) an acid,   (ii) a base.   [2 marks]
    b)   Show, by writing appropriate equations, how HSO_4^- can behave as:
         (i) a Brønsted-Lowry acid,   (ii) a Brønsted-Lowry base.   [2 marks]

2   Hydrocyanic acid (HCN) is a weak acid. Define the term 'weak acid' and write the equation for the equilibrium that occurs when HCN dissolves in water.   [4 marks]

Acids and bases — the Julie Andrews and Marilyn Manson of the chemistry world...

Don't confuse strong acids with concentrated acids, or weak acids with dilute acids. Strong and weak are to do with how much an acid ionises, whereas concentrated and dilute are to do with the number of moles of acid you've got per dm³. You can have a strong dilute acid, or a weak concentrated acid. It works the same way with bases too.

pH Calculations

Get those calculators warmed up — especially the log function key.

The **pH Scale** is a Measure of the **Hydrogen Ion Concentration**

The **concentration of hydrogen ions** can vary enormously, so those wise chemists of old decided to express the concentration on a **logarithmic scale.**

$$pH = -\log_{10}[H^+]$$

The pH scale normally goes from **0** (very acidic) to **14** (very alkaline). **pH 7** is regarded as being **neutral**.

For Strong Monoprotic Acids, **Hydrogen Ion Concentration = Acid Concentration**

Hydrochloric acid and nitric acid ($HNO_{3(aq)}$) are **strong acids** so they ionise fully. They're also **monoprotic**, so each mole of acid produces **one mole of hydrogen ions**. This means the H^+ concentration is the **same** as the acid concentration.

So for 0.1 moldm^{-3} hydrochloric acid, $[H^+]$ is **0.1 moldm^{-3}**. Its pH = $-\log_{10}[H^+]$ = $-\log_{10} 0.1$ = **1.0**.

Here are a few more examples for you:

1) Calculate the pH of 0.05 moldm^{-3} nitric acid. $[H^+] = 0.05 \Rightarrow pH = -\log_{10} 0.05 = \mathbf{1.30}$

2) Calculate the pH of 0.025 moldm^{-3} hydrochloric acid. $[H^+] = 0.025 \Rightarrow pH = -\log_{10} 0.025 = \mathbf{1.60}$

You also need to be able to work out $[H^+]$ if you're given the **pH** of a solution.
You do this by finding the **inverse log of –pH**, which is $\mathbf{10^{-pH}}$.

3) If an acid solution has a pH of 2.45, what is the hydrogen ion concentration, or $[H^+]$, of the acid?

$$[H^+] = 10^{-2.45} = \mathbf{3.55 \times 10^{-3}} \textbf{ moldm}^{-3}$$

Use K_w to Find the **pH** of a **Base**

Sodium hydroxide (NaOH) and potassium hydroxide (KOH) are **strong bases** that **fully ionise** in water.
They each have **one hydroxide per molecule**, so they donate **one mole of OH$^-$ ions** per mole of base.
This means that the concentration of OH$^-$ ions is the **same** as the **concentration of the base**.
So for 0.02 moldm^{-3} sodium hydroxide solution, $[OH^-]$ is also **0.02 moldm^{-3}**.

But to work out the **pH** you need to know $[H^+]$ — luckily this is linked to $[OH^-]$ through the **ionic product of water**, K_w:

$$K_w = [H^+][OH^-] = \mathbf{1.0 \times 10^{-14}} \text{ at 298 K}$$

So if you know K_w and $[OH^-]$ for a **strong aqueous base** at a certain temperature, you can work out $[H^+]$ and then the **pH**.

4) Find the pH of 0.1 moldm^{-3} NaOH at 298 K. $[OH^-] = 0.1$ moldm$^{-3} \Rightarrow [H^+] = \dfrac{K_w}{[OH^-]} = \dfrac{1.0 \times 10^{-14}}{0.1} = 1.0 \times 10^{-13}$ moldm^{-3}

So pH = $-\log_{10} 1.0 \times 10^{-13}$ = **13.0**

To Find the **pH** of a **Weak Acid** you Use K_a (the **Acid Dissociation Constant**)

Weak acids **don't** ionise fully in solution, so the $[H^+]$ **isn't** the same as the acid concentration.
This makes it a **bit trickier** to find their pH. You have to use yet another **equilibrium constant**, K_a.

For a weak aqueous acid, HA, you get the following equilibrium: $HA_{(aq)} \rightleftharpoons H^+_{(aq)} + A^-_{(aq)}$

As only a **tiny amount** of HA dissociates, you can assume that $[HA_{(aq)}]_{start} = [HA_{(aq)}]_{equilibrium}$.

So if you apply the equilibrium law, you get: $K_a = \dfrac{[H^+][A^-]}{[HA]}$

You can also assume that **all** the **H$^+$ ions** come from the **acid**, so $[H^+_{(aq)}] = [A^-_{(aq)}]$. So $K_a = \dfrac{[H^+]^2}{[HA]}$ ← The units of K_a are moldm^{-3}.

Here's an example of how to use K_a to find the **pH** of a weak acid:

5) Calculate the hydrogen ion concentration and the pH of a 0.02 moldm^{-3} solution of propanoic acid (CH_3CH_2COOH).
K_a for propanoic acid at this temperature is 1.30×10^{-5} moldm^{-3}.

$$K_a = \frac{[H^+]^2}{[CH_3CH_2COOH]} \Rightarrow [H^+]^2 = K_a[CH_3CH_2COOH] = 1.30 \times 10^{-5} \times 0.02 = 2.60 \times 10^{-7}$$

$$\Rightarrow [H^+] = \sqrt{2.60 \times 10^{-7}} = \mathbf{5.10 \times 10^{-4}} \textbf{ moldm}^{-3} \quad \text{So pH} = -\log_{10} 5.10 \times 10^{-4} = \mathbf{3.29}$$

pH Calculations

You Might Have to Find the Concentration or K_a of a Weak Acid

You don't need to know anything new for this type of calculation. You usually just have to find [H^+] from the pH, then fiddle around with the K_a expression to find the missing bit of information.

1) The pH of an ethanoic acid (CH_3COOH) solution was 3.02 at 298 K. Calculate the molar concentration of this solution. The K_a of ethanoic acid is 1.75×10^{-5} moldm^{-3} at 298 K.

$$[H^+] = 10^{-pH} = 10^{-3.02} = 9.55 \times 10^{-4} \text{ moldm}^{-3}$$

$$K_a = \frac{[H^+]^2}{[CH_3COOH]} \Rightarrow [CH_3COOH] = \frac{[H^+]^2}{K_a} = \frac{(9.55 \times 10^{-4})^2}{1.75 \times 10^{-5}} = 0.0521 \text{ moldm}^{-3}$$

2) A solution of 0.200 moldm^{-3} HCN has a pH of 5.05 at 298 K. What is the value of K_a for HCN at 298 K?

$$[H^+] = 10^{-pH} = 10^{-5.05} = 8.91 \times 10^{-6} \text{ moldm}^{-3} \qquad K_a = \frac{[H^+]^2}{[HCN]} = \frac{(8.91 \times 10^{-6})^2}{0.200} = 3.97 \times 10^{-10} \text{ moldm}^{-3}$$

$pK_a = -\log_{10} K_a$ and $K_a = 10^{-pK_a}$ NOT for *Nuffield*

pK_a is calculated from K_a in exactly the same way as pH is calculated from [H^+] — and vice versa. So if an acid has a K_a value of 1.50×10^{-7}, its $pK_a = -\log_{10}(1.50 \times 10^{-7}) = 6.82$. And if an acid has a pK_a value of 4.32, its $K_a = 10^{-4.32} = 4.79 \times 10^{-5}$.

Notice how pK_a values aren't annoyingly tiny like K_a values.

Just to make things that bit more complicated, there might be a pK_a value in a question. If so, you need to convert it to K_a so that you can use the K_a expression.

3) Calculate the pH of 0.050 moldm^{-3} methanoic acid (HCOOH). Methanoic acid has a pK_a of 3.75 at this temperature.

$$K_a = 10^{-pK_a} = 10^{-3.75} = 1.78 \times 10^{-4} \text{ moldm}^{-3} \quad \longleftarrow \text{ First you have to convert the } pK_a \text{ to } K_a.$$

$$K_a = \frac{[H^+]^2}{[HCOOH]} \Rightarrow [H^+]^2 = K_a[HCOOH] = 1.78 \times 10^{-4} \times 0.050 = 8.9 \times 10^{-6}$$

$$\Rightarrow [H^+] = \sqrt{8.9 \times 10^{-6}} = 2.98 \times 10^{-3} \text{ moldm}^{-3} \qquad pH = -\log 2.98 \times 10^{-3} = 2.53$$

Sometimes you have to give your answer as a pK_a value. In this case, you just work out the K_a value as usual and then convert it to pK_a — and Bob's your pet hamster.

Practice Questions

Q1 Explain how to find the pH of a strong acid.

Q2 How do you find the pH of a strong base?

Q3 Explain how to find the pH of a weak acid.

Exam Questions

1 The value of K_a for the weak acid HA, at 298 K, is 5.60×10^{-4} moldm^{-3}.
   a) Write an expression for K_a for the weak acid HA. [1 mark]
   b) Calculate the pH of a 0.280 moldm^{-3} solution of HA at 298 K. [3 marks]

2 The pH of a 0.150 moldm^{-3} solution of a weak monoprotic acid, HX, is 2.65 at 298 K. Calculate the value of K_a for the acid HX at 298 K. [4 marks]

3 a) Write an expression for the ionic product of water, K_w, and give its value at 298 K. [2 marks]
   b) Hence calculate the pH of a 0.0370 moldm^{-3} solution of sodium hydroxide at 298 K. [3 marks]

My mate had a red Ka — but she drove it into a lamppost and it fell apart...

Strong acids have high K_a values and weak acids have low K_a values. For pK_a values, it's the other way round — the stronger the acid, the lower the pK_a. If something's got p in front of it, like pH, pK_w or pK_a, it tends to mean $-\log_{10}$ of whatever. Not all calculators work the same way, so make sure you know how to work logs out on your calculator.

Titrations and pH Curves

If you add alkali to an acid, the pH changes in a squiggly sort of way. **Salters can skip these two pages.**

Use **Titration** to Find the **Concentration** of an **Acid** or **Alkali**

Titrations allow you to find out **exactly** how much alkali is needed to **neutralise** a quantity of acid.

1) You measure out some **acid** of known concentration using a pipette and put it in a flask, along with some **appropriate indicator** (see below).

2) First do a rough titration — add the **alkali** to the acid using a **burette** fairly quickly to get an approximate idea where the solution changes colour (the **end point**). Give the flask a regular **swirl**.

3) Now do an **accurate** titration. Run the alkali in to within 2 cm³ of the end point, then add it **drop by drop**. If you don't notice exactly when the solution changes colour you've **overshot** and your result won't be accurate.

4) **Record** the amount of alkali needed to **neutralise** the acid. It's best to **repeat** this process a few times, making sure you get very similar answers each time (within about 0.2 cm³ of each other).

You can also find out how much **acid** is needed to neutralise a quantity of **alkali**. It's exactly the same as above, but you add **acid to alkali** instead.

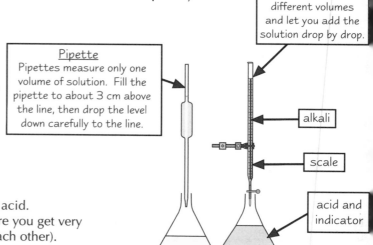

Pipette
Pipettes measure only one volume of solution. Fill the pipette to about 3 cm above the line, then drop the level down carefully to the line.

Burette
Burettes measure different volumes and let you add the solution drop by drop.

alkali

scale

acid and indicator

pH Curves Plot pH Against Volume of Acid or Alkali Added

The graphs below show the pH curves for the **different combinations** of **strong and weak** monoprotic acids and alkalis.

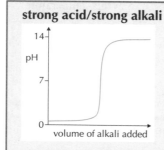

strong acid/strong alkali

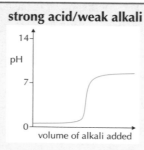

strong acid/weak alkali

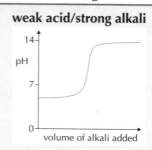

weak acid/strong alkali

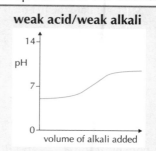

weak acid/weak alkali

All the graphs apart from the weak acid/weak alkali graph have a bit that's almost vertical — this is the **equivalence point** or **end point**. At this point, a tiny amount of alkali causes a sudden, big change in pH — it's here that all the acid is just **neutralised**.

You don't get such a sharp change in a **weak acid/weak alkali** titration. The indicator colour changes **gradually** and it's tricky to see the exact end point. You're usually better off using a **pH meter** for this type of titration.

pH Curves can Help you Decide which Indicator to Use

Methyl orange and **phenolphthalein** are **indicators** that are often used for acid-base titrations. They each change colour over a **different pH range**:

Name of indicator	Colour at low pH	Approx. pH of colour change	Colour at high pH
Methyl orange	red	3.1 – 4.4	yellow
Phenolphthalein	colourless	8.3 – 10	pink

For a **strong acid/strong alkali** titration, you can use **either** of these indicators — there's a rapid pH change over the range for **both** indicators.

For a **strong acid/weak alkali** only **methyl orange** will do. The pH changes rapidly across the range for methyl orange, but not for phenolphthalein.

For a **weak acid/strong alkali**, **phenolphthalein** is the stuff to use. The pH changes rapidly over phenolphthalein's range, but not over methyl orange's.

For **weak acid/weak alkali** titrations there's no sharp pH change, so **neither** of these indicators will work.

Titrations and pH Curves

The pH Curve for a Diprotic Acid has Two Equivalence Points
AQA only

A **diprotic acid** is one that can release **two protons** when it's in solution. **Ethanedioic acid** (HOOC-COOH) is diprotic. When **ethanedioic acid** is titrated with a **strong base**, such as sodium hydroxide, you get a pH curve with two separate near-vertical bits. These are regions where the pH increases rapidly — and the midpoints of these regions are the **equivalence points**:

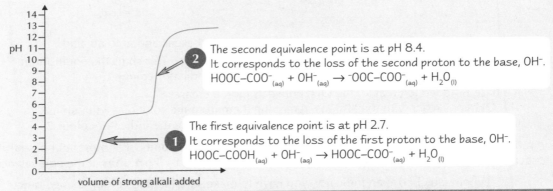

The second equivalence point is at pH 8.4. It corresponds to the loss of the second proton to the base, OH⁻.
$$HOOC-COO^-_{(aq)} + OH^-_{(aq)} \rightarrow {}^-OOC-COO^-_{(aq)} + H_2O_{(l)}$$

The first equivalence point is at pH 2.7. It corresponds to the loss of the first proton to the base, OH⁻.
$$HOOC-COOH_{(aq)} + OH^-_{(aq)} \rightarrow HOOC-COO^-_{(aq)} + H_2O_{(l)}$$

volume of strong alkali added

The pH Curve for NaCO₃²⁻ with HCl has Two Equivalence Points Too
AQA only

If you titrate **sodium carbonate** with a **strong monoprotic acid**, such as hydrochloric acid, you get another pH curve with **two equivalence points**. This is because the carbonate ion (CO_3^{2-}) can accept **two protons** from the acid.

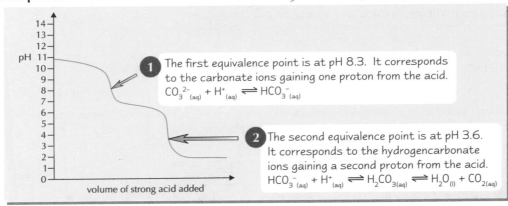

The first equivalence point is at pH 8.3. It corresponds to the carbonate ions gaining one proton from the acid.
$$CO_3^{2-}_{(aq)} + H^+_{(aq)} \rightleftharpoons HCO_3^-_{(aq)}$$

The second equivalence point is at pH 3.6. It corresponds to the hydrogencarbonate ions gaining a second proton from the acid.
$$HCO_3^-_{(aq)} + H^+_{(aq)} \rightleftharpoons H_2CO_{3(aq)} \rightleftharpoons H_2O_{(l)} + CO_{2(aq)}$$

volume of strong acid added

Practice Questions

Q1 Sketch the pH curve for a weak acid/strong alkali titration.

Q2 What indicator should you use for a strong acid/weak alkali titration — methyl orange or phenolphthalein?

Q3 Why are there two equivalence points when ethanedioic acid is titrated with sodium hydroxide?

Q4 Write the equation for the reaction that corresponds to the equivalence point at pH 8.3 when hydrochloric acid is added to sodium carbonate.

Exam Questions

1   1.0 moldm⁻³ NaOH is added separately to 25 cm³ samples of 1.0 moldm⁻³ nitric acid and 1.0 moldm⁻³ ethanoic acid. Describe two differences between the pH curves of the titrations.   [2 marks]

2   0.1 moldm⁻³ hydrochloric acid is added to 25 cm³ of 0.1 moldm⁻³ ammonia solution.
   a)   How much hydrochloric acid must be added before the equivalence point is reached?   [1 mark]
   b)   Approximately what will the pH be:
      i) at the equivalence point, ii) after 50 cm³ of acid has been added?   [2 marks]

Try learning this stuff drop by drop...
Titrations involve playing with big bits of glassware that you're told not to break as they're really expensive — so you instantly become really clumsy. If you manage not to smash the burette, you'll find it easier to get accurate results if you use a dilute acid or alkali — drops of dilute acid and alkali contain fewer particles so you're less likely to overshoot.

Indicators, pH Curves and Calculations

pH curves have more uses than toilet rolls. **Lucky old Salters can miss these two pages out.**

Indicators **Change Colour** over a **Narrow pH Range**

Indicators are **weak acids**. The key thing about them is that the acid molecules **change colour** when they dissociate.
Here's the indicator equilibrium:

$$HIn_{(aq)} \rightleftharpoons H^+_{(aq)} + In^-_{(aq)}$$
Colour 1   Colour 2

HIn is the acid indicator.

In acid-base titrations, either an acid is added to a base, or a base is added to an acid.
- If an acid is added to a base, the **H⁺** concentration increases. This shifts the equilibrium to the **left** to get rid of the extra H⁺ ions — making the indicator **colour 1**.
- If a base is added to an acid, the **OH⁻** concentration increases. The **OH⁻** ions react with the H⁺ ions, removing them from the solution — this shifts the equilibrium to the **right** to replace the H⁺ ions — making the indicator **colour 2**.

In most indicators, the colour changes over a range of about **2 pH units**. For **most** acid-base titrations (except for weak acid-weak base), there's a rapid change of at least **5 pH units** at the equivalence point.

> When you choose an indicator, you have to make sure that the **colour change range** of the indicator coincides with the **equivalence point pH range** of the titration.

Indicators have their own **Equilibrium Constants**   *Edexcel only*

The equilibrium constant for the indicator HIn is $K_{in} = \dfrac{[H^+][In^-]}{[HIn]}$

The colour at the exact end point will be an **equal mix** of colour 1 and colour 2.
At this point [HIn] = [In⁻], so **$K_{in} = [H^+]$**. ⟵ *[In⁻] and [HIn] cancel in the K_{in} expression.*

This means that the **pK_{in}** value at this point is the **same** as the **pH** — because pK_{in} and pH are both equal to **$-\log_{10}[H^+]$**.

The pK_{in} is always somewhere near the middle of the range of the indicator. In general, the range that an indicator changes colour over is approximately **1 pH unit** either side of pK_{in}.

Here's some information about some typical indicators:

Indicator	pK_{in}	pH range	HIn colour	In⁻ colour
Methyl Orange	3.7	3.1–4.4	red	yellow
Methyl red	5.1	4.2–6.3	red	yellow
Phenol red	7.9	6.8–8.4	yellow	red
Phenolphthalein	9.3	8.3–10.0	colourless	red

So if the pH is at least 1 pH unit below pK_{in}, you'd expect the indicator to have the HIn colour. If the pH is 1 or more pH units above pK_{in}, you'd expect the indicator to be the In⁻ colour.

And Another Great Use for **pH Curves** — Finding the **pK_a** of a **Weak Acid**

AQA and Edexcel only

You can work out **pK_a** of a weak acid using the pH curve for a **weak acid/strong base titration**.
It involves finding the **pH** at the **half-equivalence point**.
Half equivalence is the stage of a titration when **half** of the acid has been neutralised — it's when half of the equivalence volume of **strong base** has been added to the **weak acid**.

A weak acid, HA, dissociates like this: HA $\rightleftharpoons$ H⁺ + A⁻.
At the half-equivalence point, **[HA] = [A⁻]**.

[HA] and [A⁻] cancel.

So for the weak acid HA, $K_a = \dfrac{[H^+][A^-]}{[HA]}$ $\Rightarrow$ $K_a = [H^+]$ and $pK_a = pH$.

$$pK_a = -\log_{10}[H^+] \text{ (which is pH)}$$

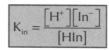

Half-equivalence point.   equivalence point

pH at this point = pK_a

volume of strong base added

Half the acid's been neutralised when this much base has been added.

So the pH at half equivalence is actually the **pK_a** value for the weak acid.
And if you know the pK_a value you can work out **K_a** ($K_a = 10^{-pK_a}$ — see page 31).

Indicators, pH Curves and Calculations

You can Calculate **Concentrations** from *Titration Data*   *Not OCR or Edexcel*

If you're doing an acid-base titration using an **indicator**, you can use the **original reading** and the reading when the **indicator changes colour** to calculate how much acid is needed to neutralise the base (or vice versa). Once you know this, you can use it to work out the **concentration** of the acid.

If you use a **pH meter** rather than an indicator, you can draw a pH curve of the titration and use it to work out how much acid or base is needed for neutralisation. You do this by finding the **equivalence point** (the mid-point of the line of rapid pH change) and drawing a **vertical line downwards** until it meets the x-axis. The value at this point on the x-axis is the volume of acid or base needed.

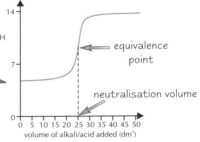

Here's an example of how you can use the neutralisation volume to calculate the concentration of the acid or base:

Example: 25 cm³ of 0.5 moldm⁻³ HCl was needed to neutralise 35 cm³ of NaOH solution. Calculate the concentration of the sodium hydroxide solution.

First write a **balanced equation** and decide **what you know** and what you **need to know**:

$$HCl + NaOH \rightarrow NaCl + H_2O$$

25 cm³   35 cm³
0.5 moldm⁻³   ?

Now work out how many **moles of HCl** you have:

$$\text{Number of moles of HCl} = \frac{\text{concentration} \times \text{volume (cm}^3)}{1000} = \frac{0.5 \times 25}{1000} = 0.0125 \text{ moles}$$

You should remember this formula from AS — you divide by 1000 to get the volume from cm³ to dm³.

From the equation, you know 1 mole of HCl neutralises 1 mole of NaOH.
So 0.0125 moles of HCl must neutralise **0.0125** moles of NaOH.

Now it's a doddle to work out the **concentration of NaOH**.

This is just the formula above, rearranged. →

$$\text{Concentration of NaOH}_{(aq)} = \frac{\text{moles of NaOH} \times 1000}{\text{volume (cm}^3)} = \frac{0.0125 \times 1000}{35} = 0.36 \text{ moldm}^{-3}$$

Practice Questions

Q1 Explain how an indicator works.
Q2 What colour is methyl orange at pH 2?
Q3 What's a half-equivalence point?

Exam Questions

1   A sample of 0.350 moldm⁻³ ethanoic acid was titrated against potassium hydroxide.

a)   Calculate the volume of 0.285 moldm⁻³ potassium hydroxide required to just neutralise with 25.0 cm³ of ethanoic acid.   **[3 marks]**

b)   From the list in the table on the right, select the best indicator for this titration and explain your choice.   **[2 marks]**

Name of indicator	pH range
bromophenol blue	3.0 – 4.6
methyl red	4.2 – 6.3
bromothymol blue	6.0 – 7.6
thymol blue	8.0 – 9.6

2   A titration curve is plotted showing the change in pH as a 0.250 moldm⁻³ solution of sodium hydroxide is added to 25.0 cm³ of a solution of ethanedioic acid, $H_2C_2O_4$. The titration curve obtained has two equivalence points.

a)   Write an equation for the reaction which is completed at the **first** equivalence point.   **[1 mark]**

b)   When the **second** equivalence point is reached, a total of 38.4 cm³ of 0.250 moldm⁻³ sodium hydroxide has been added. Calculate the concentration of the ethanedioic acid solution.   **[3 marks]**

If your stomach's rumbling, it indicates you need a biscuit...

I bet you thought you'd seen the last of those equilibrium constants — no chance. They're like The Doctor. You think he's gone for good, but then he's back, looking a bit different, but still pretty much the same. As there's no getting into your Tardis to escape — you just have to learn the stuff. At least you don't have to worry about pesky daleks though.

Buffers

I always found buffers a bit mind-boggling. How can a solution resist becoming more acidic if you add acid to it? And why would it want to? Here's where you find out...

Buffers Resist Changes in pH

A **buffer** is a solution that **resists** changes in pH when **small** amounts of acid or alkali are added.

A buffer **doesn't** stop the pH from changing completely — it does make the changes **very slight** though.
Buffers only work for small amounts of acid or alkali — put too much in and they'll go "Waah" and not be able to cope.
You get **acidic buffers** and **alkaline buffers**.

Acidic Buffers are Made from a Weak Acid and one of its Salts

Acidic buffers have a pH of less than 7 — they're made by mixing a **weak acid** with one of its **salts**.
Ethanoic acid and **sodium ethanoate** ($CH_3COO^-Na^+$) is a good example:

The salt **fully** dissociates into its ions when it dissolves: $CH_3COO^-Na^+_{(aq)} \rightarrow CH_3COO^-_{(aq)} + Na^+_{(aq)}$.
Sodium ethanoate — Ethanoate ions

The ethanoic acid is a **weak acid**, so it only **slightly** dissociates: $CH_3COOH_{(aq)} \rightleftharpoons H^+_{(aq)} + CH_3COO^-_{(aq)}$

So in the solution you've got heaps of **ethanoate ions** from the salt, and heaps of **undissociated ethanoic acid molecules**.

Le Chatelier's principle explains how buffers work:

Lots of undissociated weak acid

Addition of H⁺ (acid) ←

$$CH_3COOH_{(aq)} \rightleftharpoons H^+_{(aq)} + CH_3COO^-_{(aq)}$$

Addition of OH⁻ (alkali) →

Lots of CH_3COO^-

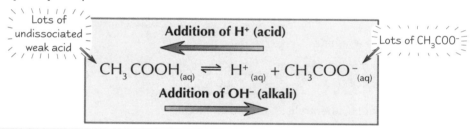

If you add a **small** amount of **acid** the **H⁺ concentration** increases. Most of the extra H⁺ ions combine with CH_3COO^- ions to form CH_3COOH. This shifts the equilibrium to the **left**, reducing the H⁺ concentration to close to its original value. So the **pH** doesn't change much.

The large number of CH_3COO^- ions make sure that the buffer can cope with the addition of acid.

If a **small** amount of **alkali** (e.g. NaOH) is added, the **OH⁻ concentration** increases. Most of the extra OH⁻ ions react with H⁺ ions to form water — removing H⁺ ions from the solution. This causes more CH_3COOH to **dissociate** to form H⁺ ions — shifting the equilibrium to the **right**. The H⁺ concentration increases until it's close to its original value, so the **pH** does not change much.

There's no problem doing this as there's absolutely loads of spare CH_3COOH molecules.

Alkaline Buffers are Made from a Weak Base and one of its Salts

A mixture of **ammonia solution** (a base) and **ammonium chloride** (a salt of ammonia) acts as an **alkaline** buffer.
It works in a similar way to acidic buffers:

The **salt** is fully ionised in solution: $NH_4Cl_{(aq)} \rightarrow NH_4^+{}_{(aq)} + Cl^-_{(aq)}$.
ammonium chloride — ammonium ions

An equilibrium is set up between the **ammonium ions** and **ammonia**:

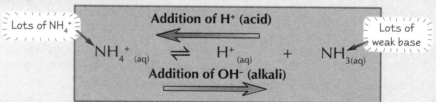

Lots of NH_4^+

Addition of H⁺ (acid) ←

$$NH_4^+{}_{(aq)} \rightleftharpoons H^+_{(aq)} + NH_{3(aq)}$$

Addition of OH⁻ (alkali) →

Lots of weak base

If a small amount of **acid** is added, the H⁺ concentration **increases** — most of the added H⁺ reacts with NH_3 and the equilibrium shifts **left**. This reduces the H⁺ concentration until it's close to its original value again, meaning that the pH **doesn't** change much. There are lots of NH_3 molecules, so they won't run out (as long as you don't add too much acid).

If a small amount of **alkali** is added, the OH⁻ concentration **increases**. OH⁻ ions react with the H⁺ ions, removing them from the solution. There's plenty of **NH_4^+** molecules around that can dissociate to generate replacement **H⁺ ions** — so the equilibrium shifts **right**, stopping the pH from changing much.

Buffers

Here's How to Calculate the pH of a Buffer Solution

Calculating the **pH** of an acidic buffer isn't too tricky. You just need to know the K_a of the weak acid and the **concentrations** of the weak acid and its salt. Here's how to go about it:

Example: A buffer solution contains 0.40 moldm^{-3} methanoic acid, HCOOH, and 0.6 moldm^{-3} sodium methanoate, HCOO$^-$ Na$^+$. For methanoic acid, $K_a = 1.6 \times 10^{-4}$ moldm^{-3}. What is the pH of this buffer?

Firstly, write the expression for K_a of the weak acid:

$$HCOOH_{(aq)} \rightleftharpoons H^+_{(aq)} + HCOO^-_{(aq)} \Rightarrow K_a = \frac{[H^+_{(aq)}] \times [HCOO^-_{(aq)}]}{[HCOOH_{(aq)}]}$$

Remember — these all have to be equilibrium concentrations.

Then rearrange the expression and stick in the data to calculate $[H^+_{(aq)}]$:

$$[H^+_{(aq)}] = K_a \times \frac{[HCOOH_{(aq)}]}{[HCOO^-_{(aq)}]}$$

$$\Rightarrow [H^+_{(aq)}] = 1.6 \times 10^{-4} \times \frac{0.4}{0.6} = 1.07 \times 10^{-4} \text{ moldm}^{-3}$$

You have to make a **few assumptions** here:
- HCOO$^-$ Na$^+$ is fully dissociated, so assume that the equilibrium concentration of HCOO$^-$ is the same as the initial concentration of HCOO$^-$ Na$^+$.
- HCOOH is only slightly dissociated, so assume that its equilibrium concentration is the same as it's initial concentration.

Finally, convert $[H^+_{(aq)}]$ to pH: $pH = -\log_{10}[H^+_{(aq)}] = -\log_{10}(1.07 \times 10^{-4}) = \mathbf{3.97}$ And that's your answer.

Buffers are Really Handy   *OCR and Salters only*

Most **shampoos** contain a pH 5.5 buffer — it counteracts the alkaline soap in the shampoo. The soap might get your hair squeaky clean, but alkalis don't make your hair look shiny.

See page 13.

Biological washing powders contain buffers too. They keep the pH at the right level for the enzymes to work.

It's vital that **blood** stays at a pH very near 7.4, so it's buffered. Here's what happens:

Carbon dioxide is formed in **respiration**, but when it's dissolved it produces **H$^+$ ions**, which lower the pH of the blood:

$$CO_{2(aq)} + H_2O_{(l)} \rightleftharpoons HCO_3^-{}_{(aq)} + H^+_{(aq)}$$

If the H$^+$ ion concentration **rises**, the equilibrium shifts **left** — this forms extra $CO_{2(aq)}$, which is breathed out from the lungs as **CO$_2$ gas**. If the H$^+$ ion concentration **falls**, the equilibrium shifts **right**. Proteins in blood with side groups such as –COO$^-$ and –NH$_3^+$ also act as buffers — they accept and donate **protons** as needed.

Practice Questions

Q1 What's a buffer solution?

Q2 How can a mixture of ethanoic acid and sodium ethanoate act as a buffer?

Q3 Describe how to make an alkaline buffer.

Exam Questions

1 A buffer solution contains 0.40 moldm^{-3} benzoic acid, C$_6$H$_5$COOH, and 0.20 moldm^{-3} sodium benzoate, C$_6$H$_5$COO$^-$Na$^+$. At 25 °C, K_a for benzoic acid is 6.4×10^{-5} moldm^{-3}.
   a) Calculate the pH of the buffer solution. [3 marks]
   b) Explain the effect on the buffer of adding a small quantity of dilute sulphuric acid. [3 marks]

2 A buffer was prepared by mixing solutions of butanoic acid, CH$_3$(CH$_2$)$_2$COOH, and sodium butanoate, CH$_3$(CH$_2$)$_2$COO$^-$Na$^+$, so that they had the **same** concentration.
   a) Write a balanced chemical equation to show butanoic acid acting as a weak acid. [1 mark]
   b) Given that K_a for butanoic acid is 1.5×10^{-5} moldm^{-3}, calculate the pH of the buffer solution. [3 marks]

Old buffers are often resistant to change...

So that's how buffers work. There's a pleasing simplicity and neatness about it that I find rather elegant. Like a fine wine with a nose of berry and undertones of... OK, I'll shut up now.

Metal-Aqua Ions

You know how it said on page 28 that acids donate protons and bases accept protons? Well, that was just Mr Brønsted and Mr Lowry's version. Mr Lewis came up with a different idea... **These two pages are not for** *Nuffield*.

A *Lewis Acid* is an *Electron Pair Acceptor*   *AQA only*

Here's the definition of Lewis acids and bases:

> If a molecule or ion can **accept** an **electron pair**, it's a **Lewis acid**.
> If it can **donate** an **electron pair**, it's a **Lewis base**.

You need to be able to work out which is the acid and which is the base in a reaction. Have a look at these examples:

$H_2O + H^+ \rightarrow H_3O^+$

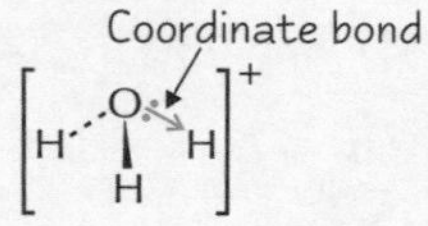

Coordinate bond

The water molecule's donating an electron pair to the hydrogen ion, and the hydrogen ion is accepting an electron pair from the water. So the water molecule is the Lewis base and the hydrogen ion's the Lewis acid.

$AlCl_3 + Cl^- \rightarrow AlCl_4^-$

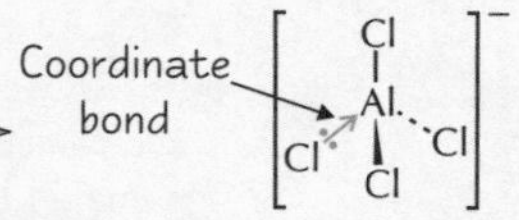

Coordinate bond

The aluminium chloride is accepting an electron pair — so it's the acid. The chloride ion is donating an electron pair — so it's the base.

Coordinate bonds have been formed in both these reactions. Coordinate bonds are **covalent** bonds in which **both** of the electrons come from just **one** of the atoms involved. So if there's a coordinate bond, there are no two ways about it — it **must** have been formed in a **Lewis acid-base reaction**.

Metal Ions Become *Hydrated* in Water

When **transition metal compounds** dissolve in water, the water molecules form **coordinate bonds** with the **metal ions**. This forms **metal-aqua complex ions**. In general, **6 water molecules** form coordinate bonds with each metal ion.

The water molecules do this by donating a **non-bonding pair of electrons** from the oxygen. The diagram on the right shows the metal-aqua ion formed by **iron** — $Fe(H_2O)_6^{2+}$.

Lots of other transition metal ions form similar complexes, e.g.: $Co(H_2O)_6^{2+}$, $Cu(H_2O)_6^{2+}$, $Fe(H_2O)_6^{3+}$, $Cr(H_2O)_6^{3+}$ and $V(H_2O)_6^{3+}$.

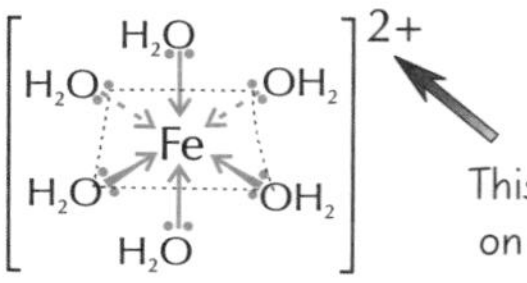

This is the charge on the metal ion.

Metal-Aqua Ions are Found in the *Solid State* too

The bonding between the metal ions and water is so **strong** that the aqua-metal ion stays intact when the solution is evaporated. For example, heating iron sulphate solution eventually leaves you with green crystals of $FeSO_4 \cdot 7H_2O$ which contain the ion $Fe(H_2O)_6^{2+}$, and heating cobalt nitrate solution leaves you with pink crystals of $Co(NO_3)_2 \cdot 6H_2O$ which contain $Co(H_2O)_6^{2+}$.

Solutions Containing *Metal Ions* are *Acidic*   *AQA only*

In a solution containing metal-aqua **2+** ions, there's a reaction between the metal-aqua ion and the water — this is a **hydrolysis** or **acidity reaction**.

E.g. $Fe(H_2O)_6^{2+}{}_{(aq)} + H_2O_{(l)} \rightleftharpoons [Fe(H_2O)_5(OH)]^+{}_{(aq)} + H_3O^+{}_{(aq)}$

The metal-aqua **2+** ions release H^+ ions, so an **acidic** solution is formed. There's only **slight** dissociation though, so the solution is only **weakly acidic**.

Metal-aqua **3+** ions react in the same way. They form **more acidic** solutions though.

E.g. $Al(H_2O)_6^{3+}{}_{(aq)} + H_2O_{(l)} \rightleftharpoons [Al(H_2O)_5(OH)]^{2+}{}_{(aq)} + H_3O^+{}_{(aq)}$

Aqua-ironing — it keeps those flat fish smooth.

Here's why 3+ metal-aqua ions form more acidic solutions than 2+ metal-aqua ions:

The metal 3+ ions are pretty **small** but have a **big charge** — so they've got a **high charge density** (otherwise known as **charge/size ratio**). The metal 2+ ions have a **much lower** charge density.

So the 3+ ions are much more **polarising** than the 2+ ions. More polarising power means that they attract the **electrons** from the oxygen atoms of the coordinated water molecules more strongly, weakening the O–H bond.

As a result, it's more likely that a **hydrogen ion** will be released...
...and more hydrogen ions released means a **more acidic** solution.

Metal-Aqua Ions

You Can *Hydrolyse* Metal-Aqua Ions *Further*

M, for example, can be Fe, Al or Cr.

This equilibrium occurs in water with **metal-aqua 3+ ions**: $M(H_2O)_6^{3+} + H_2O \rightleftharpoons [M(H_2O)_5(OH)]^{2+} + H_3O^+$.
If you add **OH⁻ ions** to the equilibrium H_3O^+ ions are removed — this shifts the equilibrium to the **right**.

Now another equilibrium's set up in the solution: $[M(H_2O)_5(OH)]^{2+} + H_2O \rightleftharpoons [M(H_2O)_4(OH)_2]^+ + H_3O^+$.
Again OH⁻ ions remove H_3O^+ ions from the solution, pushing the equilibrium to the right.

This happens one last time — now you're left with an **uncharged metal hydroxide**:
$[M(H_2O)_4(OH)_2]^+ + H_2O \rightleftharpoons [M(H_2O)_3(OH)_3] + H_3O^+$

M, for example, can be Fe, Co or Cu.

The same thing happens with **metal-aqua 2+ ions**, except this time there are only **two** steps:
$M(H_2O)_6^{2+} + H_2O \rightleftharpoons [M(H_2O)_5(OH)]^+ + H_3O^+ \longrightarrow [M(H_2O)_5(OH)]^+ + H_2O \rightleftharpoons [M(H_2O)_4(OH)_2] + H_3O^+$

These uncharged metal hydroxides are **insoluble in water** — so they form coloured **precipitates** (see the table below).

Some Metal Hydroxides are Amphoteric — they can act as both Acids and Bases
Aluminium hydroxide and **chromium(III) hydroxide** act as **acids** in an excess of OH⁻. They **donate H⁺ ions** to the OH⁻ ions and form **soluble compounds**. $\Longrightarrow$ $Al(H_2O)_3(OH)_{3(s)} + OH^-_{(aq)} \rightleftharpoons [Al(H_2O)_2(OH)_4]^-_{(aq)} + H_2O_{(l)}$
$Cr(OH)_3(H_2O)_{3(s)} + 3OH^-_{(aq)} \rightleftharpoons [Cr(OH)_6]^{3-}_{(aq)} + 3H_2O_{(l)}$
All metal hydroxides will act as **bases** and **accept H⁺ ions**. Adding H⁺ ions just **reverses** the hydrolysis reactions above.

Precipitates *Form with Ammonia Solution...*   *Not for OCR*

The obvious way of adding hydroxide ions is to use a strong alkali, like **sodium hydroxide solution** — but you can use **ammonia solution** too. When ammonia dissolves in water this equilibrium occurs: $NH_3 + H_2O \rightleftharpoons NH_4^+ + OH^-$
Because of the hydroxide ions, adding a **small** quantity of ammonia solution gives the same results as sodium hydroxide.

In some cases, a further reaction happens if you add an **excess** of ammonia solution. Check out the table below.

...and *Sodium Carbonate Too*   *AQA only*

Metal **3+** ions form **hydroxide precipitates** with **sodium carbonate** (Na_2CO_3).
Like the hydroxide ions above, the **carbonate ions** react with the H_3O^+ ions, removing them from the solution.
The equilibrium then shifts to the **right**, which leads to the hydroxide precipitate being formed.

Metal 2+ ions aren't acidic enough for this to happen.
Instead, they form **insoluble metal carbonates**, like this: $M(H_2O)_6^{2+}{}_{(aq)} + CO_3^{2-}{}_{(aq)} \rightleftharpoons MCO_{3(s)} + 6H_2O_{(l)}$

This handy table summarises all the reactions on this page:

The last column is for AQA only

Metal-aqua ion	With OH⁻$_{(aq)}$ or NH$_{3(aq)}$	With excess OH⁻$_{(aq)}$	With excess NH$_{3(aq)}$	With Na$_2$CO$_{3(aq)}$
$Co(H_2O)_6^{2+}$ — pink solution	$Co(H_2O)_4(OH)_2$ — blue-green ppt	no change	$[Co(NH_3)_6]^{2+}$ — pale yellow solution	$CoCO_3$ — pink ppt
$Cu(H_2O)_6^{2+}$ — blue solution	$Cu(H_2O)_4(OH)_2$ — blue ppt	no change	$[Cu(NH_3)_4(H_2O)_2]^{2+}$ — deep blue solution	$CuCO_3$ — green-blue ppt
$Fe(H_2O)_6^{2+}$ — green solution	$Fe(H_2O)_4(OH)_2$ — green ppt	no change	no change	$FeCO_3$ — green ppt
$Al(H_2O)_6^{3+}$ — colourless solution	$Al(H_2O)_3(OH)_3$ — white ppt	$[Al(H_2O)_2(OH)_4]^-$ — colourless solution	no change	$Al(H_2O)_3(OH)_3$ — white ppt
$Cr(H_2O)_6^{3+}$ — violet solution	$Cr(H_2O)_3(OH)_3$ — green ppt	$[Cr(OH)_6]^{3-}$ — colourless solution	$[Cr(NH_3)_6]^{3+}$ — purple solution	$Cr(H_2O)_3(OH)_3$ — green ppt
$Fe(H_2O)_6^{3+}$ — yellow solution	$Fe(H_2O)_3(OH)_3$ — brown ppt	no change	no change	$Fe(H_2O)_3(OH)_3$ — brown ppt

Practice Questions

Q1 What is: a) a Lewis base?   b) a coordinate bond?
Q2 What's the formula of cobalt hydroxide? What colour is it?

Exam Questions

1   Explain why separate solutions of iron(II) sulphate and iron(III) sulphate at equal concentrations have different pH values.   [4 marks]

2   Describe what you would see when ammonia solution is added slowly to a solution containing copper(II) sulphate until it is in excess. Write equations for all reactions described.   [8 marks]

Test-tube reactions — proper chemistry at last...

These colours might all be very pretty, but the rubbish thing is you have to learn them all. There's no rhyme or reason why a solution should decide to be pink or a precipitate should decide to be blue... or is there... (turn to page 66).

Period 3 Elements and Oxides

Period 3's the third row down on the Periodic table — that's right, the one that starts with sodium and ends with argon.

Ionic Radii Affects Polarisation and Hydration Enthalpy

1) Going from sodium ions (Na^+) to aluminium ions (Al^{3+}), the **nuclear charge** increases from 1+ to 3+. As the nuclear charge increases, the electrons are **pulled in** closer, making the **ionic radii decrease**.

2) As the cations get **smaller** and **more highly charged** from sodium to aluminium, their **polarising power increases**. This means that the bonding in their compounds becomes **less ionic** and **more covalent**.

3) Water molecules are attracted **more strongly** by smaller, more highly charged cations, so **hydration enthalpy** (see page 6) also **increases** from sodium to aluminium.

4) **Phosphorus**, **sulphur** and **chlorine** gain electrons in their outer shell to form negative **anions**. Because these anions have got **one more shell** than Na^+, Mg^{2+} and Al^{3+}, they're a lot **bigger**.

5) From P^{3-} to Cl^-, the nuclear charge **increases**. Again, this pulls the electrons in closer so the ionic radii **decrease**.

6) A **large anion** with a **large negative charge** will be more **easily polarised**. So if it's bonded with a small cation it'll make a compound with a significant amount of **covalent character**.

Reactivity of Metals with Water Decreases Across the Period

As you move from left to right across Period 3 the metallic elements become **less reactive** with **water**. This is because it gets **harder** to remove the outer electrons as **nuclear charge increases**.

1) Sodium reacts **vigorously** with water, forming **sodium hydroxide**.

$$2Na_{(s)} + 2H_2O_{(l)} \rightarrow 2NaOH_{(aq)} + H_{2(g)}$$

2) Magnesium reacts extremely **slowly** with water to form **magnesium hydroxide**.

$$Mg_{(s)} + H_2O_{(l)} \rightarrow Mg(OH)_{2(s)} + H_{2(g)}$$

It'll react rapidly with **steam** though, mainly forming **magnesium oxide** and hydrogen.

$$Mg_{(s)} + H_2O_{(g)} \rightarrow MgO_{(s)} + H_{2(g)}$$

3) Aluminium hardly reacts at all with water or steam.

Sodium hydroxide is more soluble than magnesium hydroxide. This means there are more OH^- ions in the solution, making it a stronger alkali.

The only Period 3 non-metal that reacts with water is **chlorine**. It dissolves and reacts to give an **acidic** solution of **HCl** and **HClO**.

$$Cl_{2(g)} + H_2O_{(l)} \rightarrow HClO_{(aq)} + HCl_{(aq)}$$

Period 3 Elements React with Oxygen to Form Oxides

Period 3 elements are usually oxidised to their **highest** oxidation state — which is the same as their group number. Sulphur's the exception to this — it forms SO_2, in which it's only got a +4 oxidation state. A **catalyst** is needed to make SO_3, where it's got its highest oxidation state of **+6**.

Going from left to right across the period, the structure of the oxides changes from **giant** to **simple molecular**, and the bonding changes from **ionic** to **covalent**. This affects lots of the **properties** of the oxides — as you'll see on the next page.

Element	Na	Mg	Al	Si	P	S	Cl
Formula of oxide	Na_2O	MgO	Al_2O_3	SiO_2	P_4O_{10}	SO_2 SO_3	Cl_2O
Group number	1	2	3	4	5	6	7
Oxidation state of Period 3 element	+1	+2	+3	+4	+5	+4 +6	+1
Structure of oxide	giant	giant	giant	giant	simple molecular	simple molecular	simple molecular
Bonding in oxide	ionic	ionic	ionic, some covalent	covalent	covalent	covalent	covalent
Appearance of oxide	white solid	white solid	white solid	white solid	white solid	colourless gas colourless liquid	brown-yellow gas

Chlorine also forms other oxides including Cl_2O_7, where it's got an oxidation state of +7.

Oxidation states are covered on page 46.

Period 3 Elements and Oxides

The **Structure** of Period 3 Oxides Affects **Melting Points**

1) The **metal oxides** form **giant ionic lattices**, with **strong** forces of attraction between each ion. It takes lots of energy to break these forces, so they've got **high** melting points. Magnesium forms **2+ ions**, so it bonds more strongly to oxygen than sodium ions do — so **MgO** has a higher melting point than Na_2O.

Al_2O_3 has a **lower melting point** than you'd expect — this is because Al^{3+} ions are **highly polarising** and distort oxygen's electron cloud, making the bonds **partially covalent**.

2) The **non-metal oxides** P_4O_{10}, SO_2 and Cl_2O have **simple covalent** structures. There are weak **van der Waals forces** between individual molecules, so these oxides have **low** melting points.

SiO_2 isn't your typical non-metal oxide — it's a **giant** structure, so it has a **higher** melting point than P_4O_{10} and SO_2.

Oxide	Na_2O	MgO	Al_2O_3	SiO_2	P_4O_{10}	SO_2	Cl_2O
Melting point (°C)	1275	2852	2072	1610	580	-75	-20

Acidity of Period 3 Oxides in Water **Increases** Across the Period

1) **Metal oxides** dissolve in water to form **hydroxides** — so the solution's **alkaline** (about pH 13).

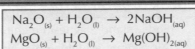

$$Na_2O_{(s)} + H_2O_{(l)} \rightarrow 2NaOH_{(aq)}$$
$$MgO_{(s)} + H_2O_{(l)} \rightarrow Mg(OH)_{2(aq)}$$

2) Al_2O_3 and SiO_2 are **insoluble** in water.

3) **Non-metal oxides** dissolve in water to form **acidic** solutions (of about pH 3).

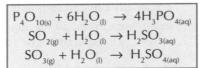

$$P_4O_{10(s)} + 6H_2O_{(l)} \rightarrow 4H_3PO_{4(aq)}$$
$$SO_{2(g)} + H_2O_{(l)} \rightarrow H_2SO_{3(aq)}$$
$$SO_{3(g)} + H_2O_{(l)} \rightarrow H_2SO_{4(aq)}$$

By the time Julie got to Period 3, she was wishing she was a boy.

Oxides React with **Acids** or **Bases**

1) **Metal oxides** are **basic** — so they'll react with **acids** to form salts: $Na_2O_{(s)} + 2HCl_{(aq)} \rightarrow 2NaCl_{(aq)} + H_2O_{(l)}$

2) **Non-metal oxides** are **acidic** — so they'll react with **bases** to form salts: $SiO_{2(s)} + 2OH^-_{(aq)} \rightarrow SiO_3^{2-}_{(aq)} + H_2O_{(l)}$

3) **Aluminium oxide** is **amphoteric** — it can react with **both** acids and alkalis.

Acting as a base: $Al_2O_{3(s)} + 3H_2SO_{4(aq)} \rightarrow Al_2(SO_4)_{3(aq)} + 3H_2O_{(l)}$
Acting as an acid: $Al_2O_{3(s)} + 2NaOH_{(aq)} + 3H_2O_{(l)} \rightarrow 2NaAl(OH)_{4(s)}$

Practice Questions

Q1 Why do metal elements become less reactive with water across Period 3?
Q2 What are the oxidation numbers of the elements from Na to S after they've been burned in oxygen?
Q3 Describe the bonding of each of the Period 3 oxides.
Q4 What is the trend in pH across Period 3 when the oxides dissolve?

Exam Question

1 Sodium, aluminium, silicon and phosphorus all form oxides when burned in oxygen.

a) Write equations for the reactions of these four elements with oxygen gas. [8 marks]

b) Name the type of structure and bonding found in the oxide of each element. [8 marks]

c) What would you expect the pH of the resulting solution to be if each of these oxides were added to water? [4 marks]

d) Explain why the melting point of phosphorus(V) oxide is much lower than that of sodium oxide. [5 marks]

This section's got more trends than a school disco...

Argon's at the end of Period 3, but it's a noble (or inert) gas, so it doesn't react with anything and you can pretty much ignore it as far as trends go. If anyone asks, it's a simple monatomic gas, so obviously it's got a low melting and boiling point.

Period 3 Chlorides and Group 4

Sorry chaps and chapesses... there's another set of trends to learn.

Period 3 Elements React with **Chlorine** to Form **Chlorides**

The **structure** and **bonding** of Period 3 chlorides follow very similar patterns to Period 3 oxides.
From left to right across Period 3 the structure changes from **giant** to **simple molecular** and the bonding changes from **ionic** to **covalent**.

Element	Na	Mg	Al	Si	P	S	Cl
Formula of chloride	NaCl	$MgCl_2$	$AlCl_3$	$SiCl_4$	PCl_3 PCl_5	S_2Cl_2	Cl_2
Oxidation state of Period 3 element in chloride	+1	+2	+3	+4	+3 (in PCl_3) +5 (in PCl_5)	+1	0
State of chloride	solid	solid	solid	liquid	liquid (in PCl_3) solid (in PCl_5)	liquid	gas
Structure of chloride	giant	giant	simple molecular	simple molecular	simple molecular	simple molecular	simple molecular
Bonding in chloride	ionic	ionic	covalent	covalent	covalent	covalent	covalent

Aluminium Chloride is **Different** From the Other Metal Chlorides

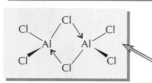

Sodium and magnesium chloride are giant ionic lattices — but aluminium chloride **isn't**.

Aluminium is **highly polarising**, so it distorts chlorine's electron cloud and ends up as a **covalently bonded simple molecular** compound.

In the liquid state aluminium chloride exists as Al_2Cl_6, which is made from two molecules of $AlCl_3$ joined together by two **coordinate** (dative covalent) bonds. This is a **dimer**.

Melting Points of Period 3 Chlorides Depend on **Structure**

1) NaCl and $MgCl_2$ have **giant ionic** structures, held together by **strong** ionic forces of attraction. A lot of energy is needed to break the ions apart, so their melting points are **high**.

2) $AlCl_3$, $SiCl_4$ and PCl_3 have **simple covalent** structures, held together by **weak** van der Waals forces. Not much energy is needed to break these attractions, so their melting points are **low**.

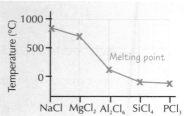

   PCl_5 has a higher melting point than PCl_3 and is solid at room temperature. This is because PCl_5 actually exists as $[PCl_6]^-[PCl_4]^+$, so it's **partially ionic** and needs more energy to break the electrostatic attractions.

Ionic Chlorides Dissolve in **Water** to Form **Neutral** or **Slightly Acidic** Solutions

Sodium chloride dissolves in water to give a **neutral** solution — there are equal amounts of H^+ and OH^- ions from the water.

$$NaCl_{(s)} \rightarrow Na^+_{(aq)} + Cl^-_{(aq)} \longrightarrow \text{pH 7}$$

Magnesium chloride dissolves in water to give a very **slightly acidic** solution.
The OH^- ions from the water are more strongly attracted to the Mg^{2+} ions, than the H^+ ions are to the Cl^- ions, so there are more **H^+ ions** than OH^- ions in the solution.

$$MgCl_{2(s)} \rightarrow Mg^{2+}_{(aq)} + 2Cl^-_{(aq)} \longrightarrow \text{pH 6.5}$$

Covalent Chlorides React with **Water** to Form **Acidic** Solutions

The Period 3 covalent chlorides are **hydrolysed** when they're added to water

These tend to be **vigorous** reactions which produce **acidic** solutions containing **hydrochloric acid**.
As you go across the period the solutions get increasingly **more acidic**.

$$AlCl_{3(s)} + 3H_2O_{(l)} \rightarrow Al(OH)_{3(s)} + 3HCl_{(aq)} \longrightarrow \text{pH 3}$$

$$SiCl_{4(l)} + 2H_2O_{(l)} \rightarrow SiO_{2(s)} + 4HCl_{(aq)} \longrightarrow \text{pH 1}$$

$$PCl_{3(s)} + 3H_2O_{(l)} \rightarrow H_3PO_{3(aq)} + 3HCl_{(aq)} \longrightarrow \text{pH 0-1}$$

$$PCl_{5(s)} + 4H_2O_{(l)} \rightarrow H_3PO_{4(aq)} + 5HCl_{(aq)} \longrightarrow \text{pH 0-1}$$

Edexcel and Nuffield only: S_2Cl_2 also forms acidic solutions on hydrolysis with water.

Period 3 Chlorides and Group 4

This stuff on Group 4 elements is just for Edexcel.

The Elements Become **More Metallic** As You Go **Down Group 4**

In Group 4, carbon and silicon are **non-metals**, germanium is a **metalloid**, and tin and lead are both **metals**. The elements nearer the bottom of the group are **more metallic** because they're larger and their outer electrons are **less tightly held**, so they're more easily lost.

A metalloid's properties are in between those of metals and non-metals.

C	Non-metals
Si	
Ge	Metalloid
Sn	Metals
Pb	

So the metallic properties **conductivity**, **ductility** and **malleability** increase as you go down the group.

As far as conductivity goes, carbon's a strange one — in graphite form it has **delocalised electrons** and conducts electricity. In its diamond form it doesn't conduct though.

Silicon and germanium are **semiconductors**, which means they're not insulators, but they don't conduct as well as your average metal. At the bottom of the table, tin and lead are **good** conductors.

Tin and **Lead** can Form **Stable +2** Oxidation States

1) The elements in Group 4 all have **four electrons** in their outer shell — two in **p-orbitals** and two in an **s-orbital**.

E.g. C's electronic configuration is [He] $2s^2 2p^2$. Si's electronic configuration is [Ne] $3s^2 3p^2$.

2) They show **+2** and **+4** oxidation states. In the +4 state they use **all** their outer electrons in bonding, and in the +2 state they just use their **p electrons**. The s electrons are **more tightly held** and less available for bonding — this is called the **inert pair effect**.

3) The **+2** state becomes **more stable** the further down the group you go:
   - Carbon, silicon and germanium form **covalent compounds**, virtually always with a +4 oxidation state.
   - Tin forms both **covalent +4 compounds** and **ionic +2 compounds**, but the **+4 state** is more stable, so tin(II) compounds, like SnO or $SnCl_2$, are **reducing agents**.
   - Lead forms mainly **ionic +2 compounds**, but also **covalent +4 compounds**. The +2 state is more stable for lead, so lead(IV) compounds, such as PbO_2, are **oxidising agents**.

The **Oxides** Become **More Basic** as You Go Down Group 4

'Amphoteric' means they react with both acids and alkalis.

The +4 oxides CO_2 and SiO_2 are **acidic**, so they react with alkalis. SnO_2 and PbO_2 are +4 oxides too — these are **amphoteric** but they tend to behave more like **acidic oxides** than basic oxides.

$$CO_{2(g)} + 2OH^-_{(aq)} \rightarrow CO_3^{2-}_{(aq)} + H_2O_{(l)}$$ $$SiO_{2(s)} + 2OH^-_{(aq)} \rightarrow SiO_3^{2-}_{(aq)} + H_2O_{(l)}$$ $$PbO_{2(s)} + 2OH^-_{(aq)} \rightarrow PbO_3^{2-}_{(aq)} + H_2O_{(l)}$$

The +2 oxides SnO and PbO are **amphoteric**, but they tend to behave more like **basic oxides**:

$$PbO_{(s)} + 2HCl_{(aq)} \rightarrow PbCl_{2(aq)} + H_2O_{(l)}$$

CCl_4 and $SiCl_4$ have a **Tetrahedral** Structure

Group 4 tetrachlorides are **simple molecular**, **covalent** structures with a **tetrahedral** shape. E.g.:

- **Silicon tetrachloride** ($SiCl_4$) hydrolyses in water — silicon uses an **empty 3d-orbital** to bond with oxygen in water.
- **Carbon tetrachloride** (CCl_4) doesn't react with water. The difference between carbon's 2p and 3d energy levels is too big to make the d-orbitals available for bonding with water.

Practice Questions

Q1 What's the trend in bond type for the chlorides as you go across Period 3?

Q2 Why does the metallic character of the elements increase down Group 4?

Exam Question

1 Sodium chloride, aluminium chloride and phosphorus(V) chloride react with water.
   a) Write an equation for each reaction. [6 marks]
   b) What is the approximate pH for each of the resulting solutions? Explain your answers. [6 marks]

Silicon chips — it'll take more than vinegar to help them slide down your throat...

You know what they say — there's an exception to every rule, and these pages prove it. Take extra care when writing formulas for compounds like PbO and $SnCl_2$. Look at the oxidation states and make sure you've got the right number of oxygens or chlorines attached. If you don't, then you're just throwing marks away. And that's just plain daft.

The Nitrogen Cycle and Water

Nitrogen and water are pretty big fish on planet Earth — the air's about 80% nitrogen and the Earth's surface is about two thirds water. Not to mention us — we're about 70% water.
This double page is mainly for Salters, but there's a bit for Edexcel folks too.

Nitrogen has lots of **Different Oxidation States** *Salters only*

Plants and animals need **protein** — and nitrogen's a vital element in proteins. Nitrogen can't be absorbed directly from the air though — it needs to be converted into a **more accessible** form in a different **oxidation state**.

The **nitrogen cycle** converts nitrogen between its different oxidation states:

Form of nitrogen	Formula	Oxidation state	What produces this form of nitrogen
Nitrogen in the air	$N_{2(g)}$	0	Denitrifying bacteria
Ammonium ions in the soil	$NH_{4}^{+}{}_{(aq)}$	−3	Bacteria and micro-organisms in the soil
Nitrates in the soil	$NO_{3}^{-}{}_{(aq)}$	+5	Nitrifying bacteria in the soil, bacteria in root nodules
Nitrites in the soil	$NO_{2}^{-}{}_{(aq)}$	+3	Nitrifying bacteria in the soil, bacteria in root nodules
Nitrogen(II) oxide	$NO_{(g)}$	+2	Thunderstorms, car engines, denitrifying bacteria in the soil
Nitrogen(IV) oxide	$NO_{2(g)}$	+4	Oxidation of NO in the atmosphere
Nitrogen(I) oxide	$N_{2}O_{(g)}$	+1	Denitrifying bacteria in the soil

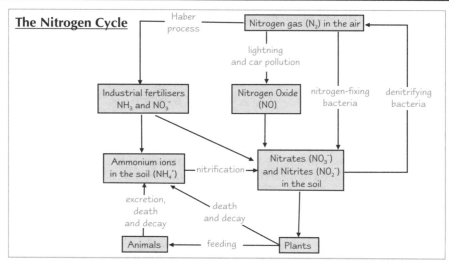

The Nitrogen Cycle

Crops Need the Right **Nutrients** and the Right **pH** to Grow Well *Salters only*

Plants won't grow well without the right **nutrients** — the main ones are **nitrogen**, **phosphorus** and **potassium**.

Plants take these nutrients from the soil, so **fertilisers** are used to add more. Plants can only absorb nutrients when they're in solution, so **ammonium ions** (NH_4^+), **nitrate(V) ions** (NO_3^-), **phosphate ions** (PO_4^{3-}) and **potassium ions** (K^+) are used. Fertilisers that contain the three main nutrients are called **NPK** fertilisers.

Most plants like to have soil of a particular **pH**. If the soil's too acidic, **lime** (calcium oxide) can be added to raise the pH.

Nitrogen Can Be Added in **Inorganic Compounds** or **Organic Compounds**

Edexcel only

INORGANIC ammonium or nitrate fertilisers are usually made through the Haber process

These are good because they're **cheap** and are **very soluble**, so they're easily spread.

The problem is, they're **easily washed out** of the soil, so you need to add lots. If fertilisers are washed into rivers or lakes they cause algae to multiply really fast. When bacteria decompose the algae, they use up all the oxygen, meaning organisms like fish die — this is **eutrophication**. Inorganic fertilisers also tend to **damage the soil structure** and are quite **corrosive** to handle.

ORGANIC fertilisers like urea, $(NH_2)_2CO$, are becoming increasingly popular with farmers

These are also **cheap** and **very soluble**, but, unlike inorganic fertilisers, they **don't damage the soil structure**. They're **less corrosive** to handle too. And if that's not enough, they also have **higher** nitrogen **concentration** — so you don't need as much. Unfortunately, they're still **washed away easily**, causing **eutrophication**.

The Nitrogen Cycle and Water

Water's a hydride of the Group 6 element, oxygen. **This page is just for Salters people.**

Water has **Strong Intermolecular** Bonds

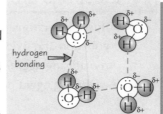

1) **Oxygen** is **very electronegative**, so it pulls the bonding electrons towards itself, creating a dipole. The bonds are **so polarised** that the **hydrogen** of one molecule forms a weak bond with the **oxygen** of **another molecule**. These intermolecular forces are called **hydrogen bonds**. None of the other Group 6 elements are electronegative enough for their hydrides to have hydrogen bonding.

2) Hydrogen bonds are strong as **intermolecular forces** go, and a fair amount of energy's needed to break them, so water's got a higher **melting** and **boiling** point than you'd expect.

Water's Got a **Higher Specific Heat Capacity** Than Other Group 6 Hydrides

1) Two **different substances** absorbing the **same** amount of energy won't necessarily change **temperature** by the same amount. **Specific heat capacity** measures how easy it is to raise the temperature of something — it usually depends on a substance's **internal structure**.

2) Water's got a fairly **high** specific heat capacity — that's why the sea heats up (and cools down) a lot slower than land. The ocean affects **climate** by absorbing energy from the Sun and carrying it in **currents** to different places.

Water's Got a **Higher Enthalpy of Vaporisation** Than Other Group 6 Hydrides

1) **Enthalpy of vaporisation** is the amount of energy needed to change a substance from its standard state to a **vapour**.

2) Water needs a lot more energy to **evaporate** than most other liquids. If it's on your skin, it takes loads of energy (in the form of heat) from your skin as it evaporates — so it cools you down quickly.

3) The water cycle acts as a **climate control**. Water in the oceans around the equator absorbs a lot of heat, and some of it **evaporates**. This water vapour then moves in air currents away from the equator, where it releases energy as it **condenses**, then falls back to the ground. This helps prevent temperatures becoming **too extreme**.

Ice is **Less Dense** Than Water

1) Substances, including water, generally get **denser** as they cool. But as water cools below 4 °C, the density **decreases**. It's because as water freezes, the molecules arrange themselves in a more **regular open structure** — it's all because of the **hydrogen bonding**. This means ice **floats** on water.

2) Ice on the top of ponds and lakes acts to **insulate** the water underneath. So there's always liquid water underneath the surface — great for fish and plants living there.

3) Other Group 6 hydrides don't have hydrogen bonding, so their density continues to **increase** as they cool. This means their solid form will **sink** in their liquid form.

Practice Questions

Q1 State the oxidation state of nitrogen in the following compounds or ions: a) N_2 b) NO_3^- c) NH_4^+

Q2 Name three compounds or ions that are added to fertilisers to improve crop production.

Q3 Give two advantages of using urea rather than inorganic ammonium fertilisers.

Q4 What type of bonding is responsible for many of the unusual properties of water?

Exam Question

1 The properties of water are unusual compared to the other Group 6 hydrides.

a) How does the boiling point of water compare with the boiling points of the other Group 6 hydrides? Explain why this is. [3 marks]

b) Why does ice float on water but frozen hydrogen sulphide sink in liquid hydrogen sulphide? [5 marks]

c) Water helps prevent climates becoming too extreme. State which property of water enables it to do this. Explain how water helps to maintain a moderate climate in both the hotter areas of the globe at the equator and in the cooler areas near the Poles. [5 marks]

It's getting hot in here — so tip a bucket of water over your head...

... rather than a bucket of hydrogen sulphide. I wonder what the world'd be like if ice was more dense than water — you'd have underwater iceskating for a start, and they'd have to wear wetsuits instead of those sparkly little skimpy numbers. Anyway, enough wondering. Make sure you know about nitrogen in its many oxidation states and why water's so special.

Oxidation and Reduction

This is where it all begins — Section 5 that is...

If Electrons are Transferred, it's a **Redox Reaction**

I couldn't find a red ox, so you'll have to make do with a multicoloured donkey instead.

1) A **loss** of electrons is called **oxidation**. A **gain** of electrons is called **reduction**.
2) Reduction and oxidation happen **simultaneously** — hence the term "redox" reaction.
3) An **oxidising agent accepts** electrons and gets reduced.
4) A **reducing agent donates** electrons and gets oxidised.

$$Na + \tfrac{1}{2}Cl_2 \longrightarrow Na^+ Cl^-$$

Na is oxidised
Cl is reduced

Sometimes it's easier to talk about **Oxidation States**

(They're also called oxidation <u>numbers</u>.)

There are lots of rules. Take a deep breath...

1) All atoms are treated as **ions** for this, even if they're covalently bonded.

2) Uncombined **atoms** have an oxidation state of **0**.

3) Atoms only bonded to **identical atoms**, like in O_2 and H_2, also have an oxidation state of **0**.

4) The oxidation state of a simple **monatomic ion**, e.g. Na^+, is the same as its **charge**.

5) Combined **oxygen** is –2, except in peroxides and F_2O, where it's –1 (and O_2, where it's 0).

In H_2O, oxidation state of O = –2, but in H_2O_2, oxidation state of H has to be +1 (an H atom can only lose one electron), so oxidation state of O = –1.

6) Combined **hydrogen** is +1, except in metal hydrides, where it's –1 (and H_2, where it's 0).

In **HF**, oxidation state of H = +1, but in **NaH**, oxidation state of H = –1.

7) In **compounds** or **compound ions**, the **overall oxidation state** is just the ion charge.

SO_4^{2-} — overall oxidation state = –2,
oxidation state of each O = –2 (total = –8),
so oxidation state of S = +6

Within an ion, the most electronegative element has a negative oxidation state (equal to its ionic charge). Other elements have more positive oxidation states.

8) The sum of the oxidation states for a **neutral compound** is 0.

Fe_2O_3 — overall oxidation state = 0, oxidation state of O = –2 (total = –6), so oxidation state of Fe = +3

9) The oxidation state of a **ligand** is equivalent to the charge on the ligand.

See page 62.

So the oxidation state of CN^- = –1 and the oxidation state of NH_3 = 0.

If you see **roman numerals** in a chemical name, it's an **oxidation number** — it applies to the atom or group immediately before it. E.g. copper has oxidation number **2** in **copper(II) sulphate**, and manganese has oxidation number **7** in a **manganate(VII) ion** (MnO_4^-).

Oxidation States go **Up** or **Down** as Electrons are **Lost** or **Gained**

1) The oxidation state for an atom will **increase by 1** for each **electron lost**.
2) The oxidation state will **decrease by 1** for each **electron gained**.
3) An element can also be **oxidised and reduced** at the same time — this is called **disproportionation**.

Oxidation No.
$$Na + \tfrac{1}{2}Cl_2 \longrightarrow Na^+ Cl^-$$
0   0   +1 –1

Example:
Chlorine and its ions undergo disproportionation reactions:

Oxidation No.
$$Cl_2 + 2OH^- \longrightarrow OCl^- + Cl^- + H_2O$$
0   +1   –1
oxidation
reduction

Oxidation and Reduction

You can separate Redox Reactions into **Half-Equations**

1) **Ionic half-equations** show oxidation or reduction.
2) An oxidation half-equation can be **combined** with a reduction half-equation to make a **full equation**.

Example: **Zinc metal** displaces **silver ions** from silver nitrate solution to form **zinc nitrate** and a deposit of **silver metal**.

The zinc atoms each lose 2 electrons (oxidation) $Zn_{(s)} \rightarrow Zn^{2+}_{(aq)} + 2e^-$
The silver ions each gain 1 electron (reduction) $Ag^+_{(aq)} + e^- \rightarrow Ag_{(s)}$

Two silver ions are needed to accept the **two electrons** released by each zinc atom.
So you need to double the silver half-equation before the two half-equations can be combined: $2Ag^+_{(aq)} + 2e^- \rightarrow 2Ag_{(s)}$

Now the number of electrons lost and gained
balance, so the half-equations can be combined: $Zn_{(s)} + 2Ag^+_{(aq)} \rightarrow Zn^{2+}_{(aq)} + 2Ag_{(s)}$

Electrons aren't included in the full equation.

H^+ *ions* May be Needed to **Reduce** Some **Oxidising Agents**

1) **Manganate(VII) ions**, MnO_4^-, contain Mn with an oxidation number of **+7**. When these ions are **reduced** they gain five electrons to become Mn^{2+} ions, with an oxidation number of **+2**.

2) In a **+2 state**, Mn can exist as simple $Mn^{2+}_{(aq)}$ ions. But in a **+7 state**, Mn has to combine with **oxygen** to form MnO_4^- ions, as $Mn^{7+}_{(aq)}$ ions wouldn't be stable.

3) MnO_4^- ions are good **oxidising agents**. The trouble is, when they get reduced to Mn^{2+} the four O^{2-} ions have to go somewhere. To solve this problem, H^+ **ions** are added. The $4O^{2-}$ now can react with $8H^+$ to form $4H_2O$. This is why manganate(VII) ions must be **acidified** to work as an oxidising agent.

Example: Acidified manganate(VII) ions can be reduced by Fe^{2+} ions.

The half-equations are: $MnO_4^-{}_{(aq)} + 8H^+_{(aq)} + 5e^- \rightarrow Mn^{2+}_{(aq)} + 4H_2O_{(l)}$
$Fe^{2+}_{(aq)} \rightarrow Fe^{3+}_{(aq)} + e^-$

To balance the electrons you have to multiply the second half-equation by 5: $5Fe^{2+}_{(aq)} \rightarrow 5Fe^{3+}_{(aq)} + 5e^-$

Now you can combine both half-equations: $MnO_4^-{}_{(aq)} + 8H^+_{(aq)} + 5Fe^{2+}_{(aq)} \rightarrow Mn^{2+}_{(aq)} + 4H_2O_{(l)} + 5Fe^{3+}_{(aq)}$

Practice Questions

Q1 What is an oxidising agent?
Q2 What is disproportionation?
Q3 Why do manganate(VII) ions have to be acidified to oxidise metals?

Exam Questions

1) What is the oxidation number of the following elements?
   a) Ti in $TiCl_4$ b) V in V_2O_5 c) Cr in CrO_4^{2-} d) Cr in $Cr_2O_7^{2-}$ [4 marks]

2) Acidified manganate(VII) ions will react with aqueous iodide ions to form iodine.
   The two half-equations for the changes that occur are $MnO_4^-{}_{(aq)} + 8H^+_{(aq)} + 5e^- \rightarrow Mn^{2+}_{(aq)} + 4H_2O_{(l)}$
   and $2I^-_{(aq)} \rightarrow I_{2(aq)} + 2e^-$

   a) Write the balanced equation to show the reaction taking place.   [2 marks]
   b) Use oxidation numbers to explain the redox processes which have occurred.   [4 marks]
   c) Suggest why a fairly reactive metal such as zinc will not react with aqueous iodide ions in a similar manner to manganate(VII) ions.   [2 marks]

Redox — relax in a lovely warm bubble bath...

The words oxidation and reduction are tossed about a lot in chemistry — so they're important.
*Don't forget, oxidation is really about electrons being lost, **not** oxygen being gained.*
I suppose you ought to learn the most famous memory aid thingy in the world — here it is...

OIL RIG
- **Oxidation Is Loss**
- **Reduction Is Gain**
  (of electrons)

Electrode Potentials

There's electrons toing and froing in redox reactions. And when electrons move, you get electricity.
OCR folks NOT doing the Transition Elements Option can skip these two pages.

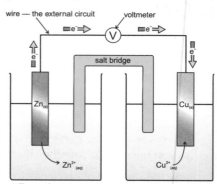
wire — the external circuit   voltmeter

Electrochemical Cells Make Electricity

Electrochemical cells can be made from **two different metals** dipped in salt solutions of their **own ions** and connected by a wire (the **external circuit**).

There are always **two** reactions within an electrochemical cell — one's an oxidation and one's a reduction — so it's a **redox process** (see page 46).

Here's what happens in the **zinc/copper** electrochemical cell on the right:

1) Zinc **loses electrons** more easily than copper. So in the half-cell on the left, zinc (from the zinc electrode) is **OXIDISED** to form $Zn^{2+}_{(aq)}$ ions. This releases electrons into the external circuit.

2) In the other half-cell, the **same number of electrons** are taken from the external circuit, **REDUCING** the Cu^{2+} ions to copper atoms.

The solutions are connected by a **salt bridge** made from filter paper soaked in $KNO_{3(aq)}$. This allows ions to flow through and balance out the charges.

So **electrons** flow through the wire from the most reactive metal to the least.

A voltmeter in the external circuit shows the **voltage** between the two half-cells. This is the **cell potential** or **emf**, E_{cell}.

The boys tested the strength of the bridge, whilst the girls just stood and watched.

You can also have half-cells involving **solutions of two aqueous ions of the same element**, such as $Fe^{2+}_{(aq)}/Fe^{3+}_{(aq)}$. The conversion from Fe^{2+} to Fe^{3+}, or vice versa, happens on the surface of the **electrode**.

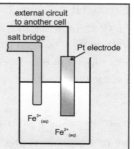
external circuit to another cell — salt bridge — Pt electrode — $Fe^{3+}_{(aq)}$ — $Fe^{2+}_{(aq)}$

There's a Convention for Drawing Electrochemical Cells   *Not for Edexcel*

It's a bit of a faff drawing electrochemical cells like the one above. There's a shorthand way of representing them though.

It's conventional to draw the **half-cell** with the **more negative** standard electrode potential, $E^{\ominus}$, on the **left**. In the case of the zinc/copper cell above, this is zinc. ⟶

Half-cell	$E^{\ominus}$/V
$Zn^{2+}_{(aq)}/Zn_{(s)}$	−0.76
$Cu^{2+}_{(aq)}/Cu_{(s)}$	+0.34

Here's the Zn/Cu cell:

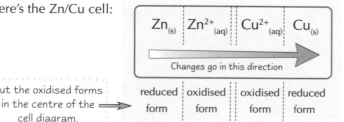
$$Zn_{(s)} \mid Zn^{2+}_{(aq)} \mid\mid Cu^{2+}_{(aq)} \mid Cu_{(s)}$$
Changes go in this direction

Put the oxidised forms in the centre of the cell diagram. ⟶

reduced form | oxidised form | oxidised form | reduced form

The metal with the more negative standard electrode potential is the one that's most eager to lose electrons and form ions. Don't panic — standard electrode potentials are covered on the next page.

Now you can **calculate** the **cell potential** by doing the calculation:

$$E^{\ominus}_{cell} = \left(E^{\ominus}_{\text{right hand side}} - E^{\ominus}_{\text{left hand side}}\right)$$

So the cell potential for the Zn/Cu cell = $+0.34 - (-0.76) = $ **+1.10 V**

The cell potential will always be a **positive voltage**, because the more negative $E^{\ominus}$ value is being subtracted from the more positive $E^{\ominus}$ value. If the positions of the half-cells were **swapped** over then the **size** of the voltage would have the **same value**, but would be **negative**.

There's a Convention for Writing Equations Too

For electrochemical cell reactions, it's the done thing to always write the **oxidised substance first**, like this:

$$Zn^{2+}_{(aq)} + 2e^- \rightleftharpoons Zn_{(s)}$$
$$Cu^{2+}_{(aq)} + 2e^- \rightleftharpoons Cu_{(s)}$$

The reversible arrow tells you the reaction can go in either direction.

So the top reaction here goes **backwards** and the bottom reaction goes **forwards**.
This gives you two half-equations $Zn_{(s)} \rightleftharpoons Zn^{2+}_{(aq)} + 2e^-$ and $Cu^{2+}_{(aq)} + 2e^- \rightleftharpoons Cu_{(s)}$.
These can be combined to give the overall equation: $Zn_{(s)} + Cu^{2+}_{(aq)} \rightleftharpoons Zn^{2+}_{(aq)} + Cu_{(s)}$

Electrode Potentials

The **Standard Electrode Potential** Tells You Which **Metal** is Oxidised

All electrode potentials are measured against a **standard hydrogen electrode**.

> The **standard electrode potential** of a half-cell is the **voltage measured** under **standard conditions** when the **half-cell** is connected to a **standard hydrogen electrode**.

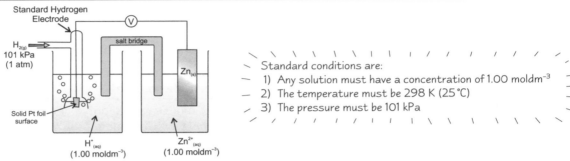

Standard conditions are:
1) Any solution must have a concentration of 1.00 moldm^{-3}
2) The temperature must be 298 K (25 °C)
3) The pressure must be 101 kPa

1) The **standard hydrogen electrode** is always shown on the **left** — it doesn't matter whether or not the other half-cell has a more positive value. The standard hydrogen electrode half-cell has a value of **0.00 V**.

3) The whole cell potential = $E^{\circ}_{\text{right-hand side}} - E^{\circ}_{\text{left-hand side}}$.

   $E^{\circ}_{\text{left-hand side}} = 0.00\,V$, so the **voltage reading** will be equal to $E^{\circ}_{\text{right-hand side}}$.
   This reading could be **positive** or **negative**, depending which way the **electrons flow**.

3) In an electrochemical cell, the half-cell with the **most negative** standard electrode potential is the one in which **oxidation** happens.

The **Standard Hydrogen Electrode** is **Tricky** to Use

The standard hydrogen electrode is a bit **awkward** to use — and it can be a trifle **dangerous** too.

Instead a **secondary standard electrode**, such as a **calomel electrode**, is used. This is easier to set up and safer to operate. It can only be used after it's been **calibrated** against a standard hydrogen electrode though.

In fact, once you know one half-cell's standard electrode potential, you can **compare** other half-cells against this known value. **The conditions must always be standard though.**

Practice Questions

1) Draw a diagram of the half-cell used for determining the standard electrode potential for the Fe^{3+}/Fe^{2+} system.
2) Fe^{3+} + e^{-} $\rightleftharpoons$ Fe^{2+}, E° = +0.77 V $\qquad$ Mn^{3+} + e^{-} $\rightleftharpoons$ Mn^{2+}, E° = +1.48 V
   Calculate the cell potential for the above system.
3) List the three standard conditions used when measuring standard electrode potentials.

Exam Question

1 An electrochemical cell containing a zinc half-cell and a silver half-cell was set up using a potassium nitrate salt bridge. The cell potential at 25 °C was measured to be 1.40 V.

$$Zn^{2+}_{(aq)} + 2e^{-} \rightleftharpoons Zn_{(s)} \qquad E^{\circ} = -0.76\,V$$
$$Ag^{+}_{(aq)} + e^{-} \rightleftharpoons Ag_{(s)} \qquad E^{\circ} = +0.80\,V$$

a) Draw a labelled diagram of this cell. [3 marks]

b) Use the standard electrode potentials given to calculate the standard cell potential for a zinc-silver cell. [1 mark]

c) Suggest two possible reasons why the actual cell potential was different from the value calculated in part (b). [2 marks]

d) Write an equation for the overall cell reaction. [1 mark]

e) Which half-cell released the electrons into the circuit? Why is this? [1 mark]

Like Greased lightning — IT'S ELECTRIFYING...

You've just got to think long and hard about this stuff. The metal on the left-hand electrode disappears off into the solution, leaving its electrons behind. This makes the left-hand electrode the negative one. So the right-hand electrode's got to be the positive one. It makes sense if you think about it. This electrode gives up electrons to turn the positive ions into atoms.

The Electrochemical Series

If you put lots of half-equations in order of their standard electrode potentials, you get the electrochemical series.
OCR folks NOT doing the Transition Elements Option can skip these two pages.

The **Electrochemical Series** Shows You What's **Reactive** and What's Not

1) The **more reactive** a **metal** is, the **more** it wants to **lose electrons** to form a **positive ion**. **More reactive metals** have **more negative standard electrode potentials**.

> **Example:** Magnesium is **more reactive** than zinc — so it's more eager to form 2+ ions than zinc is.
> The list of standard electrode potentials shows that Mg^{2+}/Mg has a **more negative** value than Zn^{2+}/Zn.
> In terms of oxidation and reduction, magnesium would **reduce** Zn^{2+} (or Zn^{2+} would **oxidise** magnesium).

2) The more reactive a **non-metal** the **more** it wants to **gain electrons** to form a **negative ion**. **More reactive non-metals** have **more positive standard electrode potentials**.

> **Example:** Chlorine is **more reactive** than bromine — so it's more eager to form a negative ion than bromine is.
> The list of standard electrode potentials shows that $Cl_2/2Cl^-$ is **more positive** than $Br_2/2Br^-$.
> In terms of oxidation and reduction, chlorine would **oxidise** Br^- (or Br^- would **reduce** chlorine).

3) Here's an **electrochemical series** showing some standard electrode potentials:

Chestnut wondered if his load was hindering his pulling potential.

Half-reaction	E°/V
$Mg^{2+}_{(aq)} + 2e^- \rightleftharpoons Mg_{(s)}$	−2.38
$Al^{3+}_{(aq)} + 3e^- \rightleftharpoons Al_{(s)}$	−1.66
$Zn^{2+}_{(aq)} + 2e^- \rightleftharpoons Zn_{(s)}$	−0.76
$Ni^{2+}_{(aq)} + 2e^- \rightleftharpoons Ni_{(s)}$	−0.25
$2H^+_{(aq)} + 2e^- \rightleftharpoons H_{2(g)}$	0.00
$Sn^{4+}_{(aq)} + 2e^- \rightleftharpoons Sn^{2+}_{(aq)}$	+0.15
$Cu^{2+}_{(aq)} + 2e^- \rightleftharpoons Cu_{(s)}$	+0.34
$Fe^{3+}_{(aq)} + e^- \rightleftharpoons Fe^{2+}_{(aq)}$	+0.77
$Ag^+_{(aq)} + e^- \rightleftharpoons Ag_{(s)}$	+0.80
$Br_{2(aq)} + 2e^- \rightleftharpoons 2Br^-_{(aq)}$	+1.07
$Cr_2O_7^{2-}{}_{(aq)} + 14H^+_{(aq)} + 6e^- \rightleftharpoons 2Cr^{3+}_{(aq)} + 7H_2O_{(l)}$	+1.33
$Cl_{2(aq)} + 2e^- \rightleftharpoons 2Cl^-_{(aq)}$	+1.36
$MnO_4^-{}_{(aq)} + 8H^+_{(aq)} + 5e^- \rightleftharpoons Mn^{2+}_{(aq)} + 4H_2O_{(l)}$	+1.52

More positive electrode potentials mean that:
1. The left-hand substances are more easily reduced.
2. The right-hand substances are more stable.

More negative electrode potentials mean that:
1. The right-hand substances are more easily oxidised.
2. The left-hand substances are more stable.

The **Anticlockwise Rule** Predicts Whether a Reaction Will Happen

To figure out if a metal will react with the aqueous ions of another metal, you can use the **anticlockwise rule**.

For example, will zinc react with aqueous copper ions?
First you write the two half-equations down, putting the one with the **more negative** standard electrode potential on **top**. Then you draw on some **anticlockwise arrows** — these give you the **direction** of each half-reaction.

$Zn^{2+}_{(aq)} + 2e^- \rightleftharpoons Zn_{(s)}$ $E^\circ = -0.76\,V$
$Cu^{2+}_{(aq)} + 2e^- \rightleftharpoons Cu_{(s)}$ $E^\circ = +0.34\,V$

The **half-equations** are: $Zn_{(s)} \rightleftharpoons Zn^{2+}_{(aq)} + 2e^-$
$Cu^{2+}_{(aq)} + 2e^- \rightleftharpoons Cu_{(s)}$

Which combine to give: $Zn_{(s)} + Cu^{2+}_{(aq)} \rightleftharpoons Zn^{2+}_{(aq)} + Cu_{(s)}$
So zinc **does** react with aqueous copper ions.

To find the **cell potential** you always do $E^\circ_{bottom} - E^\circ_{top}$, so the cell potential for this reaction is $+0.34 - (-0.76) = +1.10\,V$.

You can also draw an **electrode potential chart**. It's the same sort of idea.
You draw an 'upside-down y-axis' with the more negative number at the top.
Then you put both half-reactions on the chart and draw on your **anticlockwise** arrows which give you the **direction** of each half-reaction.

The **difference** between the values is the **cell potential** — in this case it's **+1.10 V**.

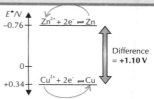

The Electrochemical Series

High $E^\ominus$ Values Suggest a High K_c and a Positive Entropy Change

1) Reactions will only happen spontaneously in the direction that results in a **positive cell potential**.
   If the cell potential is **positive** and **greater than about 0.4 V**, the reaction will go to **completion**. If the value is **between 0.0 V and about 0.4 V** then the reaction will still happen, but it'll be in **equilibrium** and will be **reversible**.

2) If you get a negative E° value then the reaction will go in **reverse**.

3) A **highly positive** cell potential means that the equilibrium lies far to the **right**. So the value of the equilibrium constant, K_c, will be **high**.

See page 20 for K_c and pages 8-9 for entropy change.

4) A **highly positive** cell potential also means that the forward reaction is likely to happen **spontaneously**. This suggests a **positive entropy change**.

5) But this isn't a definite rule — both predictions **may be wrong**. The **rate** may be so extremely **slow** that the reaction doesn't appear to happen. Alternatively, a feasible reaction might not happen at all because the **activation energy** is too high.

6) If you use non-standard **concentrations** (or temperatures) the reaction might not occur as expected.

> For example, in the zinc/copper cell these equilibria are set up: $Zn_{(s)} \rightleftharpoons Zn^{2+}_{(aq)} + 2e^-$ and $Cu^{2+}_{(aq)} + 2e^- \rightleftharpoons Cu_{(s)}$
>
> If the Zn^{2+} concentration is **increased** then the **zinc equilibrium shifts to the left-hand side**.
> This **reduces the ease of electron loss**, so the whole cell potential would be lower.
>
> If the Cu^{2+} concentration is **increased** then the **copper equilibrium shifts to the right-hand side**.
> This **increases the ease of electron gain**, so the cell potential would be higher.

$E^\ominus$ Values Can be Used to Predict the Stability of Oxidation States

OCR (Transition Elements Option) and Salters only

1) Both **iron** and **cobalt** form simple **+2** and **+3** ions, which are in equilibrium:

$$Fe^{3+}_{(aq)} + e^- \rightleftharpoons Fe^{2+}_{(aq)} \qquad E^\circ = +0.77 \text{ V}$$
$$Co^{3+}_{(aq)} + e^- \rightleftharpoons Co^{2+}_{(aq)} \qquad E^\circ = +1.82 \text{ V}$$

2) The cobalt half-equation has a **more positive** E° value than the iron half-equation.
   This suggests that cobalt +3 ions are **less stable** than iron +3 ions.
   In fact $Co^{3+}_{(aq)}$ ions are **incredibly unstable** in water, **reducing** very readily to $Co^{2+}_{(aq)}$.

Practice Questions

Q1 Cu is less reactive than Pb. Which half-reaction below would have a more negative standard electrode potential?

$$Pb^{2+} + 2e^- \rightleftharpoons Pb$$
$$\text{or} \quad Cu^{2+} + 2e^- \rightleftharpoons Cu$$

Q2 What is the overall E° value which a reaction must have for it to proceed?

Q3 Give two reasons why a reaction that is expected to be spontaneous might not happen.

Q4 Use the table on the opposite page to predict whether or not Zn^{2+} ions can oxidise Fe^{2+} ions to Fe^{3+} ions.

Exam Question

Q1 Use E° values quoted on the opposite page to determine the outcome of mixing the following solutions.

If there is a reaction, determine the E° value and write the equation. If there isn't a reaction, state this and explain why.

a) Zinc metal and Ni^{2+} ions   [2 marks]
b) Acidified MnO_4^- ions and Sn^{2+} ions   [2 marks]
c) $Br_{2(aq)}$ and acidified $Cr_2O_7^{2-}$ ions   [2 marks]
d) Silver ions and Fe^{2+} ions   [2 marks]

The forward reaction that happens is the one with the most positive $E^\ominus$ value...

To see if a reaction will happen, you basically find the two half-equations in the electrochemical series and check whether you can draw anticlockwise arrows on them to get from your reactants to your products. If you can — great. The reaction will have a positive electrode potential, so it should happen. If you can't — well, it ain't gonna work.

Applying Electrochemistry

Copper(I) solutions are about as common as hens' teeth. Read on and I'll tell you why. Oooh, bet you're excited now.
AQA people and OCR people NOT doing the Transition Elements Option can skip these two pages.

Copper(I) Ions are Unstable in Water — *Only for Edexcel and OCR (Transition Elements Option)*

1) Copper(I) compounds, such as CuI, CuCl and Cu_2SO_4, do exist as white **solids**, but they **can't** exist in **solution** because Cu^+ ions undergo **disproportionation**. This means the ions are **both** reduced and oxidised **simultaneously**.

2) The disproportionation happens because one Cu^+ ion **oxidises** another Cu^+ ion into a Cu^{2+} ion. The oxidising Cu^+ ion is itself **reduced** to a Cu atom.

$$2Cu^+_{(aq)} \longrightarrow Cu^{2+}_{(aq)} + Cu_{(s)}$$

3) The Cu^{2+} ions give the solution a **blue** colour and the solid copper forms a **red-brown precipitate**.

4) The disproportionation can be explained by looking at the E° values for the two half-equations and applying the anticlockwise rule (see page 50):

$$Cu^{2+}_{(aq)} + e^- \rightleftharpoons Cu^+_{(aq)} \quad E^\circ = +0.15 \text{ V}$$
$$Cu^+_{(aq)} + e^- \rightleftharpoons Cu_{(s)} \quad E^\circ = +0.52 \text{ V}$$

Combining the equations gives:

$$Cu^+_{(aq)} + Cu^+_{(aq)} \rightleftharpoons Cu_{(s)} + Cu^{2+}_{(aq)} \quad E^\circ = +0.37 \text{ V}$$

So half the Cu^+ ions (the ones shown in blue) are reduced to Cu. And the other half (shown in red) are oxidised to Cu^{2+}.

Standard Electrode Potentials are Used as the Basis of **Battery** Making

Edexcel and Nuffield only.

The batteries that are used for everyday objects, like torches, clocks, and Duracell bunnies, are a type of **portable cell**. The contents of these batteries are '**dry**' — the half-cells aren't solid/solution mixtures.

A typical battery is shown below.

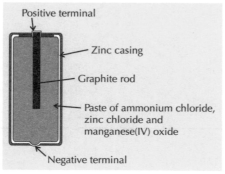

Positive terminal
— Zinc casing
— Graphite rod
— Paste of ammonium chloride, zinc chloride and manganese(IV) oxide
Negative terminal

9 Volts
battery hen
(deceased)

1) The **negative terminal** is the zinc casing where Zn atoms are **oxidised** to Zn^{2+} ions, releasing electrons to the external circuit.

$$Zn \longrightarrow Zn^{2+} + 2e^-$$

2) The **positive terminal** is made from graphite within a **paste of solid ammonium chloride**, **solid zinc chloride** and **solid manganese(IV) oxide**.

The MnO_2 within the paste is **reduced** to MnO(OH) by the electrons taken in from the external circuit.

$$MnO_2 + NH_4^+ + e^- \longrightarrow MnO(OH) + NH_3$$

3) The **ammonia** released in the reduction forms a **complex** with the Zn^{2+} **ions** formed in the oxidation.

$$Zn^{2+} + 4NH_3 \longrightarrow [Zn(NH_3)_4]^{2+}$$

When all the reactants have been used up, the battery won't work any more.

Rusting All Comes Down to Electrochemical Processes Too — *Edexcel and Salters only*

If iron's exposed to **oxygen** and **water**, it'll turn into crumbly, flaky stuff called **rust**. Here's how:

1) There are two half-equations involved:

$$Fe^{2+}_{(aq)} + 2e^- \rightleftharpoons Fe_{(s)} \quad E^\circ = -0.44 \text{ V}$$
$$2H_2O_{(l)} + O_{2(g)} + 4e^- \rightleftharpoons 4OH^-_{(aq)} \quad E^\circ = +1.23 \text{ V}$$

So the overall reaction is:

$$2H_2O_{(l)} + O_{2(g)} + 2Fe_{(s)} \longrightarrow 2Fe^{2+}_{(aq)} + 4OH^-_{(aq)} \quad E^\circ = +1.67 \text{ V}$$

2) The $Fe^{2+}_{(aq)}$ and $OH^-_{(aq)}$ ions produced combine to form a precipitate of iron(II) hydroxide, $Fe(OH)_2$.

$$Fe^{2+}_{(aq)} + 2OH^-_{(aq)} \longrightarrow Fe(OH)_{2(s)}$$

3) The $Fe(OH)_2$ is further oxidised to $Fe(OH)_3$ by oxygen and water.

$$2H_2O_{(l)} + O_{2(g)} + 4Fe(OH)_{2(s)} \longrightarrow 4Fe(OH)_{3(s)}$$

4) Iron(III) hydroxide gradually turns into hydrated iron(III) oxide, $Fe_2O_3.xH_2O$ — this is **rust**.

Applying Electrochemistry

There are Lots of Ways to Prevent Rusting
Edexcel and Salters only

1) The obvious way to prevent rusting is to coat the iron with a **barrier** to keep out either the oxygen, the water or both. Barrier methods include:

Painting	This is ideal for large and small structures alike. It can also be decorative.
Oiling/Greasing	This has to be used when moving parts are involved, like on bike chains.
Galvanising	A coat of zinc is sprayed onto the object. E.g. buckets, corrugated roofing.
Tinning	A coat of tin is applied to the object. E.g. food cans.
Chrome plating	A coat of chromium is electroplated onto the iron.

2) The other way is the **sacrificial method**. This involves placing a **more reactive metal** with the iron. The water and oxygen then react with this **sacrificial metal** instead of with the iron. **Galvanising** and **chrome plating** are examples of the sacrificial method.

3) Another sacrificial method involves bolting **blocks of zinc or magnesium** to the iron. This is used on the hulls of ships, or on underground iron pipes.

   Both Zn / Zn^{2+} and Mg / Mg^{2+} systems have **more negative** E° values than the Fe / Fe^{2+} system. This means the **zinc** will be **oxidised** to Zn^{2+} ions and the **magnesium** to Mg^{2+} ions in preference to the iron.

$$Mg^{2+}_{(aq)} + 2e^- \rightleftharpoons Mg_{(s)} \qquad E^\circ = -2.38\,V$$
$$Zn^{2+}_{(aq)} + 2e^- \rightleftharpoons Zn_{(s)} \qquad E^\circ = -0.76\,V$$
$$Fe^{2+}_{(aq)} + 2e^- \rightleftharpoons Fe_{(s)} \qquad E^\circ = -0.44\,V$$

4) Tinning **isn't** a sacrificial method. It only works as long as the coat of tin remains **intact**. If the iron is **exposed**, then it will rust in preference to the tin, because the Fe / Fe^{2+} system has a more negative E° value than the Sn / Sn^{2+} system

$$Sn^{2+}_{(aq)} + 2e^- \rightleftharpoons Sn_{(s)} \qquad E^\circ = -0.14\,V$$

This is why discarded cans go very rusty very quickly.

Don't Just Leave Iron to Rust — Recycle It
Salters only

Iron ore is a **finite resource** — it's going to **run out** one day. It also takes **lots of energy** to extract new iron from the ore.

Scrap iron and steel is recycled by **adding it to the melt** in the **steelmaking process**.

This helps because steelmaking is very **exothermic** and produces heaps of energy — which leads to the **temperature soaring** if it's left uncontrolled.

The best way of preventing a dangerous rise in temperature is to allow the excess energy to melt scrap steel.

Practice Questions

Q1 What is disproportionation?

Q2 What's the chemical name for rust?

Q3 Name two metals which could be used in the sacrificial method of rust prevention.

Exam Question

1 a) Rust-proofing of iron takes many forms. Explain the chemistry involved for each of the methods below.
   (i) Placing magnesium blocks on an object.
   (ii) Coating an object with a layer of zinc.
   (iii) Coating an object with a layer of tin. [8 marks]
   b) Name an object suitable for each method. [2 marks]

Rust — the crumbliest, flakiest hydrated oxide in the world...

So if the iron man thought he could stand next to the tin man for protection, he'd be wrong — he'd still go rusty first. The iron man needs to hope someone invents the zinc man. Being the iron man definitely wouldn't be ideal — he's also in danger from Magneto. He needs to get himself an adamantium skeleton like Wolverine, then he won't crumble completely.

Transition Metals — The Basics

The rest of this section's about transition metals — and there's a lot of it left.
So it's obviously important stuff in the A2 chemistry world.

Transition Elements are Found in the d-Block

The **d-block** is the block of elements in the middle of the periodic table.

Most of the elements in the d-block are **transition elements** (or transition metals). You mainly need to know about the ones in the first row of the d-block. These are the elements from **titanium to copper**.

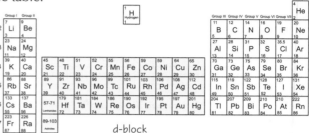

d-block

You Have to Know the Electronic Configurations of the Transition Metals

The transition metals in the **first row** of the d-block all have their highest energy electrons in the **4s** and **3d** subshells.

To find the **total number of 4s and 3d electrons**, you take **18** away from the element's **atomic number**.
E.g., the atomic number of **cobalt** is 27, so it has 27 − 18 = **9** 4s and 3d electrons.

The **4s** subshell is always filled **first**, **EXCEPT** for in **chromium** and **copper**. See below for why.

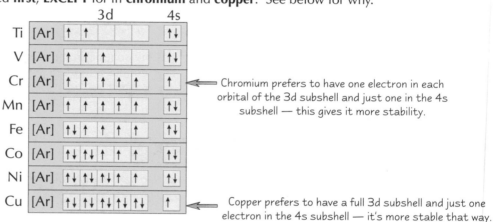

The 3d orbitals are occupied singly at first. They only double up when they have to.

Chromium prefers to have one electron in each orbital of the 3d subshell and just one in the 4s subshell — this gives it more stability.

Copper prefers to have a full 3d subshell and just one electron in the 4s subshell — it's more stable that way.

Don't forget you can also use **subshell notation** to show electronic configurations.
In subshell notation, **vanadium** would have the electronic configuration $1s^2 2s^2 2p^6 3s^2 3p^6 3d^3 4s^2$, or $[Ar]3d^3 4s^2$.
Check back to your AS notes if this makes less than perfect sense to you.

Sc and Zn Ain't Transition Metals

Here's the definition of a transition metal:

A **transition metal** is one that can form **one or more stable ions** with an **partially filled d-subshell**.

d-orbitals can fit **10** electrons in. So transition metals must form **at least one ion** that has **between 1 and 9 electrons** in the d-orbital. All the Period 4 d-block elements are transition metals apart from **scandium** and **zinc**.

Scandium only forms one ion, Sc^{3+}, which has an **empty d-subshell**. Scandium has the electronic configuration $[Ar]3d^1 4s^2$, so when it loses three electrons to form Sc^{3+}, it ends up with the electronic configuration $[Ar]$.

Zinc only forms one ion, Zn^{2+}, which has a **full d-subshell**. Zinc has the electronic configuration $[Ar]3d^{10} 4s^2$.
When it forms Zn^{2+} it loses 2 electrons, both from the 4s subshell. This means it keeps its full 3d subshell.

When Ions are Formed, the s Electrons are Removed First

When the atoms form **positive** ions, the **s electrons** are removed **first**, **then** d electrons.

For example, iron forms Fe^{2+} ions and Fe^{3+} ions.
When it forms 2+ ions, it loses **both its 4s electrons**. $Fe = [Ar]3d^6\ 4s^2 \rightarrow Fe^{2+} = [Ar]3d^6$
Only once the 4s electrons are removed can a **3d electron** be removed. E.g. $Fe^{2+} = [Ar]3d^6 \rightarrow Fe^{3+} = [Ar]3d^5$

Transition Metals — The Basics

The Transition Metals all Have Similar Physical Properties

The transition elements don't gradually change across the periodic table like you might expect.
They're all typical metals and have **similar physical properties**:

> 1) They all have a **high density**.
> 2) They all have **high melting** and **high boiling points**.
> 3) Their **ionic radii** are more or less the same.

Transition Metals have Special Chemical Properties

1) They can form **complex ions** — see pages 62-65. E.g. iron forms a **complex ion with water** — $[Fe(H_2O)_6]^{2+}$.

2) They form **coloured ions** — see pages 66-67. E.g. Fe^{2+} ions are **pale green** and Fe^{3+} ions are **yellow**.

3) They can act as **good catalysts** — see page 68. E.g. iron is the catalyst used in the **Haber process**.

4) They can exist in **variable oxidation states**. E.g. iron can exist in the **+2** oxidation state as Fe^{2+} ions and in the **+3** oxidation state as Fe^{3+} ions.

Some common **coloured** ions and **oxidation states** are shown below. The colours refer to the **aqueous ions**.

oxidation state +7	+6	+5	+4	+3	+2
		VO_2^+ (yellow)	VO^{2+} (blue)	V^{3+} (green)	V^{2+} (violet)
	$Cr_2O_7^{2-}$ (orange)			Cr^{3+} (green/violet)	
MnO_4^- (purple)				*See page 58.*	Mn^{2+} (pale pink)
				Fe^{3+} (yellow)	Fe^{2+} (pale green)
				Co^{3+} (brown)	Co^{2+} (pink)
					Ni^{2+} (green)
					Cu^{2+} (blue)

These elements show **variable** oxidation states because the **energy levels** of the 4s and the 3d subshells are **very close** to one another. So different numbers of electrons can be gained or lost using fairly **similar** amounts of energy.

Practice Questions

Q1 Which two d-block elements from Period 4 aren't transition elements?
Q2 Give the electronic arrangement of: (a) a vanadium atom, (b) a V^{2+} ion.
Q3 Do transition metals have high or low boiling points?
Q4 State four chemical properties which are characteristic of transition elements.

Exam Question

1 When solid copper(I) sulphate is added to water, a blue solution forms with a red-brown precipitate of copper metal.

   a) Give the electron configuration of copper(I) ions. [1 mark]

   b) Does the formation of copper(I) ions show copper acting as a transition metal? Explain your answer. [2 marks]

   c) Identify the blue solution. [1 mark]

   d) Explain why transition metals such as copper exist in more than one oxidation state. [2 marks]

s electrons — like rats leaving a sinking ship...

Definitely have a quick read of the electronic configuration stuff in your AS notes if it's been pushed to a little corner of your mind labelled, "Well, I won't be needing that again in a hurry". It should come flooding back pretty quickly. This page is just an overview of transition metal properties. Don't worry — they're all looked at in lots more detail in the coming pages...

Transition Metals — Vanadium and Cobalt

Vanadium ions come in just about as many colours as Bertie Bott's Every-Flavour Beans.
OCR people who are NOT doing the Transition Metals Option and Salters can skip these two pages.

Vanadium has **Four Different Oxidation States** and **Four Different Colours**

You need to know the **formula** and **colour** of each oxidation state — so here they are:

Oxidation state	Formula of ion	Colour of ion
+2	$V^{2+}_{(aq)}$	Violet
+3	$V^{3+}_{(aq)}$	Green
+4	$VO^{2+}_{(aq)}$	Blue
+5	$VO_2^{+}{}_{(aq)}$	Yellow

Vanadium can exist as simple ions in the +2 and +3 oxidation states. V^{4+} and V^{5+} wouldn't be stable though, so vanadium exists as oxocations in these higher oxidation states.

Vanadium(V) can be **Reduced** to Vanadium(IV), Vanadium(III) or Vanadium(II)

What it's reduced to depends on how **strong** a **reducing agent** you use.

ZINC (IN ACID) REDUCES VANADIUM(V) ALL THE WAY TO VANADIUM(II)

When a bit of zinc is added to a solution of acidified VO_2^+ the solution turns from **yellow**, to **blue** (VO^{2+}), to **green** (V^{3+}) and finally to **violet** (V^{2+}).

This can be explained by looking at E° values. Here are the half-equations (and **standard electrode potentials**) for the conversions between vanadium's oxidation states:

half-equation 1	$VO_2^+{}_{(aq)} + 2H^+{}_{(aq)} + e^- \rightleftharpoons VO^{2+}{}_{(aq)} + H_2O_{(l)}$	$E^{\circ} = +1.00\ V$
half-equation 2	$VO^{2+}{}_{(aq)} + 2H^+{}_{(aq)} + e^- \rightleftharpoons V^{3+}{}_{(aq)} + H_2O_{(l)}$	$E^{\circ} = +0.34\ V$
half-equation 3	$V^{3+}{}_{(aq)} + e^- \rightleftharpoons V^{2+}{}_{(aq)}$	$E^{\circ} = -0.26\ V$

And here's the half-equation for the reducing agent, zinc: $Zn^{2+}{}_{(aq)} + 2e^- \rightleftharpoons Zn_{(s)}$ $E^{\circ} = -0.76\ V$

You can combine this half-equation with **any** of the half-equations above to show that zinc will reduce VO_2^+ ions, VO^{2+} ions, and V^{3+} ions. E.g.

See page 50 for how to combine half-equations.

$Zn^{2+}{}_{(aq)} + 2e^- \rightleftharpoons Zn_{(s)}$ $E^{\circ} = -0.76\ V$
$VO_2^+{}_{(aq)} + 2H^+{}_{(aq)} + e^- \rightleftharpoons VO^{2+}{}_{(aq)} + H_2O_{(l)}$ $E^{\circ} = +1.00\ V$

Combined equation: $Zn_{(s)} + 2VO_2^+{}_{(aq)} + 4H^+{}_{(aq)} \rightleftharpoons 2VO^{2+}{}_{(aq)} + 2H_2O_{(l)} + Zn^{2+}{}_{(aq)}$ $E^{\circ} = +1.76\ V$

A positive value means the reaction can happen.

SULPHUR DIOXIDE REDUCES VANADIUM(V) TO VANADIUM(III)   *Not AQA*

Sulphur dioxide gas has **less reducing power** than acidified zinc. When sulphur dioxide gas is bubbled through a solution of acidified vanadate(V) ions, the solution turns from **yellow** (VO_2^+), to **blue** (VO^{2+}) and then to **green** (V^{3+}). It **doesn't** reduce the green (V^{3+}) solution to violet (V^{2+}) though.

The half-equation for the reducing agent **sulphur dioxide** is: $SO_4^{2-}{}_{(aq)} + 4H^+{}_{(aq)} + 2e^- \rightleftharpoons SO_{2(g)} + 2H_2O_{(l)}$ $E^{\circ} = +0.17\ V$.

This can be combined with half-equations 1 and 2 to show that VO_2^+ can be **reduced** to $VO^{2+}{}_{(aq)}$, and that this can then be reduced to V^{3+}.

The E° value of $+0.17\ V$ is **too positive** to allow SO_2 to reduce the V^{3+} to V^{2+}. You **can't** combine the SO_2 half-equation with half-equation 3 so that V^{3+} is reduced to V^{2+} and still get a positive E° value.

A positive E° value means the reaction can happen.

IRON(II) IONS ONLY REDUCE VANADIUM(V) TO VANADIUM(IV)   *Not AQA*

Iron(II) ions are the wimps of the reducing agents. They can turn a solution of acidified vanadate(V) ions from **yellow** (VO_2^+) to **blue** (VO^{2+}), but **can't reduce** these ions any **further**.

The half-equation for Fe^{2+}/Fe^{3+} is: $Fe^{3+}{}_{(aq)} + e^- \rightleftharpoons Fe^{2+}{}_{(aq)}$ $E^{\circ} = +0.77\ V$

This can be combined with half-equation 1 to show that Fe^{2+} **reduces** yellow VO_2^+ to blue VO^{2+}.
The E° value of $+0.77\ V$ is **too positive** to allow Fe^{2+} to reduce the blue VO^{2+} ions any further though.

Transition Metals — Vanadium and Cobalt

Cobalt can exist as Co²⁺ and Co³⁺   OCR (Transition Elements Option) and Nuffield only

Cobalt can exist in two oxidation states — **+2** as **Co²⁺**, and **+3** as **Co³⁺**. It much prefers to be in the **+2** state though.

Co²⁺ ions are hydrated in water to $[Co(H_2O)_6]^{2+}$ — which is a **stable** complex. You can oxidise $[Co(H_2O)_6]^{2+}$ to $[Co(H_2O)_6]^{3+}$, but $[Co(H_2O)_6]^{3+}$ is very **unstable**, so it's **reduced really easily** back to $[Co(H_2O)_6]^{2+}$. This makes $[Co(H_2O)_6]^{3+}$ ions **very good oxidising agents**.

The $E°$ value for the half-equation $[Co(H_2O)_6]^{3+} + e^- \rightleftharpoons [Co(H_2O)_6]^{2+}$ is **highly positive** (+1.82). This means the right-hand ion is likely to be **more stable** than the left-hand ion, and the equilibrium will lie **far to the right**.

Even **water reduces** $[Co(H_2O)_6]^{3+}$ ions:

$$O_{2(g)} + 4H^+_{(aq)} + 4e^- \rightleftharpoons 2H_2O_{(l)} \qquad E° = +1.23 \text{ V}$$

$$[Co(H_2O)_6]^{3+}_{(aq)} + e^- \rightleftharpoons [Co(H_2O)_6]^{2+}_{(aq)} \qquad E° = +1.82 \text{ V}$$

Combined equation: $2H_2O_{(l)} + 4[Co(H_2O)_6]^{3+}_{(aq)} \rightleftharpoons 4[Co(H_2O)_6]^{2+}_{(aq)} + O_{2(g)} + 4H^+_{(aq)} \qquad E° = +0.59 \text{ V}$

Co²⁺ can be Oxidised by Alkaline H₂O₂ or by Air with Ammonia Solution

AQA and OCR (Transition Elements Option) only

1) Co³⁺ can be made by oxidising Co²⁺₍ₐq₎ with **alkaline hydrogen peroxide**.

$$2Co^{2+}_{(aq)} + H_2O_{2(aq)} + 4OH^-_{(aq)} \rightarrow 2Co^{3+}_{(aq)} + 6OH^-_{(aq)}$$

2) You can also oxidise Co²⁺ with **ammonia solution and air**. You place Co²⁺ ions in an excess of **ammonia solution**, which causes **$[Co(NH_3)_6]^{2+}$ ions** to form. If these complex ions are left to stand in **air**, oxygen oxidises them to **$[Co(NH_3)_6]^{3+}$**, which is **stable**.

The NH₃ ligands stabilise the Co³⁺ ions.

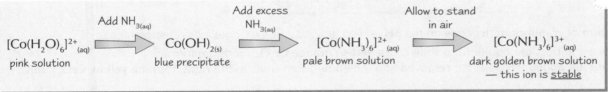

$[Co(H_2O)_6]^{2+}_{(aq)}$ — pink solution → Add NH₃₍ₐq₎ → Co(OH)₂₍ₛ₎ blue precipitate → Add excess NH₃₍ₐq₎ → $[Co(NH_3)_6]^{2+}_{(aq)}$ pale brown solution → Allow to stand in air → $[Co(NH_3)_6]^{3+}_{(aq)}$ dark golden brown solution — this ion is stable

$[Co(NH_3)_6]^{2+}_{(aq)}$ complex ions are far **easier to oxidise** than aqueous Co²⁺ ions because the $E°$ value is **far less positive**.

So this equilibrium lies further to the left. → $[Co(NH_3)_6]^{3+} + e^- \rightleftharpoons [Co(NH_3)_6]^{2+} \qquad E° = +0.11 \text{ V}$

Practice Questions

Q1 What are the common oxidation states shown by vanadium?

Q2 What is the colour of the aqueous V³⁺ ion?

Q3 Why are Co³⁺ ions so unstable in water?

Q4 Name two different solutions which allow the formation of the +3 state of cobalt.

Exam Question

1 Acidified zinc is a good reducing agent.

   a) Describe the colour changes seen if acidified zinc is added to a yellow solution of vanadate(V) ions. [3 marks]

   b) Use equations and oxidation numbers to explain any colour changes seen. [7 marks]

All reducing agents are not the same...

Lots of transition metal chemistry comes back to electrode potentials — that's why they've been bunged in the same section. Remember — you can tell if things'll react by looking at the half-equations and using that anticlockwise rule from page 50. It only works if you put the most electronegative reaction on top though. It's bottom minus top to get the overall E° value.

Transition Metals — Chromium and Copper

With all these transition metals that begin with C, it's hard to keep them straight...
OCR people who are NOT doing the Transition Metals Option and Salters can skip this page.

Chromium Exists in the +2, +3 and +6 Oxidation States

Chromium is used to make **stainless steel**. It's also added to steel to make it **harder**.
It mainly exists in the **+3** and **+6** oxidation states though, as the +2 oxidation state is **unstable**.

Chromium forms two oxoanions in the +6 oxidation state — chromate(VI) ions, CrO_4^{2-}, and dichromate(VI) ions, $Cr_2O_7^{2-}$. Both these ions are good oxidising agents because they can easily be reduced to Cr^{3+}.

Oxidation state	Formula of ion	Colour of ion
+6	$Cr_2O_7^{2-}{}_{(aq)}$	Orange
+6	$CrO_4^{2-}{}_{(aq)}$	Yellow
+3	$Cr^{3+}{}_{(aq)}$	Green (Violet)
+2	$Cr^{2+}{}_{(aq)}$	Blue

When Cr^{3+} ions are surrounded by 6 water ligands they're violet. The water ligands are usually substituted though, so the colour is usually green.

Chromate(VI) Ions, CrO_4^{2-}, and Dichromate(VI) Ions, $Cr_2O_7^{2-}$, Exist in Equilibrium

AQA and OCR (Transition Elements Option) only

1) When an **alkali** (OH⁻ ions) is added to aqueous **dichromate(VI) ions**, the orange colour turns **yellow**, because aqueous **chromate(VI) ions** form.

$$Cr_2O_7^{2-}{}_{(aq)} + OH^-{}_{(aq)} \rightarrow 2CrO_4^{2-}{}_{(aq)} + H^+{}_{(aq)}$$
orange   **yellow**

2) When an **acid** (H⁺ ions) is added to aqueous **chromate(VI) ions**, the yellow colour turns **orange**, because aqueous **dichromate(VI) ions** form.

$$2CrO_4^{2-}{}_{(aq)} + H^+{}_{(aq)} \rightarrow Cr_2O_7^{2-}{}_{(aq)} + OH^-{}_{(aq)}$$
yellow   **orange**

3) These are **opposite processes** and the two ions exist in **equilibrium**.

$$Cr_2O_7^{2-}{}_{(aq)} + H_2O_{(l)} \rightleftharpoons 2CrO_4^{2-}{}_{(aq)} + 2H^+{}_{(aq)}$$

This isn't a redox process because chromium stays in the +6 oxidation state.

The **position** of equilibrium depends on the **pH** — yep, it's good ol' Le Chatelier's principle again.
If **H⁺ ions** are added, the equilibrium shifts to the **left** so orange $Cr_2O_7^{2-}$ ions are formed.
If **OH⁻ ions** are added, H+ ions are **removed** and the equilibrium shifts to the **right**, forming **yellow CrO_4^{2-} ions**.

Chromium Ions can be Oxidised and Reduced   **AQA only**

1) Dichromate(VI) ions can be **reduced** using a good reducing agent, such as **zinc and dilute acid**.

Oxidation states: +6   0   +2   +3
$$Cr_2O_7^{2-}{}_{(aq)} + 14H^+{}_{(aq)} + 3Zn_{(s)} \rightarrow 3Zn^{2+}{}_{(aq)} + 2Cr^{3+}{}_{(aq)} + 7H_2O_{(l)}$$
orange   **green**

2) Zinc will **reduce** Cr^{3+} further to Cr^{2+} —

$$2Cr^{3+}{}_{(aq)} + Zn_{(s)} \rightarrow Zn^{2+}{}_{(aq)} + 2Cr^{2+}{}_{(aq)}$$
green   **blue**

But unless you use an inert atmosphere, you're wasting your time — Cr^{2+} is so **unstable** that it oxidises straight back to Cr^{3+} in air.

3) You can oxidise Cr^{3+} to chromate(VI) ions with **hydrogen peroxide** in an **alkaline** solution.

Oxidation states: +3   +6
$$2Cr^{3+}{}_{(aq)} + 10OH^-{}_{(aq)} + 3H_2O_{2(aq)} \rightarrow 2CrO_4^{2-}{}_{(aq)} + 8H_2O_{(l)}$$
green   **yellow**

Here's a summary of all the chromium reactions you need to know:

$$Cr_2O_7^{2-}{}_{(aq)} \underset{OH^-{}_{(aq)}}{\overset{H^+{}_{(aq)}}{\rightleftarrows}} CrO_4^{2-}{}_{(aq)}$$

REDUCTION
$H^+{}_{(aq)}/Zn_{(s)}$

OXIDATION
$OH^-{}_{(aq)}/H_2O_{2(aq)}$

$Cr^{3+}{}_{(aq)}$

REDUCTION
$H^+{}_{(aq)}/Zn_{(s)}$
(inert atmosphere)

OXIDATION
air

$Cr^{2+}{}_{(aq)}$

Transition Metals — Chromium and Copper

The stuff on copper is only for OCR (Transition Elements Option), Salters, Edexcel and Nuffield.

Copper Ions Show Two Oxidation States — +1 and +2

Oxidation state	Formula	Electronic configuration	Colour
0	Cu	$1s^2\ 2s^2\ 2p^6\ 3s^2\ 3p^6\ 3d^{10}\ 4s^1$	Shiny brown metal (or pink if in powder form)
+1	Cu^+	$1s^2\ 2s^2\ 2p^6\ 3s^2\ 3p^6\ 3d^{10}$	White solid
+2	Cu^{2+}	$1s^2\ 2s^2\ 2p^6\ 3s^2\ 3p^6\ 3d^9$	Blue in aqueous solution

1) Cu^{2+} ions are **stable** in aqueous solution. They're **reduced** to Cu metal by more **electropositive metals**, like zinc or nickel. E.g. $Cu^{2+} + Zn \rightarrow Cu + Zn^{2+}$. This type of reaction is known as a **displacement** reaction.

2) **Solid** copper(I) compounds are **stable**, but **aqueous** Cu^+ ions are **unstable** and **disproportionate** in aqueous solution (see page 52).

$$2Cu^+_{(aq)} \rightarrow Cu^{2+}_{(aq)} + Cu_{(s)}$$

Aqueous copper(I) complexes are also stable.

3) Copper(I) compounds **aren't coloured** because copper(I) has a full **3d subshell** ($3d^{10}$) — see page 66.

Copper(II) Ions are Reduced to Copper(I) Ions by Potassium Iodide Solution

OCR (Transition Elements Option) only
$Cu^{2+}_{(aq)}$ ions can be **reduced** to **copper(I)** compounds.
This can be done by adding **aqueous potassium iodide** to **aqueous copper(II) ions**.

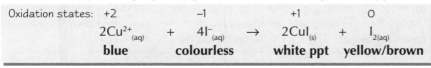

Oxidation states: +2   −1   +1   0

$$2Cu^{2+}_{(aq)} + 4I^-_{(aq)} \rightarrow 2CuI_{(s)} + I_{2(aq)}$$

blue   colourless   white ppt   yellow/brown

Copper iodide (CuI) is a white solid, but it looks like a mustard coloured precipitate due to masking by aqueous iodine.

Copper Metal is Used to Make Alloys

OCR (Transition Elements Option) only
When you mix two or more metals you get an **alloy**. Alloys often have **more useful properties** than any of the metals do by themselves.

Copper + zinc = brass Brass is **attractive** and resistant to corrosion. It's **harder** and **more ductile** than copper.

Copper + tin = bronze Bronze is **harder** than copper. It's good for making medals and statues from.

Copper + tin + zinc = coinage bronze This stuff's nice and hard and doesn't corrode.

Practice Questions

Q1 What colour change is seen when aqueous chromate(VI) ions have acid added to them? Name the new chromium ion.
Q2 Name two reagents needed to oxidise chromium(III) to chromium(VI).
Q3 Give the electronic configurations of :   (a) Cu^{2+} ions,   (b) Cu^+ ions.

Exam Questions

1   Half-equation 1: $Fe^{3+}_{(aq)} + e^- \rightleftharpoons Fe^{2+}_{(aq)}$ Half-equation 2: $Cr_2O_7^{2-}_{(aq)} + 14H^+_{(aq)} + 6e^- \rightleftharpoons 2Cr^{3+}_{(aq)} + 7H_2O_{(l)}$

   a)   Use the two half-equations to produce a balanced ionic equation for the reaction between acidified potassium dichromate and a solution of iron(II) sulphate.   [2 marks]
   b)   Describe, using oxidation numbers, the redox processes which have taken place.   [4 marks]

2   Copper(I) sulphate, Cu_2SO_4, is a white powder. It reacts with water to form a blue solution, A, and a pinky brown solid, B.
   a) Identify A and B.   [2 marks]
   b) Write an equation for the reaction.   [1 mark]
   c) Name the **type** of reaction taking place   [1 mark]

Whether you give a brass monkey or not — you've still got to learn it...

Here's an interesting fact to brighten up your dreary pre-exam existence — a brass monkey was something used to hold cannon balls. If it got really cold then it'd shrink and the cannon balls would roll off. So that's where the saying "cold enough to freeze the balls off a brass monkey" comes from. It's got nowt to do with monkey-shaped brass ornaments.

Transition Metals — Titrations and Calculations

These titrations are redox titrations. They're like acid-base titrations, but different. You don't need an indicator for a start.

Titrations Using Transition Element Ions are Redox Titrations

Titrations using transition element ions let you find out how much **oxidising agent** is needed to **exactly** react with a quantity of **reducing agent**. If you know the **concentration** of either the oxidising agent or the reducing agent, you can use the titration results to work out the concentration of the other.

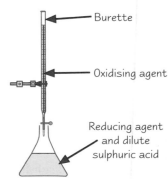

Burette

Oxidising agent

Reducing agent and dilute sulphuric acid

1) First you measure out a quantity of **reducing agent**, e.g. aqueous Fe^{2+} ions, using a pipette, and put it in a conical flask.

2) Using a **measuring cylinder**, you add about **20 cm³ of dilute sulphuric acid** to the flask — this is an excess, so you don't have to be too exact.

3) Now you add the **oxidising agent**, e.g. aqueous potassium manganate(VII), to the reducing agent using a **burette**, **swirling** the conical flask as you do so.

4) You stop when the mixture in the flask **just** becomes tainted with the colour of the oxidising agent (the **end point**) and record the volume of the oxidising agent added. This is the **rough titration**.

5) Now you do some **accurate titrations**. You need to do a few until you get **two or more** readings that are **within 0.10 cm³** of each other.

You can also do titrations the **other way round** — adding the reducing agent to the oxidising agent.

The Sharp Colour Change Tells You when the Reaction's Just Been Completed

When the **coloured oxidising agent** is added to the reducing agent, they start reacting and the reducing agent starts to **lose its colour**. The reaction continues until **all** of the reducing agent has reacted. The **very next drop** into the flask will give the mixture the **colour of the oxidising agent**. The trick is to spot **exactly** when this happens.

> The two main **oxidising agents** used are:
>
> 1) **Manganate(VII) ions** (MnO_4^-) in **aqueous potassium manganate(VII)** ($KMnO_4$) — these are **purple**.
> 2) **Dichromate(VI) ions** ($Cr_2O_7^{2-}$) in **aqueous potassium dichromate(VI)** ($K_2Cr_2O_7$) — these are **orange**.

The **acid** is added to make sure there are plenty of H^+ **ions** to allow the oxidising agent to be reduced — see page 47.

You Can Calculate the Concentration of a Reagent from the Titration Results

EXAMPLE: 27.50 cm³ of 0.0200 moldm⁻³ aqueous potassium manganate(VII) reacted with 25.0 cm³ of acidified iron(II) sulphate solution. Calculate the concentration of Fe^{2+} ions in the solution.

$$MnO_{4\ (aq)}^- + 8H^+_{(aq)} + 5Fe^{2+}_{(aq)} \rightarrow Mn^{2+}_{(aq)} + 4H_2O_{(l)} + 5Fe^{3+}_{(aq)}$$

1) Work out the number of **moles of MnO_4^- ions** added to the flask.

> Number of moles MnO_4^- added $= \dfrac{\text{concentration} \times \text{volume}}{1000} = \dfrac{0.0200 \times 27.50}{1000} = 5.50 \times 10^{-4}$ moles

2) Look at the balanced equation to find how many moles of Fe^{2+} react with **every mole** of MnO_4^-. Then you can work out the **number of moles of Fe^{2+}** in the flask.

> 5 moles of Fe^{2+} react with 1 mole of MnO_4^-. So moles of $Fe^{2+} = 5.50 \times 10^{-4} \times 5 = 2.75 \times 10^{-3}$ moles.

3) Work out the **number of moles of Fe^{2+}** that would be in 1000 cm³ (1 dm) of solution — this is the **concentration**.

> 25.0 cm³ of solution contained 2.75×10^{-3} moles of Fe^{2+}.
>
> 1000 cm³ of solution would contain $\dfrac{(2.75 \times 10^{-3}) \times 1000}{25.0} = 0.11$ moles of Fe^{2+}.
>
> So the concentration of Fe^{2+} is **0.11 moldm⁻³**.

Manganate 007 licensed to oxidise.

The same type of calculation could be done using **acidified potassium dichromate(VI)** and **iron(II) sulphate**. This time **6 moles** of Fe^{2+} react with **each mole** of $Cr_2O_7^{2-}$. $Cr_2O_7^{2-}{}_{(aq)} + 14H^+_{(aq)} + 6Fe^{2+}_{(aq)} \rightarrow 2Cr^{3+}_{(aq)} + 7H_2O_{(l)} + 6Fe^{3+}_{(aq)}$

Transition Metals — Titrations and Calculations

This stuff is just for OCR (Transition Elements Option) and Edexcel.

Here's Another Way to Find Out the Concentration of an Oxidising Agent

Edexcel only

To find out the concentration of an **oxidising agent**, such as aqueous **potassium dichromate**, you can use this method:

1) First measure out some of the **oxidising agent** using a pipette, and put it in a conical flask.
   Then add about **20 cm³** of **aqueous potassium iodide** to the oxidising agent — this is an **excess**.
   Some of the colourless I⁻ ions are oxidised to I_2, making the mixture **dark brown**.

$$2I^-_{(aq)} \rightarrow I_{2(aq)} + 2e^-$$

The more oxidising agent molecules there are, the more iodine will be produced.

2) Then you titrate the mixture in the conical flask with **sodium thiosulphate solution** of **known concentration**.
   You add the sodium thiosulphate solution from a **burette**, **swirling** the conical flask as you do so.

   This begins to **decolourise** the mixture in the conical flask. What's happening is that the I_2 is being **reduced back to I⁻** by the thiosulphate ions:

$$I_{2(aq)} + 2S_2O_3^{2-}_{(aq)} \rightarrow 2I^-_{(aq)} + S_4O_6^{2-}_{(aq)}$$

3) It's dead hard to see the end point, so when there's **very little colour** left you add a few drops of **starch**.
   This produces a **dark blue colour** with any remaining I_2. When all the I_2 has gone, the mixture changes sharply from dark blue to colourless. This is the **end point** — so record how much **sodium thiosulphate solution** you've added.

4) Now you know roughly where the end point is, you have to **repeat** the titration accurately a few times.

5) From the amount of **sodium thiosulphate solution** you've added, you can work out how many moles of I_2 there were.
   And from the number of moles of I_2, you can figure out how much **oxidising agent** there was.

You can Analyse the Percentage of Copper in Brass Using Iodide

OCR (Transition Elements Option) only

Here's how you can use titration to find out how much copper there is in brass.

1) Dissolve a small measured mass of **brass** in excess warm **nitric acid**. This converts all the **copper** into **copper ions**.

2) Add an **excess** of aqueous **potassium iodide**. A **white precipitate of copper(I) iodide** forms, which actually looks **mustard** coloured because **iodine** is formed at the same time.

3) Now titrate this mixture with **thiosulphate ions**, as described in stages 2-4 above. The end point is when there's **no colouration due to iodine** and just the **white precipitate of copper(I) iodide remains** in the conical flask.

4) From the amount of **thiosulphate ions** you've added, you can work out how many moles of I_2 there were.
   And from the number of moles of I_2, you can figure out how much **copper** there was in the brass.

Practice Questions

Q1 What colour change would you expect to see in a conical flask containing acidified aqueous iron(II) sulphate if just enough manganate(VII) ions were added to react with the mixture?

Q2 What substance is added to a mixture to highlight the presence of iodine?

Exam Question

1 0.500 g of brass was added to excess warm nitric acid. On cooling, excess aqueous potassium iodide was added.
  The iodine formed in this mixture needed 30.00 cm³ of thiosulphate ions of concentration 0.200 moldm⁻³ to fully react.

a) Calculate the number of moles of thiosulphate ions that were added. [1 mark]

b) $2S_2O_3^{2-}_{(aq)} + I_{2(aq)} \rightarrow S_4O_6^{2-}_{(aq)} + 2I^-_{(aq)}$
   Use this equation to calculate the number of moles of iodine that reacted with the thiosulphate ions. [1 mark]

c) $2Cu^{2+}_{(aq)} + 4I^-_{(aq)} \rightarrow 2CuI_{(s)} + I_{2(aq)}$
   Use the equation above to calculate the number of moles of copper(II) ions that were present. [1 mark]

d) Calculate the percentage of copper in brass. (A_r of Cu = 63.5) [2 mark]

And how many moles does it take to change a light bulb....

You might remember the iodine/starch reaction from biology. In biology you normally use the iodine to test for starch, but here you're using the starch to test for iodine. It makes no odds really — you get a dark blue solution whichever way round you do it. Make sure you're happy with the methods for the calculations, and have mastered the one in the exam question.

Complex Ions — The Basics

Transition metals are always forming complex ions. These aren't as complicated as they sound, though. Honest.

Complex Ions are Metal Ions Surrounded by Ligands

1) A **complex ion** is a **metal ion** surrounded by **coordinately** bonded **ligands**.

2) The **coordination number** is the **number** of **coordinate bonds** that are formed with the central metal ion. The usual coordination numbers are **6** and **4**. If the ligands are **small**, like H_2O, CN^- or NH_3, **6** can fit around the central metal ion. But if the ligands are **larger**, like Cl^-, only **4** can fit around the central metal ion.

A coordinate bond (or dative covalent bond) is a covalent bond in which both electrons in the shared pair come from the same atom — see page 38.

6 COORDINATE BONDS MEAN AN OCTAHEDRAL SHAPE

There were lots of examples of complex ions with 6 coordinate bonds on page 39. The table on the right has a few more too.

The different types of bond arrow show that the complex is 3-D. The wedge-shaped arrows represent bonds coming towards you and the dashed arrows represent bonds sticking out behing the molecule.

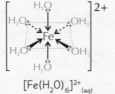

$[Fe(H_2O)_6]^{2+}_{(aq)}$

Complex	Oxidation State of Metal	Colour
$[Cr(H_2O)_6]^{2+}_{(aq)}$	+2	Blue
$[Cr(H_2O)_6]^{3+}_{(aq)}$	+3	Violet
$[Fe(CN)_6]^{4-}_{(aq)}$	+2	Yellow
$[Fe(CN)_6]^{3-}_{(aq)}$	+3	Orange brown
$[Co(NH_3)_6]^{2+}_{(aq)}$	+2	Pale brown
$[Co(NH_3)_6]^{3+}_{(aq)}$	+3	Golden brown
$[Cu(NH_3)_4(H_2O)_2]^{2+}_{(aq)}$	+2	Deep blue

The ligands don't always have to be the same.

4 COORDINATE BONDS USUALLY MEAN A TETRAHEDRAL SHAPE...

E.g. $[CuCl_4]^{2-}$, which is yellow and shown below, and $[CoCl_4]^{2-}$, which is blue.

...BUT NOT ALWAYS

In a **few** complexes, **4 coordinate bonds** form a **square planar** shape (see the cis-trans isomerism diagrams on the opposite page).

SOME SILVER COMPLEXES HAVE 2 COORDINATE BONDS AND FORM A LINEAR SHAPE
Not Salters or Edexcel

And some of these silver complexes are mighty handy:

1) $[Ag(NH_3)_2]^+$ is the complex ion found in **Tollens' reagent**. This is used for testing for aldehydes — see page 76.

2) $[Ag(S_2O_3)_2]^{3-}$ is formed in **photography**. To stop the whole film turning black in the light any undeveloped silver bromide has to be removed. This is done by adding thiosulphate ions — these form the soluble complex $[Ag(S_2O_3)_2]^{3-}$ with any silver ions: $AgBr_{(s)} + 2S_2O_3^{2-} \rightarrow [Ag(S_2O_3)_2]^{3-}_{(aq)} + Br^-_{(aq)}$

3) $[Ag(CN)_2]^-$ is used in **silver plating**. It decomposes under **electrolysis** and deposits Ag^+ ions on the **cathode**. If a metal object is used as a cathode, the object becomes **silver plated**.

Complex Ions Have an Overall Charge or Total Oxidation State

The **overall charge** on the complex ion is its **total oxidation state**. It's put **outside** the **square** brackets. For example:

$[Cu(H_2O)_6]^{2+}_{(aq)}$ ⟵ *Overall charge is 2+.*

You have to know how to work out the **oxidation state of the metal**:

The oxidation state of the metal ion = the total oxidation state – the sum of the oxidation states of the ligands

E.g. $[Fe(CN)_6]^{4-}_{(aq)}$ The total oxidation state is **–4** and each CN^- ligand has an oxidation state of **–1**. So in this complex, iron's oxidation state = $-4 - (6 \times -1) = +2$.

A Ligand Must Have at Least One Lone Pair of Electrons

A ligand must have **at least one lone pair of electrons**, or it won't have anything to form a **coordinate bond** with.

- Ligands with **one lone pair** are called **monodentate** (or **unidentate**) — e.g. $H_2\ddot{O}$, $\ddot{N}H_3$, $\ddot{C}\ddot{l}^-$, $\ddot{C}N^-$.

- Ligands with **two lone pairs** are called **bidentate** — e.g. ethane-1,2-diamine: $\ddot{N}H_2CH_2CH_2\ddot{N}H_2$. Bidentate ligands can each form **two coordinate bonds** with a metal ion.

- Ligands with **more than two lone pairs** are called **polydentate** (or **multidentate**) — e.g. **EDTA** has six lone pairs (it's **hexadentate** to be precise). It can form **six coordinate bonds** with a metal ion. Haemoglobin contains a molecule with **four nitrogens** that each form a coordinate bond with Fe^{2+} — so this is a multidentate ligand too. See page 69.

Complex Ions — The Basics

Those of you not doing AQA *or the* OCR Transition Element Option *can jump straight to the questions.*

Complex Ions show Geometric Isomerism

Geometric (or **cis-trans**) isomerism is a type of **stereoisomerism**. In stereoisomerism, the atoms are attached together in the same order, with the same types of bonds, but they're **arranged differently in space**.

With **complex ions**, geometric isomerism occurs when **square planar** ions have **two pairs** of ligands. **[NiCl₂(NH₃)₂]** is an example of this:

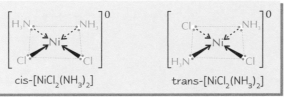

cis-[NiCl₂(NH₃)₂]   trans-[NiCl₂(NH₃)₂]

See page 70 for more on isomerism.

AQA only   **Platin**, $PtCl_2(NH_3)_2$ is another square planar molecule which exists as **two geometric isomers**. The **cis** form is a very effective **anti-cancer drug** used a lot in chemotherapy.

They show Optical Isomerism Too   *Not AQA*

Optical isomerism is another type of **stereoisomerism**. With complex ions, it happens when an ion can exist in **two non-superimposable mirror images**.

This happens when **three bidentate ligands**, such as ethane-1,2-diamine, $H_2NCH_2CH_2NH_2$, use the lone pairs on **both** nitrogen atoms to coordinately bond with **nickel**.

You might see ethane-1,2-diamine abbreviated to "en".

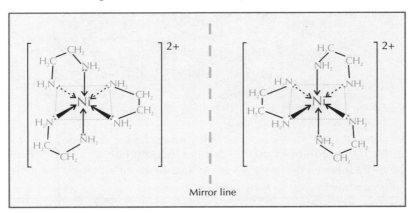

Mirror line

Practice Questions

Q1 What's meant by the term 'complex ion'?

Q2 Explain how a ligand, such as ammonia, bonds to a central metal ion.

Q3 What is meant by the term 'bidentate ligand'?

Q4 Draw the shape of the complex ion $[Co(NH_3)_6]^{3+}$. Name the shape.

Exam Question

1   Solution A is aqueous copper(II) sulphate and is pale blue in colour.
     When concentrated hydrochloric acid is added, solution B forms. B is yellow in colour.
     When excess concentrated aqueous ammonia is added, solution C forms. C is deep blue in colour.

   a) Identify the formula of the complex ion causing the colour of each solution.   [3 marks]

   b) Write the equation to show the conversion of solution A to solution B.   [1 mark]

   c) Draw and name the shapes of the complex ions in solutions A, B and C.   [6 marks]

Put your hands up — we've got you surrounded...

You'll never get transition metal ions floating round by themselves in a solution — they'll always be surrounded by other molecules. It's kind of like what'd happen if you put a dish of sweets in a room of eight (or eighteen) year-olds. When you're drawing complex ions, you should always include some wedge-shaped bonds to show that it's 3-D.

Complex Ions – Ligand Exchange

There are more equations on this page than the number of elephants you can fit in a Mini.

Ligands can Exchange Places with One Another

One ligand can be **swapped** for another ligand — this is **ligand exchange**. It pretty much always causes a **colour change**.

1) If the ligands are of **similar size**, e.g. H_2O and NH_3, then the **coordination number** of the complex ion doesn't change, and neither does the **shape**.

$$[Co(H_2O)_6]^{2+}{}_{(aq)} + 6NH_{3(aq)} \rightarrow [Co(NH_3)_6]^{2+}{}_{(aq)} + 6H_2O_{(l)}$$
octahedral   octahedral
pink   pale brown

$$[Cr(H_2O)_6]^{3+}{}_{(aq)} + 6OH^-{}_{(aq)} \rightarrow [Cr(OH)_6]^{3-}{}_{(aq)} + 6H_2O_{(l)}$$
octahedral   octahedral
violet   pale green

2) If the ligands are **different sizes**, e.g. H_2O and Cl^-, there's a **change of coordination number** and a **change of shape**.

$$[Cu(H_2O)_6]^{2+}{}_{(aq)} + 4Cl^-{}_{(aq)} \rightleftharpoons [CuCl_4]^{2-}{}_{(aq)} + 6H_2O_{(l)}$$
octahedral   tetrahedral
pale blue   yellow

$$[Co(H_2O)_6]^{2+}{}_{(aq)} + 4Cl^-{}_{(aq)} \rightleftharpoons [CoCl_4]^{2-}{}_{(aq)} + 6H_2O_{(l)}$$
octahedral   tetrahedral
pink   blue

> The forward reaction is endothermic, so the equilibrium can be shifted to the right-hand side by heating. The equilibrium will also shift to the right if you add more concentrated hydrochloric acid. Adding water to this equilibrium shifts it back to the left.

3) Sometimes the substitution is only **partial**.

$$[Cu(H_2O)_6]^{2+}{}_{(aq)} + 4NH_{3(aq)} \rightarrow [Cu(NH_3)_4(H_2O)_2]^{2+}{}_{(aq)} + 4H_2O_{(l)}$$
octahedral   elongated octahedral
pale blue   deep blue

$$[Fe(H_2O)_6]^{3+}{}_{(aq)} + SCN^-{}_{(aq)} \rightarrow [Fe(H_2O)_5SCN]^{2+}{}_{(aq)} + H_2O_{(l)}$$
octahedral   distorted octahedral
pale violet when pure   blood red

but this usually looks yellow ⟹ pale violet when pure

Have a quick peek back at the reactions on page 39 — these were **ligand exchange reactions** too. In these, **hydroxide precipitates** were formed when a little bit of **sodium hydroxide** or **ammonia solution** was added to metal-aqua ions. The hydroxide precipitates sometimes **dissolved** when excess sodium hydroxide or ammonia solution was added.

Edexcel only: Here are a few more reactions for you to learn:

Aqua-complex	With a little $OH^-{}_{(aq)}$ or $NH_{3(aq)}$	With excess $OH^-{}_{(aq)}$	With excess $NH_{3(aq)}$
$Mn(H_2O)_6^{2+}$	$Mn(H_2O)_4(OH)_{2(s)}$ — light brown ppt	no change	no change
$Ni(H_2O)_6^{2+}$	$Ni(H_2O)_4(OH)_{2(s)}$ — green ppt	no change	precipitate dissolves to form a blue-green solution $[Ni(NH_3)_6]^{2+}{}_{(aq)}$
$Zn(H_2O)_6^{2+}$	$Zn(H_2O)_4(OH)_{2(s)}$ — white ppt	precipitate dissolves to form a colourless solution $Zn(H_2O)_2(OH)_4^{2-}{}_{(aq)}$	precipitate dissolves to form a colourless solution $[Zn(H_2O)_2(NH_3)_4]^{2+}{}_{(aq)}$

Different Ligands Form Different Strength Bonds   *AQA and Nuffield only*

Ligand exchange reactions can be easily **reversed**, **EXCEPT** if the new complex ion is much **more stable** than the old one.

1) If the new ligands form **stronger** bonds with the central metal ion than the old ligands did, the change is less easy to reverse. E.g. **CN^- ions** form stronger coordinate bonds with Fe^{3+} ions than H_2O molecules did, so it's hard to reverse this reaction:
$$[Fe(H_2O)_6]^{3+}{}_{(aq)} + 6CN^-{}_{(aq)} \rightarrow [Fe(CN)_6]^{3-}{}_{(aq)} + 6H_2O_{(l)}$$

2) **Bidentate** ligands form more stable complexes than monodentate ligands, so a change like the one below is hard to reverse:
$$[Ni(H_2O)_6]^{2+}{}_{(aq)} + 3H_2NCH_2CH_2NH_{2(aq)} \rightarrow [Ni(H_2NCH_2CH_2NH_2)_3]^{2+}{}_{(aq)} + 6H_2O_{(l)}$$

3) A **hexadentate** ligand, like EDTA, forms even more stable complexes.
So a change like this one is **very** hard to reverse:
$$[Cu(H_2O)_6]^{2+}{}_{(aq)} + EDTA^{4-}{}_{(aq)} \rightarrow [Cu(EDTA)]^{2-}{}_{(aq)} + 6H_2O_{(l)}$$

When **bidentate** ligands take the place of **monodentate** ligands there are generally **more product** molecules than **reactant** molecules. This means the **total entropy change** for the reaction is likely to be **positive**, so the reaction would be expected to "go". The same happens when a hexadentate ligand replaces the monodentate ligands. *See pages 8-9 for entropy changes.*

Complex Ions – Ligand Exchange

You Can Measure the Stability of Complex Ions by Using Stability Constants

Salters and Nuffield only

1) The **stability constant, K_{stab}**, is a **measure** of how **stable** a particular **complex ion** is, compared to the **aqua complex** of the metal ion. It's the **equilibrium constant** for the ligand exchange reaction taking place.

E.g.

$$[Cu(H_2O)_6]^{2+}_{(aq)} + 4Cl^-_{(aq)} \rightleftharpoons CuCl_4^{2-}_{(aq)} + 6H_2O_{(l)}$$

$$K_{stab} = \frac{\left[CuCl_4^{2-}_{(aq)}\right]}{\left[Cu(H_2O)_6^{2+}_{(aq)}\right]\left[Cl^-_{(aq)}\right]^4} = 4.2 \times 10^5 \text{ dm}^{12}\text{mol}^{-4}$$

The concentrations of reactants and products are measured at equilibrium. You don't include water in the expression — its concentration doesn't change.

2) The **higher the K_{stab}**, the further the equilibrium lies to the **right**. The K_{stab} indicates which complex ion is more stable, and whether one ligand will **replace** another. A **high** K_{stab} indicates that ligand exchange **will** happen.

You Need to Know How to Prepare a Complex Compound   *Nuffield only*

Here are the steps you have to go through to prepare a particular complex:

1) First look at the **balanced equation** to find the **ratio** of **moles of central metal ion** to **moles of ligand**.
2) Weigh a small number of moles of the **central metal ion** and **dissolve** to form a solution.
3) Weigh the required number of **moles of ligand** and **dissolve** to form a solution.
4) **Mix** both solutions and allow the mixture to **crystallise**.

For example:

How to prepare $K_3Fe(C_2O_4)_3.3H_2O$ (which contains the complex ion $[Fe(C_2O_4)_3]^{3-}$) from iron(III) chloride.

1) The equation is: $FeCl_3.6H_2O_{(aq)} + 3K_2C_2O_4.H_2O_{(aq)} \rightarrow K_3Fe(C_2O_4)_3.3H_2O_{(aq)} + 3KCl_{(aq)} + 6H_2O_{(l)}$
   So there's 3 moles of $K_2C_2O_4.H_2O_{(aq)}$ for every 1 mole of $FeCl_3.6H_2O_{(aq)}$.

2) Weigh 0.0100 moles (or 2.70 g) of hydrated iron(III) chloride, $FeCl_3.6H_2O$. Dissolve this in 5 cm³ of water.

3) Weigh 3 × 0.0100 moles (or 5.53 g) of hydrated potassium ethanedioate, $K_2C_2O_4.H_2O$.
   Dissolve this in 10 cm³ of warm water.

4) **Mix both solutions** and allow the mixture to **cool**. The complex $K_3Fe(C_2O_4)_3.3H_2O$ will **crystallise** out on cooling.

Practice Questions

Q1 What's meant by ligand exchange?

Q2 Why is it much easier to exchange monodentate ligands for bidentate ligands rather than the other way around?

Q3 Write the expression for K_{stab} for the reaction: $[Cu(H_2O)_6]^{2+}_{(aq)} + 4NH_{3(aq)} \rightarrow [Cu(NH_3)_4(H_2O)_2]^{2+}_{(aq)} + 4H_2O_{(l)}$

Exam Questions

1 Solid anhydrous cobalt(II) chloride can be used to test for the presence of water because of the distinct pink colour which is formed. The substance will also dissolve in dilute hydrochloric acid to form a solution with the same pink colour. This solution can be made to turn blue.
   a) Identify the complex ion responsible for the pink colour. [1 mark]
   b) Suggest the identity of the complex ion responsible for the blue colour. [1 mark]
   c) Write an equation to show the formation of the blue solution. [2 marks]
   d) State two methods which could be used to turn the original cobalt(II) chloride solution blue. [2 marks]

2 Aqueous silver(I) solutions contain $[Ag(H_2O)_2]^+_{(aq)}$ ions. If aqueous ammonia is added, these ions are converted to $[Ag(NH_3)_2]^+_{(aq)}$ ions. Write an equation to show this change and write the expression for the stability constant of the $[Ag(NH_3)_2]^+_{(aq)}$ ion, giving units. [3 marks]

Ligand exchange — the musical chairs of the molecular world...

Ligands generally don't mind swapping with other ligands, so long as they're not too tightly attached to the central metal ion. They also won't fancy changing if it means forming fewer molecules and having less entropy. It's kind of like you wouldn't want to dump Brad Pitt for Mr Bean or Rachel Stevens for Vicky Pollard.

Transition Metal Ion Colour

Transition metal complex ions have distinctive colours, which is handy when it comes to identifying them.
This page explains why they're so colourful. **Nuffield people can skip these two pages.**

Ligands **Split** the 3d Subshell into **Two Energy Levels**

Normally the 3d orbitals of transition element ions **all** have the
same energy. But when **ligands** come along and bond to
the ions some of the orbitals are given more energy than others.
This splits the 3d orbitals into **two different energy levels**.

Electrons tend to **occupy the lower orbitals** — to jump up
to the higher orbitals they need **energy** equal to the energy
gap, ΔE. They get this energy from **visible light**.

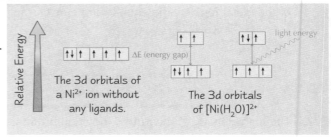

The 3d orbitals of
a Ni^{2+} ion without
any ligands.

The 3d orbitals
of $[Ni(H_2O)]^{2+}$

The energy **absorbed** when electrons jump up can be worked out using this formula:

$$\Delta E = hf$$

where f = frequency (hertz/Hz) and h = Planck's constant (6.63×10^{-34} Js)

The amount of energy needed to make electrons jump depends upon the **central metal ion**, the **ligands** and the
coordination number, as these cause the **size of the energy gap** to alter.

The **Colours** of Compounds are the **Complement** of Those That are **Absorbed**

When **visible light** hits a transition metal ion some frequencies are **absorbed** when electrons jump up to the higher
orbitals. The frequencies absorbed depend on the size of the **energy gap**.

The rest of the frequencies are **reflected**. It's the **colour** which corresponds to the **reflected** frequencies that you see.

frequency increases $\Longrightarrow$

If all the colours except blue are absorbed, then the solution
appears **blue**. This is what happens in $[Cu(H_2O)_6]^{2+}$.

The **central metal ion**, the **ligands** and the **coordination number**
affect the size of the energy gap, so they'll also affect the **colour**.

If there are **no** 3d electrons, there are no electrons to jump and so **no energy** is absorbed. Also, if the 3d subshell
is **full**, there's no room in the upper orbitals for any electrons to jump to, so again, **no energy** will be absorbed.
If there's no energy absorbed, the compound will look **white** or **colourless**.

Transition Elements Make Good **Pigments** for **Paints**   *OCR Transition Elements Option only*

1) Titanium oxide contains **Ti^{4+} ions**. These have a **$3d^0$** electronic configuration and so are **colourless**. They don't absorb
   any light from the visible spectrum and reflect it all as **white** light. So TiO_2 makes an **excellent white paint pigment**.

2) **Monastral blue** is a **copper(II)** complex in which the Cu^{2+} ion is bonded to 4 nitrogen atoms which are part of a large
   organic ring structure. The colour is a **bright blue** pigment. The **multidentate** organic ligand of Monastral blue can be
   tweaked a little by **substituting in halide atoms** — this causes the pigment to be **different** shades of blue or green.

Colorimeters can be used to Find **Concentrations** of Transition Metal Ions   *Not Edexcel*

A colorimeter measures how much visible light is **absorbed** by a solution.
The more **concentrated** a coloured solution is, the more light it'll absorb.
Colorimeters are handy for working out **how concentrated** a solution of
transition metal ions is.

The light passes through a **filter**, which **only** lets the colour of light through
which is absorbed by the sample. After passing through the sample, the
remaining light hits a **detector** — this shows how much light was absorbed.

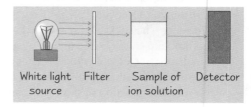

White light   Filter   Sample of   Detector
source   ion solution

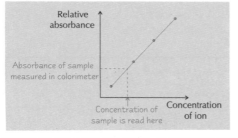

Before you can use a colorimeter to find the unknown concentration of a sample,
you have to produce a **calibration graph**. This involves measuring the **absorbance**
of solutions of the ion at **known concentrations** and plotting them on a graph.
Then you can measure the absorbance of your sample and read its **concentration**
off the graph.

If the concentration of the transition metal ion solution is too **low** to absorb much
light you can add ligands such as **EDTA**. This'll **intensify** the colour and make the
solution absorb more light.

Transition Metal Ion Colour

Colorimeters can be Used to Determine the Formulas of Complexes

AQA, OCR and Salters only

$[Fe(H_2O)_6]^{3+}_{(aq)}$ ions will undergo **ligand exchange** to produce a **blood-red** complex, $[Fe(H_2O)_5SCN]^{2+}_{(aq)}$.

$$[Fe(H_2O)_6]^{3+}_{(aq)} + SCN^-_{(aq)} \rightarrow [Fe(H_2O)_5SCN]^{2+}_{(aq)} + H_2O_{(l)}$$

If you didn't know that the ratio of $SCN^-_{(aq)}$ to $[Fe(H_2O)_6]^{3+}_{(aq)}$ was **1 : 1**, you could use the method below to work it out.

1) Take several test tubes and put a measured amount of the **metal ion solution**, $Fe^{3+}_{(aq)}$ into each. Then add a different amount of the **ligand solution**, e.g. $KSCN_{(aq)}$, to each test tube. You have to know the concentration of **both** of these solutions so that you can work out the **number of moles** of each that are in the test tubes.

2) Top up all the test tubes to the same level with **water**.

3) Now use the **colorimeter** to measure the absorbance for each test tube. You'll get a set of results that looks something like this:

Test tube	1	2	3	4	5	6
Volume of 0.1 moldm^{-3} $Fe^{3+}_{(aq)}$ (cm^3)	5.0	5.0	5.0	5.0	5.0	5.0
Volume of 0.1 moldm^{-3} $SCN^-_{(aq)}$ (cm^3)	2.0	4.0	6.0	8.0	10.0	12.0
Relative absorbance	0.32	0.64	0.80	0.80	0.80	0.80

4) Now you draw a **graph** of **relative absorbance** against volume of **ligand solution**. **Extrapolate** the two straight portions of your graph so that they **intersect**.

5) At this intersection point, find the number of moles of **metal ion** and the number of moles of **ligand** — this gives you the ratio of ions reacting.

In this example, there's **5.0 cm^3** of **0.1 moldm^{-3} SCN$^-$** needed to produce maximum absorbance with **5.0 cm^3** of **0.1 moldm^{-3} Fe^{3+}**. These contain equal numbers of moles, so the ratio of Fe^{3+} : SCN$^-$ is **1:1**, and the formula is **$[Fe(H_2O)_5SCN]^{2+}$**.

Maximum absorbance is first reached here.

This is the volume you're interested in.

You'd have to actually work out the numbers of moles if the concentrations or volumes were different.

Another variation on this is to change the number of moles of **both** the metal ion **and** the ligands. If you do it this way you'll end up with a graph like the one on the right.

Just like before, you **extrapolate** the straight portions of the graph until they intersect. Then you find the **ratio** of ions at this point — the point where **maximum absorbance** is reached.

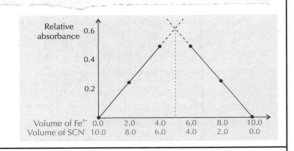

Practice Questions

Q1 Explain why complex ions are coloured.

Q2 Work out the energy associated with a single electron transition made by light of frequency 4.50×10^{14} Hz.

Exam Question

1 The formula of the complex ion formed between Cu^{2+} ions and Cl^- ions can be determined by colorimetry. Samples of the mixtures in the table on the right were placed into a colorimeter and the absorbance figures shown were obtained.

Test tube	1	2	3	4	5	6
Volume of 0.1 moldm^{-3} $Cu^{2+}_{(aq)}$ (cm^3)	5.0	5.0	5.0	5.0	5.0	5.0
Volume of 0.2 moldm^{-3} $HCl_{(aq)}$ (cm^3)	0.0	4.0	8.0	12.0	16.0	20.0
Relative absorbance	0.00	0.36	0.72	0.90	0.90	0.90

a) Explain what happens to the 3d orbitals of copper ions when $[Cu(H_2O)_6]^{2+}_{(aq)}$ forms. [1 mark]

b) Explain why the complexes formed by copper with water and chloride ions are different colours. [4 marks]

c) Plot a graph of the relative absorbance against volume of $HCl_{(aq)}$. [3 marks]

d) How many moles of Cu^{2+} ions are present in 5.0 cm^3 of 0.1 moldm^{-3} solution? [1 mark]

e) Work out the formula and charge of the complex ion. [4 marks]

Blue's not my complementary colour — it clashes with my hair...

*So that's why transition metal ions are pretty colours. Remember, it's the colours from the visible light that **aren't** absorbed that you see. The colorimeter stuff looks yukky, but it's just a matter of finding out how many moles of each thing there are when the point of maximum absorbance is reached, then figuring out the ratio — then you're almost laughin'.*

Uses of Transition Metals

Transition metals aren't just good — they're greeeeat.
If you're doing OCR you can miss these two pages, as long as you're not doing the Transition Elements Option.

Transition Metals Make **Good Catalysts**

Transition metals and their compounds make good catalysts because they can **change oxidation states** by gaining or losing electrons within their **d orbitals**. This means they can **transfer electrons** to **speed up** reactions.

They can act as **Heterogeneous Catalysts**...

Heterogeneous catalysts are in a **different state** to the reactants (see page 12).

For example: • iron catalyses the Haber process (see page 25)
• nickel is a catalyst used to harden margarine
• vanadium(V) oxide is used in the Contact process (see below)

The **Contact process** is used to make **sulphuric acid**. *AQA, OCR (Transition Elements*
During the process SO_2 gas must be **oxidised** to SO_3 gas. $SO_{2(g)} + \frac{1}{2}O_{2(g)} \rightleftharpoons SO_{3(g)}$ *Option) and Edexcel only*

This is a really **slow** reaction, but vanadium(V) oxide speeds it up by providing a route with a **lower activation energy**.

1) First of all the V_2O_5 oxidises SO_2 to SO_3. The vanadium is **reduced** from +5 to +4, forming an **intermediate**, V_2O_4.

Oxidation state of V: +5   +4
$$V_2O_{5(s)} + SO_{2(g)} \rightarrow V_2O_{4(s)} + SO_{3(g)}$$

2) Then the V_2O_4 is **oxidised** back to its **original state** by **oxygen**.

Oxidation state of V: +4   +5
$$V_2O_{4(s)} + \frac{1}{2}O_{2(g)} \rightarrow V_2O_{5(s)}$$

...and **Homogeneous Catalysts**   *AQA only*

Homogeneous catalysts are in the **same physical state** as the reactants.
Usually a **homogeneous** catalyst is an **aqueous catalyst** for a reaction between two **aqueous solutions**.

A homogeneous catalyst works by forming an **intermediate ion**. This causes the enthalpy profile for a **homogeneously catalysed** reaction to have **two humps** in it.

The activation energy needed to form the **intermediates** (and to form the products from the intermediates) is **lower** than that needed to make the products directly from the reactants.

The catalyst is always **reformed** so it can carry on catalysing the reaction.

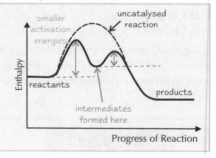

Fe^{2+} Catalyses the Reaction Between $S_2O_8^{2-}$ and I^- *AQA only*

The **redox** reaction between iodide ions and peroxodisulphate ($S_2O_8^{2-}$) ions takes place **annoyingly slowly** because there's a **high activation energy**.

$$S_2O_8^{2-}{}_{(aq)} + 2I^-{}_{(aq)} \rightarrow I_{2(aq)} + 2SO_4^{2-}{}_{(aq)}$$

One reason for this high activation energy is that both reactant ions are negatively charged, so repulsion reduces the number of collisions.

But if Fe^{2+} ions are added, things are really speeded up.
First, the Fe^{2+} ions are **oxidised** to Fe^{3+} ions by the $S_2O_8^{2-}$ ions. $\quad S_2O_8^{2-}{}_{(aq)} + 2Fe^{2+}{}_{(aq)} \rightarrow 2Fe^{3+}{}_{(aq)} + 2SO_4^{2-}{}_{(aq)}$

The newly formed intermediate Fe^{3+} ions now **easily oxidise** the I^- ions to iodine.

$$2Fe^{3+}{}_{(aq)} + 2I^-{}_{(aq)} \rightarrow I_{2(aq)} + 2Fe^{2+}{}_{(aq)}$$

Fe^{3+} will also catalyse the reaction. It'll mean the two steps happen the other way round though.

Autocatalysis is When a **Product** Catalyses the Reaction   *AQA and Nuffield only*

And here's an example of an **autocatalysed** reaction: $\quad 2MnO_4^-{}_{(aq)} + 16H^+{}_{(aq)} + 5C_2O_4^{2-}{}_{(aq)} \rightarrow 2Mn^{2+}{}_{(aq)} + 8H_2O_{(l)} + 10CO_{2(g)}$
The initial rate of this reaction is **slow**, but it's catalysed by **Mn^{2+} ions**. As these ions form, the reaction **speeds up**.

Uses of Transition Metals

Fe²⁺ in Haemoglobin Allows Oxygen to be Carried in the Blood

AQA only

1) **Haemoglobin** contains Fe^{2+} ions. The Fe^{2+} ions are **hexa-coordinated**.
Four of the **lone pairs** come from nitrogen atoms within a circular part of a molecule called '**haem**'. *Haem is a multidentate ligand.*
A fifth lone pair comes from a nitrogen atom on a protein. The last position
is the important one — this has a **water ligand** attached to the **iron**.

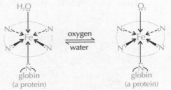

2) In the lungs the oxygen concentration is high, so the water ligand
is **substituted** for an **oxygen molecule**, forming **oxyhaemoglobin**.
This is carried around the body and when it gets to a place where oxygen
is needed the oxygen molecules are exchanged for a water molecule again.

3) If **carbon monoxide** is inhaled, the **haemoglobin** swaps its **water** ligand for a **carbon monoxide** ligand, forming
carboxyhaemoglobin. This is bad news because carbon monoxide is a **strong** ligand and **doesn't** readily
exchange with oxygen or water ligands, meaning the haemoglobin **can't transport oxygen** any more.

Iron is Alloyed with Carbon and Often Other Metals to Make Steel

Salters only

The **properties** of steel depend on the **amount** of carbon and other
metals you add, so you can design steel to meet a **specific need**.

Type of steel	Properties	Uses
Low-carbon	easily shaped	car bodies
High-carbon	strong, inflexible	bridges
Chromium	hard	ball bearings
Chromium/nickel	rust-resisting (stainless)	cutlery
Tungsten	tough, hard-wearing	tools
Titanium	high melting temperature	spacecraft

Any exposed chromium in the steel quickly forms a very tough oxide layer protecting the underlying iron.

The iron used to make steel comes from a blast furnace and contains lots of **impurities**.
The main impurities are **carbon, sulphur, phosphorus** and **silicon**, as well as small amounts
of **other metals** like manganese. These impurities are all removed in redox reactions:

1) **Sulphur** is removed from molten impure iron by adding **magnesium** — this forms molten **magnesium sulphide**,
which **floats** above the molten iron.

2) The rest of the impurities are converted to **oxides** by blowing **pure oxygen** into the molten mix.
There are three types of oxides produced:

   a) **Gaseous non-metal oxides** such as CO, CO_2 and a little SO_2. These escape from the mixture.

   b) **Other non-metal oxides** such as SiO_2 and P_4O_{10}. These are **acidic**, so they're removed by adding a **base** such as
   lime (CaO). The product of this reaction is a **slag** which **floats** on the molten metal and can be **skimmed off**.

   c) **Basic metal oxides**. These **float** above the molten metal and can be **skimmed off**.

Practice Questions

Q1 What's autocatalysis? Give an example of a reaction in which this happens.

Q2 What is the effect of adding chromium to steel?

Exam Questions

1 When aqueous potassium peroxodisulphate is added to a solution of potassium iodide, a small amount of sodium
thiosulphate and starch, there is a delay before the solution becomes dark blue. The appearance of this dark blue
coloration is accelerated by the addition of either aqueous iron(III) chloride or aqueous copper(II) sulphate.
   a) Give the ionic equation for the reaction between potassium peroxodisulphate and potassium iodide. [1 mark]
   b) Explain how the thiosulphate ions cause the appearance of the dark blue coloration to be delayed. [2 marks]
   c) Explain by the use of equations the role of the iron(III) chloride. [4 marks]
   d) Which copper ion intermediate would form when copper(II) sulphate is added? [1 mark]

2 Iron is an important part of our blood.
Outline the role of iron in the blood system and state why carbon monoxide acts as a poison. [4 marks]

Section 5 — Take 1 — and cut...

*There's been a few more uses of transition metals and other d-block elements splattered throughout this section.
There was cis-platin, the anti-cancer drug on page 63, the uses of linear silver complexes on page 62, and the stuff about
how TiO_2 and copper(II) complexes are used in paints on page 66. See if you can scribble a quick list of all the uses.*

Isomerism

Well, you've got this far — but there's still a long trek ahead of you.
The best thing to do is get a couple of matchsticks and prop your eyelids open with them.

Structural Isomers have Different Structural Arrangements of Atoms

Structural isomers have the same **molecular formula**, but their atoms are **arranged** in different ways. The molecules below are all structural isomers of each other — they all have the **molecular formula** $C_4H_{10}O$:

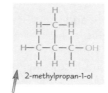

butan-1-ol　　2-methylpropan-1-ol　　butan-2-ol　　diethyl ether

Structural isomers can have...　　...different arrangements of the carbon skeleton...　　...different positions of the functional groups...　　...or different functional groups.

Stereoisomers are Arranged Differently in Space

Stereoisomers have the same molecular formula and their atoms are arranged in the same way. The only difference is the **orientation** of the bonds in **space**. There are two types of stereoisomerism — **geometric** and **optical**.

Geometric Isomers Happen Because there's no Rotation about the Double Bond

Geometric isomers **only** happen if —
* there's a C=C **double bond**, like in alkenes. C=C double bonds **can't rotate**.
* two **different** groups are attached to **each** of the double bond carbon atoms.

But-2-ene shows geometric isomerism:

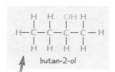

cis-but-2-ene　　trans-but-2-ene

But **but-1-ene** doesn't:

identical groups

but-1-ene

Cis isomers have similar groups on the **same side** of the double bond.

Trans isomers have similar groups going diagonally across.

If there are two identical groups attached to a double bond carbon atom, then geometric isomerism won't happen.

Optical Isomers are Mirror Images of Each Other

A **chiral** (or **asymmetric**) carbon atom is one which has **four different** groups attached to it. It's possible to arrange the groups in two different ways around the carbon atom so that two different molecules are made — these molecules are called **enantiomers** or **optical isomers**.

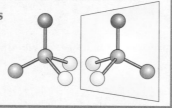

The enantiomers are **mirror images** and no matter which way you turn them, they can't be **superimposed**.

If the molecules can be superimposed, they're <u>achiral</u> — and there's no optical isomerism.

One enantiomer's always labelled **D** and one **L** — luckily you don't have to worry about which is which. Chiral compounds are very common in nature, but you usually only find **one** of the enantiomers — for example, all naturally occurring amino acids are **L–amino acids** (except glycine which isn't chiral).

You have to be able to identify any chiral centres in a molecule and draw optical isomers...

Example

Locating the chiral centre:
Look for the carbon atom with four different groups attached. Here it's the carbon with the four groups H, OH, CHO and CH₃ attached.

chiral centre

2-hydroxypropanoic acid

Drawing isomers:
Once you know the chiral carbon, draw one enantiomer in a tetrahedral shape. Don't try to draw the full structure of each group — it gets confusing. Then draw a mirror image beside it.

enantiomers of 2-hydroxypropanoic acid

Isomerism

Optical Isomers **Rotate Plane-Polarised Light**

Optical isomers are **optically active** — they **rotate plane-polarised light**.
One enantiomer rotates it in a **clockwise** direction; the other rotates it
in an **anticlockwise** direction.

Normal light vibrates in all directions, but plane-polarised light only vibrates in one direction.

Most Optical Isomers are Produced as **Racemic Mixtures** in the **Laboratory**

A **racemic mixture** (or **racemate**) contains **equal quantities** of each enantiomer of an optically active compound.

Racemic mixtures **don't** show any optical activity — the two enantiomers **cancel** each other's light-rotating effect.
Chemists often react two **achiral** things together and get a **racemic** mixture of a **chiral** product.
You can get a single enantiomer using chemical methods, but it's tricky.

Pharmaceutical Drugs Must Only Contain a **Single** Optical Isomer   *OCR only*

Enzymes control chemical reactions in our cells. Drugs work by fitting into the **active sites** of specific enzymes and changing the reactions.

A drug must be **exactly** the right **shape** to fit into the active site of the correct enzyme — only one enantiomer will do.
The other enantiomer could fit into a different enzyme, and might cause **harmful side-effects**. So chiral drugs have to be made so that they only contain one enantiomer.

Another good thing about using a single enantiomer is that you can use a **smaller dose** because it'll all be the correct drug, rather than just half of it.

Before 1963 a drug called thalidomide was used to reduce morning sickness. This drug was a chiral compound — one enantiomer helped morning sickness but the other caused serious birth defects. Although thalidomide only contained the good enantiomer, some of this was converted into the bad enantiomer inside the body.

Practice Questions

Q1 What two things are needed for geometrical isomerism to happen?

Q2 What's a chiral molecule?

Q3 What's a racemic mixture?

Exam Questions

1   There are sixteen possible structural isomers of the compound $C_3H_6O_2$, four of which show stereoisomerism.

a)   Explain the meaning of the term *stereoisomerism*.   [2 marks]

b)   Draw a pair of geometrical isomers of $C_3H_6O_2$, with hydroxyl groups. Label them *cis* and *trans*.   [3 marks]

c)   i)   There are two chiral isomers of $C_3H_6O_2$. Draw the enantiomers of one of the chiral isomers.   [2 marks]
     ii) What structural feature in the molecule gives rise to optical isomerism?   [1 mark]
     iii) State how you could distinguish between the enantiomers.   [2 marks]

2   Parkinson's disease involves a deficiency of dopamine. It is treated by giving patients L-DOPA (dihydroxyphenylalanine), a naturally occurring amino acid, which is converted to dopamine in the brain.

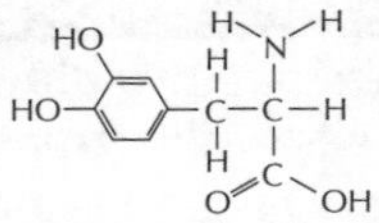

a)   DOPA is a chiral molecule. Its structure is shown on the right. Mark the structure's chiral centre.   [1 mark]

b)   A D,L-DOPA racemate was synthesised in 1911, but today natural L-DOPA is isolated from fava beans for use as a pharmaceutical.
     i)   Explain the meaning of the term *racemate*.   [1 mark]
     ii) Suggest two reasons why L-DOPA is used in preference to the D,L-DOPA racemate.   [2 marks]

Time for some quiet reflection...

This isomer stuff's not all bad — you get to draw little pretty pictures of molecules. If you're having difficulty picturing them as 3D shapes, you could always make models with blu-tack and those matchsticks that're propping your eyelids open. Blu-tack's very therapeutic anyway and squishing it about'll help relieve all that revision stress. It's great stuff.

Reaction Mechanisms

Now we're straight into the really gruesome, hardcore organic chemistry... **These two pages are for** *Edexcel* **and** *Nuffield*.

Homolytic Fission *Produces* Free Radicals *Only* Edexcel

A **free-radical substitution** reaction is where a chemical group is **substituted** (replaced) by a **free radical**.
A free radical is a particle with an **unpaired electron**. You show this unpaired electron with a dot — e.g. Cl·

Chlorine and **methane** react with a bit of a bang in U.V. light to form **chloromethane.**
The reaction mechanism has three stages:

$$CH_4 + Cl_2 \xrightarrow{\text{U.V.}} CH_3Cl + HCl$$

1 **Initiation reactions** — free radicals are produced.
1) Sunlight provides energy to break the Cl–Cl bond.

$$Cl\!-\!Cl \xrightarrow{\text{U.V.}} Cl· + Cl·$$

2) The bond splits **equally** and each atom gets to keep one electron — **homolytic fission**. The atom becomes a highly reactive **free radical**, Cl·, due to its **unpaired electron**.

Single-headed curly arrows show the movement of one electron only.

2 **Propagation reactions** — free radicals are used up and created in a **chain reaction**.

1) Cl· attacks a methane molecule (CH_4). $H_3C\!-\!H \quad Cl· \longrightarrow H_3C· + H\!-\!Cl$

2) The new methyl free radical, CH_3·, can attack another Cl_2 molecule. $H_3C· \quad Cl\!-\!Cl \longrightarrow H_3C\!-\!Cl + Cl·$

3) The new Cl· can attack another CH_4 molecule, and so on, until all the Cl_2 or CH_4 molecules are wiped out.

3 **Termination reactions** — free radicals are mopped up.
1) If any two free radicals join together, they form a **stable molecule**.
2) There are heaps of possible termination reactions. Here are a few examples:

$H_3C· \quad Cl· \longrightarrow H_3C\!-\!Cl$
$H_3C· \quad CH_3· \longrightarrow H_3C\!-\!CH_3$

Some products formed will be trace impurities in the final sample.

More substitutions
1) If **methane's** in excess, then the product will be mostly **chloromethane**, CH_3Cl.
2) But if **chlorine's** in excess, Cl· free radicals start attacking chloromethane, producing **dichloromethane**, CH_2Cl_2, **trichloromethane**, $CHCl_3$, and **tetrachloromethane**, CCl_4.

Alkenes Undergo *Electrophilic Addition* Reactions *Only* Edexcel

An **electrophilic addition** reaction is where a **carbon double bond** opens up and atoms are **added** to the carbon atoms. The double bond's got a high electron density, so it attracts **electrophiles**.
Here are two electrophilic addition reactions that you need to know:

Electrophiles attack molecules where there are high electron densities.

Chlorine and iodine do this with alkenes too.

A carbocation is an ion containing a positively charged carbon atom.

The easy one — adding halogens to alkenes

| The double bond repels the electrons in Br_2, polarising Br–Br. | Heterolytic (unequal) fission of Br_2: the closer Br gives up the bonding electrons to the other Br and sticks to the C atom. | You get a positively charged carbocation intermediate. The Br^- now zooms over... | ...and bonds to the other C atom, forming 1, 2-dibromoethane |

A slightly more tricky one — adding hydrogen bromide to an unsymmetrical alkene like propene

This time there are two **different** products — the amount of each depends on how **stable** the **carbocation intermediate** is. **Alkyl** groups push **electrons towards** the positive charge, making the carbocation **more stable**. So the **more** alkyl groups a carbocation has, the **more** stable it is, and the **more** often it forms.

Alkyl groups are alkanes with an H removed. E.g. methyl, -CH_3.

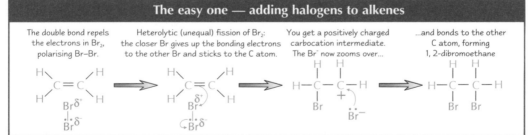

This carbocation's relatively stable — it's got two alkyl groups pushing electrons towards the carbon. So this carbocation forms most of the time.

This carbocation's less stable — it's only got one alkyl group pushing electrons towards the carbon. So this carbocation forms less often.

2-bromopropane's the major product — you get lots of this.

1-bromopropane's the minor product — you only get a bit of this.

Reaction Mechanisms

Haloalkanes Undergo *Nucleophilic Substitution*

Nucleophilic substitution is when a nucleophile attacks another molecule and is **swapped** for one of the attached groups.

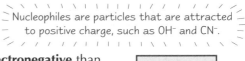
Nucleophiles are particles that are attracted to positive charge, such as OH⁻ and CN⁻.

The C–halogen bond in haloalkanes is **polar** — halogens are much more **electronegative** than carbon, so they draw the electrons **towards** themselves. The carbon is **partially positive,** so it's easily attacked by nucleophiles.

$C^{\delta+}-Br^{\delta-}$

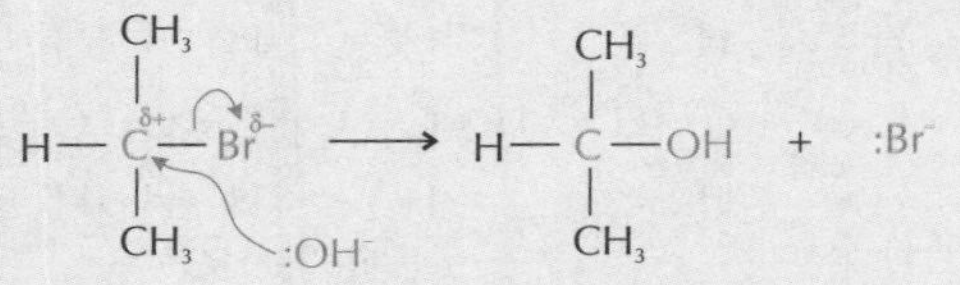

1) OH⁻ is the **nucleophile** — it provides a pair of electrons for the $C^{\delta+}$.
2) The C–Br bond breaks **heterolytically** — both electrons from the bond are taken by Br⁻.
3) Br⁻ comes away as OH⁻ bonds to the carbon.

There are two different mechanisms for nucleophilic substitution:

The S$_N$1 Reaction

This is how **tertiary haloalkanes** react. The reaction happens this way because there's very **little space** around the carbon (it's surrounded by alkyl groups).

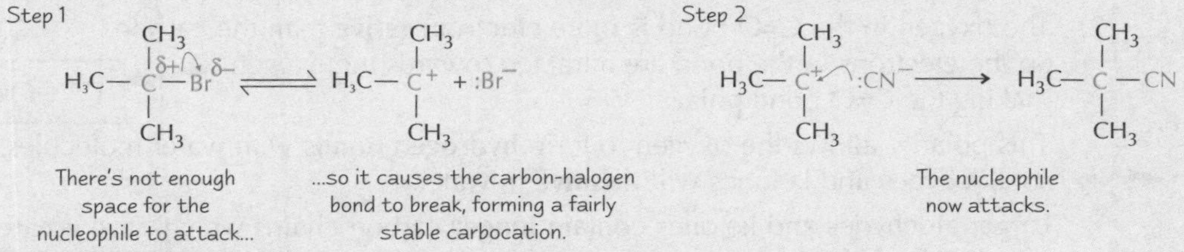

Step 1

There's not enough space for the nucleophile to attack...

...so it causes the carbon-halogen bond to break, forming a fairly stable carbocation.

Step 2

The nucleophile now attacks.

In primary haloalkanes, the halogen is joined to a carbon with just one alkyl group attached. In secondary haloalkanes the halogen is joined to a carbon with two alkyl groups attached. In tertiary haloalkanes, the halogen is attached to a carbon with three alkyl groups attached.

The S$_N$2 Reaction

This happens when there's **lots of space** around the carbon, like in **primary haloalkanes** which are surrounded mostly by H groups.

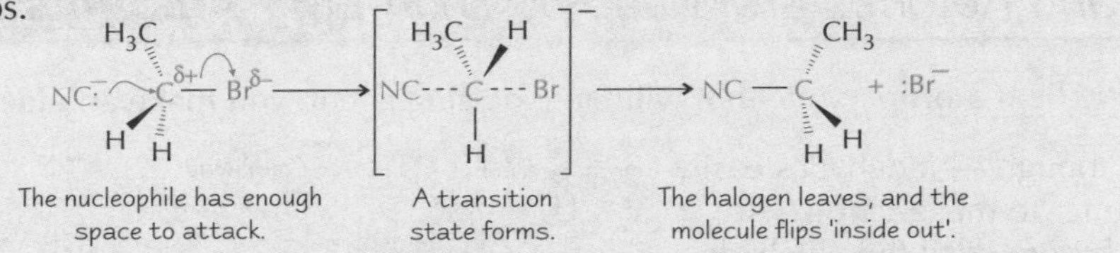

The nucleophile has enough space to attack.

A transition state forms.

The halogen leaves, and the molecule flips 'inside out'.

Secondary haloalkanes can react by **both** the S$_N$1 and S$_N$2 mechanisms — it all depends what mood they're in.

Practice Questions

Q1 What's a free radical? How are they made?

Q2 What's a carbocation?

Q3 What's a nucleophile? Give an example.

Exam Questions

1   a)   Give the conditions necessary for the formation of chloromethane as the major product from chlorine
         and methane.   [2 marks]
    b)   What kind of reaction is this?   [2 marks]

2   Consider the following reaction scheme: 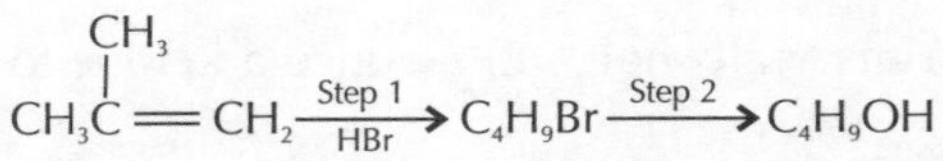
    a)   Step 1 is electrophilic addition.

    $$CH_3\overset{\overset{\textstyle CH_3}{|}}{C}=CH_2 \xrightarrow[\text{HBr}]{\text{Step 1}} C_4H_9Br \xrightarrow{\text{Step 2}} C_4H_9OH$$

    i)   Explain the term *electrophilic*.   [1 mark]
    ii) Draw the structure of the major product of the electrophilic addition reaction in Step 1.
         Explain why it is the major product.   [3 marks]
    b)   Draw the mechanism for Step 2, including the structure of the final product.   [3 marks]

Dude — this stuff's just... so... like... radical...

By now I'm sure you'd rather be chewing off your own hand than reading this. But you're going to need to know all the reactions, and be able to draw the mechanisms. Look out for these types of reactions throughout this section — examiners will expect you to recognise them. And be careful where those curly arrows are going to and from...

Aldehydes and Ketones

Aldehydes and ketones are both carbonyl compounds. They've got their carbonyl groups in different positions though.

Aldehydes and Ketones contain a Carbonyl Group

Aldehydes and ketones are **carbonyl compounds** — they contain the **carbonyl** functional group, **C=O**.

Aldehydes have their carbonyl group at the **end** of the carbon chain. Their names end in **–al**.

'R' represents a carbon chain of any length.

methanal   propanal

Ketones have their carbonyl group in the middle of the carbon chain. Their names end in **–one**, and often have a number to show which **carbon** the carbonyl group is on.

propanone   pentan-2-one

Smaller Aldehydes and Ketones will Combust and Dissolve in Water   *Nuffield only*

1) Aldehydes and ketones **burn completely** in **oxygen**, releasing CO_2 and water. The smaller the molecule, the easier it burns.

$$2CH_3CHO_{(l)} + 5O_{2(g)} \rightarrow 4CO_{2(g)} + 4H_2O_{(g)}$$

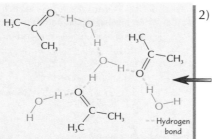

2) The oxygen in the C=O bond is more **electronegative** than the carbon. So the electrons in the bond are attracted towards the oxygen, making the C=O bond **polar**.

$$C^{\delta+}=O^{\delta-}$$

This polarity allows the oxygen to form **hydrogen bonds** with water molecules, so aldehydes and ketones will **dissolve** in water.

Larger aldehydes and ketones contain **longer** carbon chains which have **greater van der Waals** forces between them. If the van der Waals forces are stronger than the hydrogen bonds would be, the compound won't dissolve.

Hydrogen bond

Aldehydes and Ketones are Made by Oxidising Alcohols

The oxidising agent [O] used is often potassium dichromate(VI) with dilute sulphuric acid.

1) If you **gently heat** a **primary alcohol** with an oxidising agent, you make an **aldehyde**.

Be careful though — aldehydes easily oxidise further to make **carboxylic acids**. It's best to distil the aldehyde out of the reaction as soon as it's formed to stop it oxidising further.

primary alcohol   aldehyde   carboxylic acid

R and R' are two, possibly different, carbon chains.

See page 106 for primary and secondary alcohols.

2) You can reflux (see page 126) a **secondary alcohol** with an oxidising agent to make a **ketone**.

Ketones are very difficult to oxidise further — so you don't have to worry about oxidising it too far.

secondary alcohol   ketone

You can Reduce Aldehydes and Ketones Back to Alcohols

Using a **reducing agent** [H] you can:

1) reduce an **aldehyde** to a **primary alcohol**.   2) reduce a **ketone** to a **secondary alcohol**.

These are <u>nucleophilic addition</u> reactions — the reducing agent supplies an H^- that acts as a nucleophile and attacks the $\delta+$carbon.

Here are two **reducing agents** that you could use:

1) **NaBH₄** (sodium tetrahydridoborate(III) or sodium borohydride) dissolved in water with methanol — this is what you'd usually use, as it's pretty harmless.

2) **LiAlH₄** (lithium tetrahydridoaluminate(III) or lithium aluminium hydride) in **dry diethyl ether**. This is a more powerful reducing agent — it reacts violently with water, bursting into flames. Eeek.

Aldehydes and Ketones

Hydrogen Cyanide will React with Carbonyls by Nucleophilic Addition

Hydrogen cyanide reacts with carbonyl compounds to produce **hydroxynitriles** (molecules with a CN and an OH group). It's a **nucleophilic addition reaction** — a **nucleophile** attacks the molecule, causing an extra group to be **added**.

Hydrogen cyanide's a **weak acid** — it partially dissociates in water to form **H^+** ions and **CN^-** ions. $HCN \rightleftharpoons H^+ + CN^-$

1) The CN^- group **attacks** the partially positive carbon atom and **donates** a pair of electrons. Both electrons from the double bond transfer to the oxygen.

2) H^+ (from either hydrogen cyanide or water) bonds to the oxygen to form the **hydroxyl group** (OH).

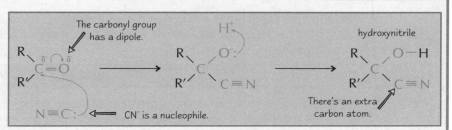

The carbonyl group has a dipole.

CN^- is a nucleophile.

hydroxynitrile

There's an extra carbon atom.

Watch out — hydrogen cyanide is a **highly poisonous gas**, so the reaction needs to be carried out in a **fume cupboard**.

Lactic Acid is formed from Ethanal   OCR only

Lactic acid (2-hydroxypropanoic acid) can be made in two steps. The first step is the **addition** of **hydrogen cyanide** to ethanal — it's exactly the same **nucleophilic addition** mechanism as above.

ethanal   2-hydroxypropanenitrile

$$H_3C{-}C{=}O + HC{\equiv}N \longrightarrow H_3C{-}\underset{H}{\overset{OH}{C}}{-}C{\equiv}N$$

This reaction's great for adding an extra carbon to the chain — dead useful when you want to make a longer compound.

This is followed by **acid hydrolysis**.

2-hydroxypropanenitrile   2-hydroxypropanoic acid

$$H_3C{-}\underset{H}{\overset{OH}{C}}{-}C{\equiv}N + 2H_2O + H^+ \longrightarrow H_3C{-}\underset{H}{\overset{OH}{C}}{-}C\overset{O}{\underset{OH}{{\Large\diagup}}} + NH_4^+$$

Practice Questions

Q1 Why are small carbonyls soluble in water? Why are large carbonyls less soluble?

Q2 How could you make butanone?

Q3 What do you get if you reduce ethanal?

Exam Question

1   The compound C_3H_6O can exist as an aldehyde and a ketone.

   a)   Draw and name the carbonyl isomers of C_3H_6O.   [4 marks]

   b)   Hydrogen cyanide, HCN, will react with C_3H_6O carbonyl compounds to form a compound with the molecular formula C_4H_6ON.
   i) Name the type of reaction that occurs.   [1 mark]
   ii) Draw the products produced from the aldehyde and the ketone.
        Indicate which product will be produced as a racemic mixture and explain why.   [4 marks]

   c)   The aldehyde C_3H_6O can be reduced to alcohol, C_3H_7OH.
   i) Write an equation for the reaction. Make sure the structures of the compounds are clear.   [1 mark]
   ii) Suggest suitable reagent(s) and conditions for the reaction.   [2 marks]

They go together as shoo-bop sha whada whada yippidy boom da boom...

You've got to be a dab hand at recognising different functional groups from a mile off. The carbonyl group's just the first of many. Make sure you know how aldehydes differ from ketones and what you get when you oxidise them both. The lactic acid reaction is a bit of a pain to learn, but if you're a hard-done-by OCR person then you've just got to do it.

Aldehydes, Ketones and Grignard Reagents

Knowing what aldehyde and ketone molecules look like won't help you decide which is which if you've got a test tube of each. There are a few chemical tests that will though. You can also just test for a carbonyl group.
Salters can skip these two pages.

Brady's Reagent Tests for a Carbonyl Group   *Not for AQA*

Brady's reagent is **2,4-dinitrophenylhydrazine** (2,4-DNPH) dissolved in methanol and concentrated sulphuric acid.

The **2,4-dinitrophenylhydrazine** forms a **bright orange precipitate** if a carbonyl group is present.

This only happens with **C=O groups**, not with more complicated ones like COOH, so it only tests for **aldehydes** and **ketones**.

The Melting Point of the Precipitate Identifies the Carbonyl Compound

The orange precipitate is a **derivative** of the carbonyl compound which can be purified by **recrystallisation** (see page 126). Each different carbonyl compound produces a crystalline derivative with a **different melting point**.

So if you measure the melting point of the crystals and compare it against the **known** melting points of the derivatives, you can **identify** the carbonyl compound.

And There are a Few Ways of Testing for Aldehydes

These tests let you distinguish between an aldehyde and a ketone.
They all work on the idea that an **aldehyde** can be **easily oxidised** to a carboxylic acid, but a ketone can't.
As an aldehyde is oxidised, another compound is **reduced** — so a reagent is used that **changes colour** as it's reduced.

TOLLENS' REAGENT   *Not for Nuffield*

Tollens' reagent is a **colourless** solution of **silver nitrate** dissolved in **aqueous ammonia**.

If it's heated in a test tube with an aldehyde, a **silver mirror** forms after a few minutes.

You should heat the test tube in a beaker of hot water, rather than directly over a flame.

$$\underset{\text{colourless}}{Ag(NH_3)_2^+{}_{(aq)}} + e^- \longrightarrow \underset{\text{silver}}{Ag_{(s)}} + 2NH_{3(aq)}$$

FEHLING'S SOLUTION OR BENEDICT'S SOLUTION   *Not for OCR*

Fehling's solution is a **blue** solution of complexed **copper(II) ions** dissolved in **sodium hydroxide**.

If it's heated with an aldehyde the copper(II) ions are reduced to a **brick-red precipitate** of **copper(I) oxide**.

$$\underset{\text{blue}}{Cu^{2+}{}_{(aq)}} + e^- \longrightarrow \underset{\text{brick-red}}{Cu^+{}_{(s)}}$$

Again, make sure you heat the mixture in a test tube in a beaker of hot water, rather than directly over a flame.

Benedict's solution is exactly the same as Fehling's solution except the copper(II) ions are dissolved in **sodium carbonate** instead. You still get a **brick-red precipitate** of copper(I) oxide though.

Some Carbonyls will react with Iodine   *This bit's just for Edexcel*

Carbonyls that contain a **methyl carbonyl** group react when heated with **iodine** in the presence of an alkali.
If there's a methyl carbonyl group you'll get a **straw-yellow precipitate** and an antiseptic smell.

This is a methyl carbonyl group.

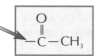

You can also use potassium iodide solution and sodium chlorate(I).

If something contains a **methyl carbonyl** group, it's got to be either...

....ethanal,

or a ketone with at least one methyl group,

Aldehydes, Ketones and Grignard Reagents

Haloalkanes React with Magnesium to form Grignard Reagents

Edexcel only

Grignard reagents have the general formula RMgX, where X is a halogen (usually bromine).

They're prepared by **refluxing** a **haloalkane** with **magnesium** in dry diethyl ether.

$$RX + Mg \xrightarrow[\text{reflux}]{\text{dry diethyl ether}} RMgX$$

The carbon (in the R-group) is more electronegative than the magnesium, so the C–Mg bond is **polarised**. As the carbon's at the negative end it acts as a **nucleophile** and will take part in nucleophilic addition or nucleophilic substitution reactions to form C–C bonds.

$$\overset{\delta-}{R}\!-\!\overset{\delta+}{MgX}$$

Here are a few reactions of Grignard reagents that you need to know:

1) **Grignard reagents** are hydrolysed using dilute acid to form **alkanes**. E.g.

$$CH_3CH_2MgBr + H_2O \xrightarrow{H^+} CH_3CH_3 + Mg(OH)Br$$
alkane

2) They react with **carbon dioxide** to form **carboxylic acids.**
An intermediate is formed which has to be hydrolysed by dilute hydrochloric acid.
E.g.

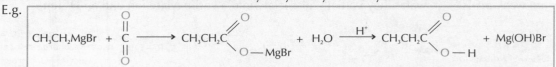

The carboxylic acid has one more carbon than the original Grignard reagent.

3) They react with **carbonyl compounds** to form **alcohols.**
Again an intermediate is formed which has to be hydrolysed by dilute hydrochloric acid.

The **type** of alcohol produced depends on the carbonyl compound used. E.g.

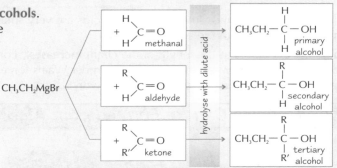

Practice Questions

Q1 What's Brady's reagent? What does it test for?

Q2 Why do you get a 'silver mirror' when an aldehyde is heated with Tollens reagent?

Q3 Why is a Grignard reagent a nucleophile?

Exam Questions

1   A student has two compounds, A and B.
   a) When tested with 2,4-dinitrophenylhydrazine, both compounds gave an orange precipitate.
       Explain what this shows.   [1 mark]
   b) Compounds A and B were also heated with Fehling's solution:
       i) The mixture containing A turned brick-red. What does this tell the student?   [1 mark]
       ii) The mixture containing B remains blue. What does this tell the student?   [1 mark]
   c) Finally, the student adds potassium iodide solution and sodium hydroxide solution until the solution
       becomes colourless. On heating, a straw-yellow precipitate forms in both mixtures.
       i) Identify A.   [1 mark]   ii) Draw the structure that must be present in B.   [1 mark]
   d) How could the student use the precipitate in (a) to identify B?   [2 marks]

2   a) Give the conditions and write an equation for the formation of a Grignard reagent from bromomethane.   [4 marks]
   b) i) Draw and name the product of the reaction of the Grignard reagent formed in (a) with propanone.   [2 marks]
       ii) Why must acid be added?   [1 mark]

It's all crystal clear — unless it's a precipitate or a silver mirror...

This is chemistry at its best — test tubes, Bunsen burners and things changing colour. The fun's spoilt a bit for the Edexcel folks by the Grignard stuff, which gets a bit yukky. The best thing to do is shut the book and try to write down the reactions from memory, then check if you're right. If you're not, then have another go. Keep doing this till you're perfect.

Carboxylic Acids

Carboxylic acids are much more interesting than cardboard boxes — as you're about to discover...

Carboxylic Acids contain –COOH

A <u>carbox</u>yl group contains a <u>carbon</u>yl group and a <u>hydrox</u>yl group.

Carboxylic acids contain the **carboxyl** functional group **–COOH**.

To name them, you find and name the longest alkane chain, take off the 'e' and add '**–oic acid**'.

The carboxyl group's always at the **end** of the molecule and when naming it's more important than other functional groups — so all the other functional groups in the molecule are numbered starting from this carbon.

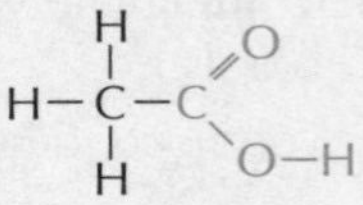

ethanoic acid 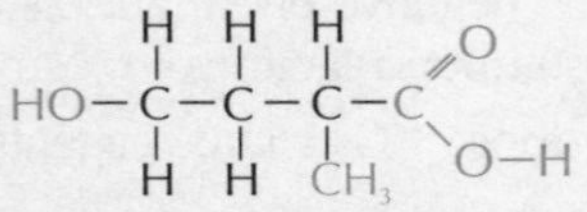   4-hydroxyl-2-methylbutanoic acid 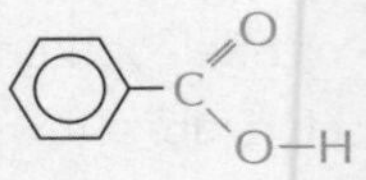   benzoic acid

Carboxylic Acids are Weak Acids

Carboxylic acids are **weak acids** — in water they partially dissociate into a **carboxylate ion** and an **H⁺ ion**.

The equilibrium lies to the left because most of the molecules don't dissociate.

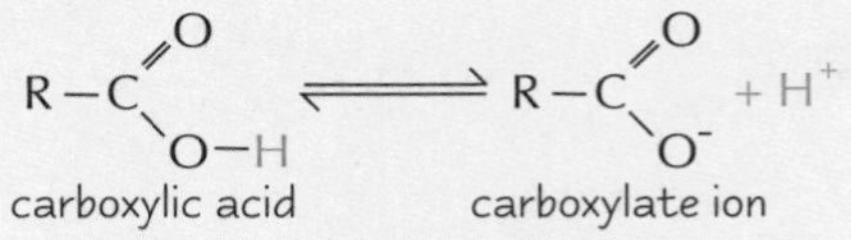

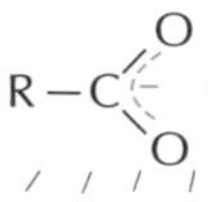

carboxylic acid ⇌ carboxylate ion + H⁺

Nuffield only — the negative charge delocalises over the COO⁻ group.

Like alcohols and carbonyls, small carboxylic acids are **very soluble** in water as they can form **hydrogen bonds** with the water molecules.

The solubility **decreases** as the length of the chain **increases**. Longer chains have stronger **van der Waals** forces between them — if the van der Waals forces are stronger than the hydrogen bonds would be, the carboxylic acid won't dissolve.

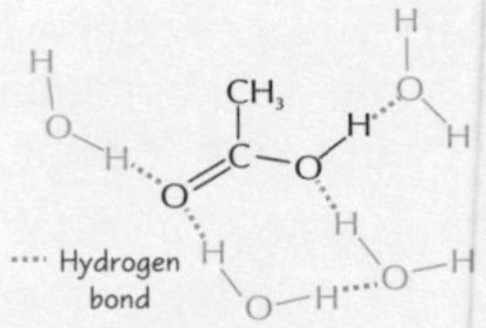

···· Hydrogen bond

Carboxylic Acids Can Be Formed from Alcohols, Aldehydes and Nitriles

Not for OCR or Salters

OXIDATION OF PRIMARY ALCOHOLS AND ALDEHYDES

You can make a carboxylic acid by **oxidising a primary alcohol** to an **aldehyde**, and then to a carboxylic acid. You'll need to reflux the alcohol with an **oxidising agent** (such as acidified potassium dichromate).

Remember, ketones **can't** be oxidised to carboxylic acids.

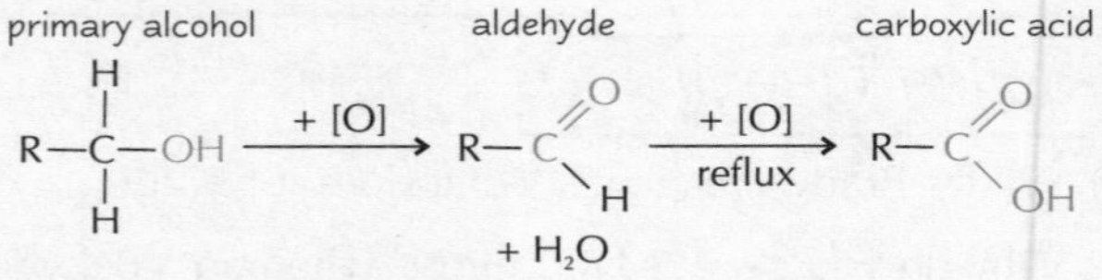

primary alcohol → aldehyde → carboxylic acid

HYDROLYSIS OF NITRILES   **Not for AQA**

Carboxylic acids can also be made by **hydrolysing a nitrile**. You vigorously reflux the nitrile with dilute hydrochloric acid, and then distill off the carboxylic acid formed.

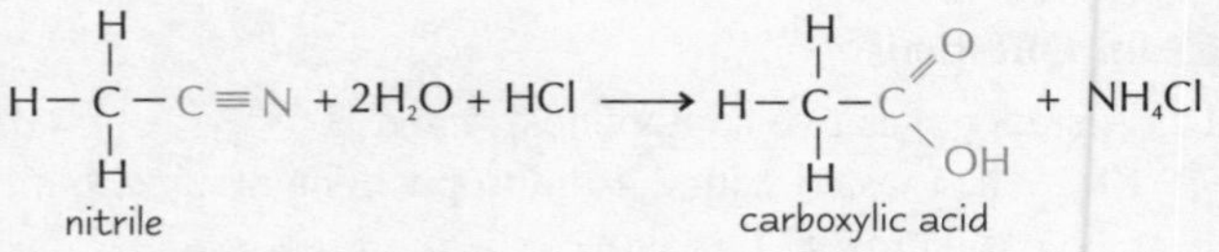

nitrile + 2H₂O + HCl ⟶ carboxylic acid + NH₄Cl

Alcohols React with Carboxylic Acids to form Esters

See pages 80-81 for esters.

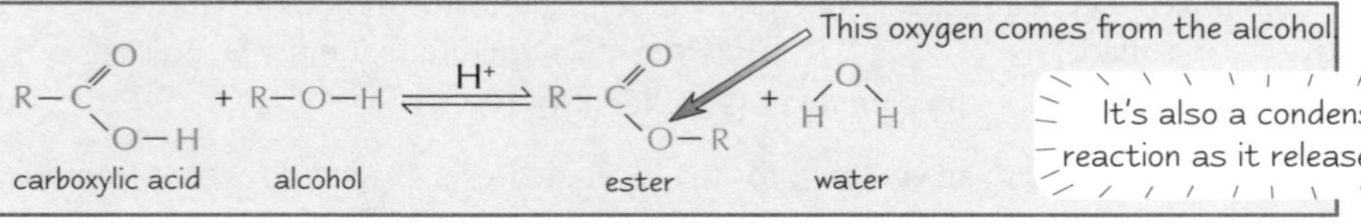

If you heat a **carboxylic acid** with an **alcohol** in the presence of an **acid catalyst**, you get an ester. It's called an **esterification** reaction. Concentrated sulphuric acid is usually used as the acid catalyst.

carboxylic acid + alcohol ⇌ ester + water

This oxygen comes from the alcohol

It's also a condensation reaction as it releases water.

Here's how ethanoic acid reacts with ethanol to make the ester, ethyl ethanoate:

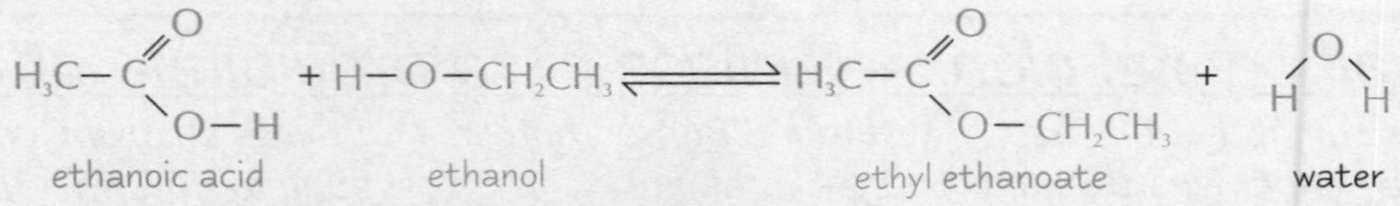

ethanoic acid + ethanol ⇌ ethyl ethanoate + water

Carboxylic Acids

If you're doing Salters you can skip this page.

Carboxylic Acids React with **Alkalis** and **Carbonates** to Form **Salts**

1) Carboxylic acids are **neutralised** by **aqueous alkalis** to form **salts and water**.

ethanoic acid   sodium ethanoate
$$CH_3COOH + NaOH \rightarrow CH_3COONa + H_2O$$

Salts of carboxylic acids are called carboxylates and their names end with –oate.

2) Carboxylic acids react with **carbonates CO_3^{2-}** or **hydrogencarbonates HCO_3^-** to form a **salt, carbon dioxide and water**.

In these reactions, carbon dioxide fizzes out of the solution.

ethanoic acid   sodium ethanoate
$$2CH_3COOH_{(aq)} + Na_2CO_{3(s)} \rightarrow 2CH_3COONa_{(aq)} + H_2O_{(l)} + CO_{2(g)}$$
$$CH_3COOH_{(aq)} + NaHCO_{3(s)} \rightarrow CH_3COONa_{(aq)} + H_2O_{(l)} + CO_{2(g)}$$

Other Reactions You'll Need to Know   *This bit's just for Edexcel and OCR*

It's quite **hard** to reduce a carboxylic acid, so you have to use a **powerful reducing agent** like **$LiAlH_4$**. It reduces the carboxylic acid right down to an **alcohol** in one go — you can't get the reduction to stop at the aldehyde.

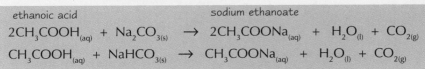

carboxylic acid   primary alcohol
$$R-C\overset{O}{\underset{OH}{}} \xrightarrow{[H^+]} R-CH_2-OH \Longleftarrow \textit{Edexcel only}$$

Acyl chlorides are covered on pages 82-83.

Mix a carboxylic acid with **phosphorus pentachloride** and you'll get an **acyl chloride**.

$$R-C\overset{O}{\underset{OH}{}} + PCl_5 \longrightarrow R-C\overset{O}{\underset{Cl}{}} + POCl_3 + HCl$$
carboxylic acid   acyl chloride

Practice Questions

Q1 How can you make methanoic acid from methanol?

Q2 What do you get if you react ethanoic acid with ethanol?

Q3 What will carboxylic acids react with to release CO_2?

Exam Questions

1 Tartaric acid is found in unripe grapes.

a) Its systematic name is 2,3-dihydroxybutanedioic acid. Draw its structure. [1 mark]

b) Draw and name the structure of the compound formed if tartaric acid is reduced. Suggest a reducing agent for the reaction. [4 marks]

c) Potassium bitartrate crystals form in wine casks during fermentation. Suggest a reagent to make potassium bitartrate from tartaric acid in the laboratory. [1 mark]

2 Propanol and propanoic acid both contain three carbon atoms.

a) Explain why propanoic acid behaves as an acid, but propanol does not. [3 marks]

b) Describe a simple test-tube reaction to distinguish between propanol and propanoic acid. Give the reagent(s) and state the observations expected. [3 marks]

c) A student refluxes propanol with propanoic acid and a little concentrated sulphuric acid.
   i) Write an equation for the reaction, showing the structures clearly. [2 marks]
   ii) How could the student increase her yield? [1 mark]

January sales — my kind of powerful reducing agent...

I'm sure this page is the least fun you've had since... well, the last page really. Carboxylic acids are just like any old acid — react them with an alkali and you get a salt and water; react them with a carbonate and you get a salt, carbon dioxide and water. Make sure you balance that acid + carbonate equation though — you don't want to go throwing away marks.

Esters

There are not that many reactions to learn on these pages — just a little bit of this and a little bit of that.

Esters have the Functional Group –COO–

Esters are made from carboxylic acids and alcohols (see page 78). Their names have two parts —
the first bit comes from the **alcohol** and the second bit from the **carboxylic acid**.

Just to confuse you, the name's written the opposite way round from the structural and displayed formulae.

ethyl ethanoate
$CH_3COO\ CH_2CH_3$

methyl benzoate
$C_6H_5COOCH_3$

1-methylpropyl methanoate
$HCOO\ CH(CH_3)CH_2CH_3$

When numbering the carbons to name the attached groups, count out from the ester link in the middle. So here, the methyl group's on carbon 1.

They're Used as Food Flavourings, Solvents and Plasticisers

1) Esters have a **sweet smell** — it varies from gluey sweet for smaller esters to a fruity 'pear drop' smell for the larger ones. The **nice fragrances** and **flavours** of lots of flowers and fruits come from esters. The food industry uses esters to **flavour** things like drinks and sweets too.

2) Esters are **polar** liquids so they dissolve lots of **polar organic compounds** (see page 7). They've also got quite **low boiling** points so they **easily evaporate** from the mixtures. This makes them good solvents in **glues** and **printing inks**.

3) Esters are used as **plasticisers** — they're added to plastics during polymerisation to make the plastic more **flexible**. Over time, the plasticiser molecules escape though, and the plastic becomes brittle and stiff.

Esters are Hydrolysed to Form Alcohols   *Not for Salters*

There are two types of hydrolysis of esters — **acid hydrolysis** and **base hydrolysis**.
With both types you get an **alcohol**, but the second product in each case is different.

ACID HYDROLYSIS

Acid hydrolysis splits the ester into an **acid** and an **alcohol** — it's just the **reverse** of the condensation reaction at the bottom of page 78.

You have to **reflux** the ester with a **dilute acid**, such as hydrochloric or sulphuric. For example:

ethyl ethanoate + H_2O ⇌ (H^+, reflux) ethanoic acid + ethanol

As it's a reversible reaction, you need to use lots of water to push the equilibrium over to the right.

BASE HYDROLYSIS

This time you have to **reflux** the ester with a **dilute alkali**, such as sodium hydroxide. You get a carboxylate ion and an alcohol. For example:

ethyl ethanoate + OH^- ⇌ (reflux) ethanoate + ethanol

Esters

The stuff on this page is just for AQA, Salters and Edexcel.

Fats and Oils are Esters of Glycerol and Fatty Acids

Fatty acids are large **carboxylic acids**. They combine with glycerol (propane-1,2,3-triol) to make fats and oils.
The fatty acids can be **saturated** (no double bonds) or **unsaturated** (with C=C double bonds).

Most of a fat or oil is made from fatty acid chains — so it's these that
give them many of their properties.

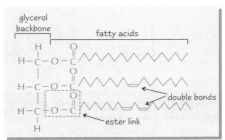

- **'Fats'** have mainly **saturated** hydrocarbon chains — they fit neatly together,
  increasing the van der Waals forces between them. This means higher
  temperatures are needed to melt them and they're **solid** at room temperature.

- **'Oils'** have unsaturated hydrocarbon chains — these don't pack
  well, decreasing the effect of the van der Waals forces.
  So they're easier to melt and are **liquids** at room temperature.

Oils and Fats Have Heaps of Uses

1) **Margarine** is made from oil — **unsaturated** oil is used to give it a **lower** melting point.

2) **Essential oils** — these are used in perfumes and were used in early medicines. They're **steam-distilled** from plants.

3) **Soaps** — if you hydrolyse a fat with a base, you get fatty acid salts that are **soaps** — see page 132.

And just for Salters —

4) **Oil paints** — these gradually harden by **oxidative cross-linking** between unsaturated fatty acid hydrocarbon chains.

   What happens is that U.V. light splits oxygen molecules into **free radicals**. The oxygen free radicals react with the
   unsaturated hydrocarbon chains and link them together. This produces new radicals which go on to attack other
   molecules. This turns the paint into a **tough layer**.

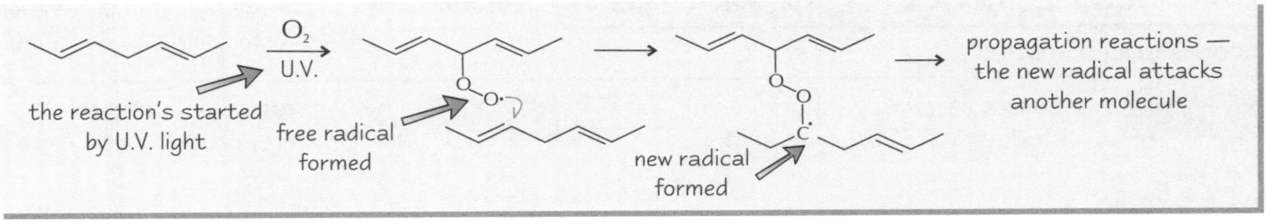

the reaction's started by U.V. light   free radical formed   new radical formed   propagation reactions — the new radical attacks another molecule

Practice Questions

Q1 What do you get if you hydrolyse an ester with a base?

Q2 What's a plasticiser?

Q3 Name three uses of fats and oils.

Exam Questions

1 Compound C, shown on the right, is found in raspberries.

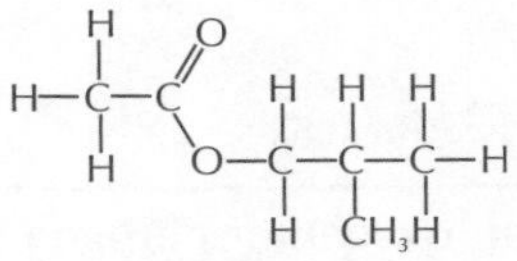

   a) Name compound C.   [1 mark]

   b) Suggest a use for compound C.   [1 mark]

   c) Draw and name the structures of the products formed when compound C is refluxed with dilute sulphuric acid.
      What kind of reaction is this?   [5 marks]

   d) If compound C is refluxed with excess sodium hydroxide, a similar reaction occurs.
      Give a difference between this reaction and the reaction described in (c).   [1 mark]

2 Many people eat margarine in preference to butter.
   Butter contains saturated fatty acids in propane-1,2,3-triol tri-ester compounds.

   a) Explain the term 'saturated'.   [1 mark]

   b) How does the proportion of saturated fatty acids affect the melting point of butter?   [1 mark]

Ahh... the sweet smell of success...

*I bet your apple flavoured sweets have never been near a nice rosy apple — it's all the work of esters. And for that matter,
I reckon prawn cocktail crisps have never met a prawn, or a cocktail either. None of it's real. And as for potatoes...*

Acyl Chlorides

Acyl chlorides are easy to make and a handy starting point for making other types of molecules.

Acyl Chlorides have the Functional Group –COCl

Acyl (or acid) chlorides have the functional group **COCl** — their general formula is $C_nH_{2n-1}OCl$. All their names end in **–oyl chloride**.

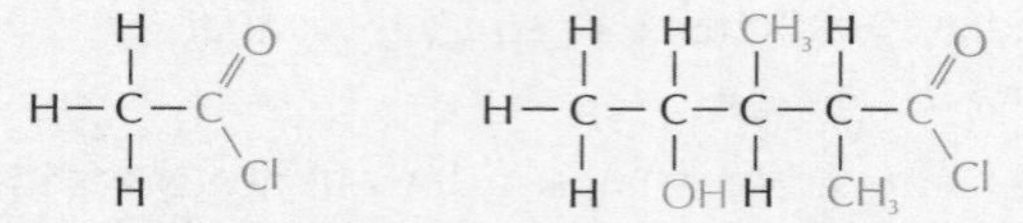

ethanoyl chloride   4-hydroxy-2,3-dimethyl**pentan**oyl chloride

Like carboxylic acids and esters, the carbon atoms are numbered from the end with the acyl functional group.

Acyl Chlorides are Made from Carboxylic Acids   *Only OCR and Edexcel*

Acyl chlorides are made by reacting a **carboxylic acid** with:

Phosphorus pentachloride (PCl₅)
PCl₅ is liquid, so you need to use **fractional distillation** to separate the acyl chloride from the mixture.

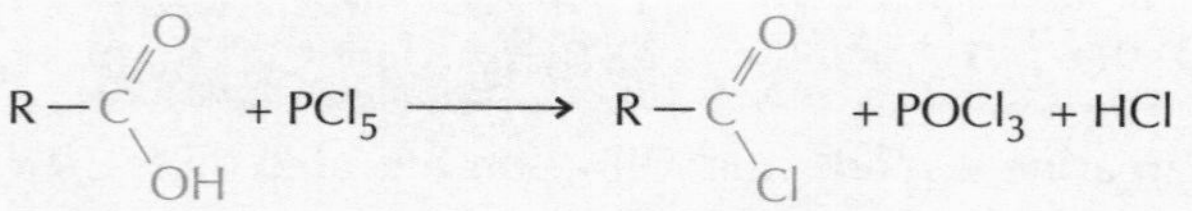

$$R-C(OH)O + PCl_5 \longrightarrow R-C(Cl)O + POCl_3 + HCl$$

carboxylic acid   acyl chloride

or... **Sulphur dichloride oxide (SOCl₂)**   *Just for OCR*
This reaction's more useful — both SO₂ and HCl are **gases**, so there's no need to separate out the mixture.

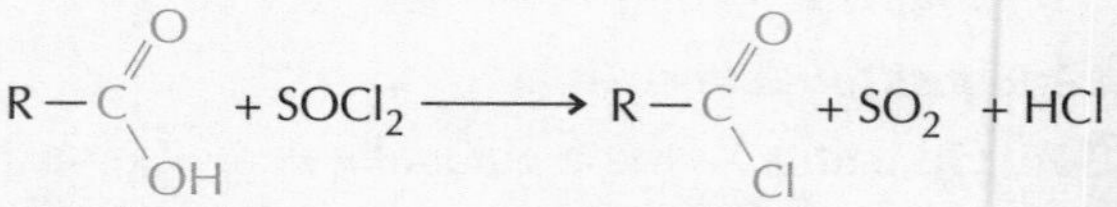

$$R-C(OH)O + SOCl_2 \longrightarrow R-C(Cl)O + SO_2 + HCl$$

carboxylic acid   acyl chloride

Acyl Chlorides Easily Lose Their Chlorine   *Not for Salters*

This irreversible reaction is a much easier, faster way to produce an ester than esterification (page 78).

Acyl chlorides react with...

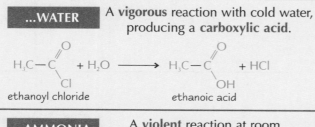

| ...WATER | A **vigorous** reaction with cold water, producing a **carboxylic acid**. |

$$H_3C-C(Cl)O + H_2O \longrightarrow H_3C-C(OH)O + HCl$$
ethanoyl chloride   ethanoic acid

| ...ALCOHOLS | A **vigorous** reaction at room temperature, producing an **ester**. |

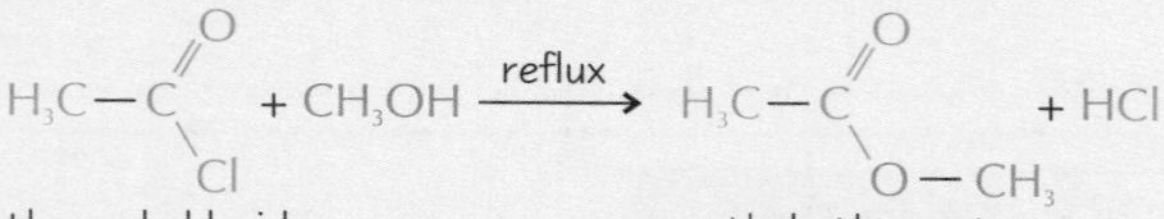

$$H_3C-C(Cl)O + CH_3OH \xrightarrow{reflux} H_3C-C(O-CH_3)O + HCl$$
ethanoyl chloride   ethyl ethanoate

| ...AMMONIA | A **violent** reaction at room temperature, producing an **amide**. |

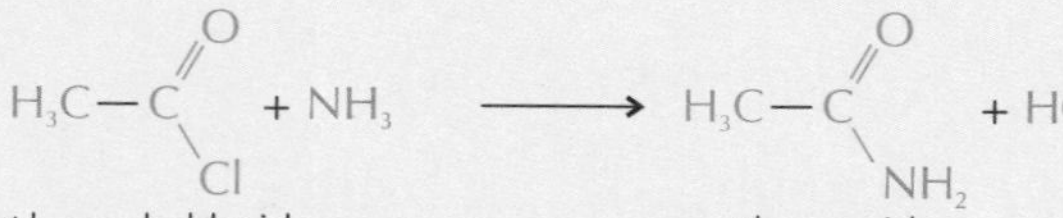

$$H_3C-C(Cl)O + NH_3 \longrightarrow H_3C-C(NH_2)O + HCl$$
ethanoyl chloride   ethanamide

| ...AMINES | A **violent** reaction at room temperature, producing an **N-substituted amide**. |

See pages 92-93 for more on amides.

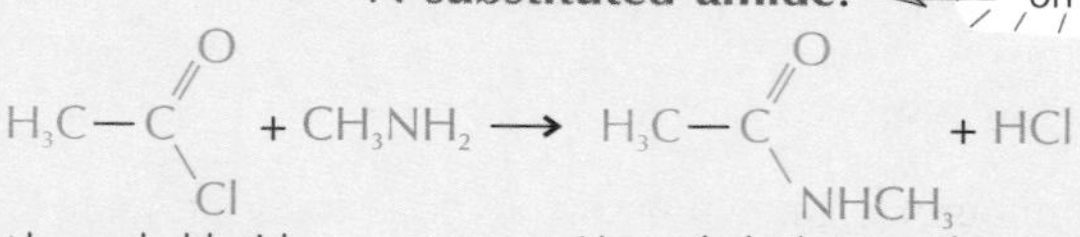

$$H_3C-C(Cl)O + CH_3NH_2 \longrightarrow H_3C-C(NHCH_3)O + HCl$$
ethanoyl chloride   N-methylethanamide

Each time, **Cl** is **substituted** by an oxygen or nitrogen group and **hydrogen chloride** fumes are given off.

Acyl Chlorides and Acid Anhydrides React in the Same Way   *AQA only*

An **acid anhydride** is made from two identical carboxylic acid molecules. If you know the name of the carboxylic acid, they're easy to name — just take away '**acid**' and add '**anhydride**'.

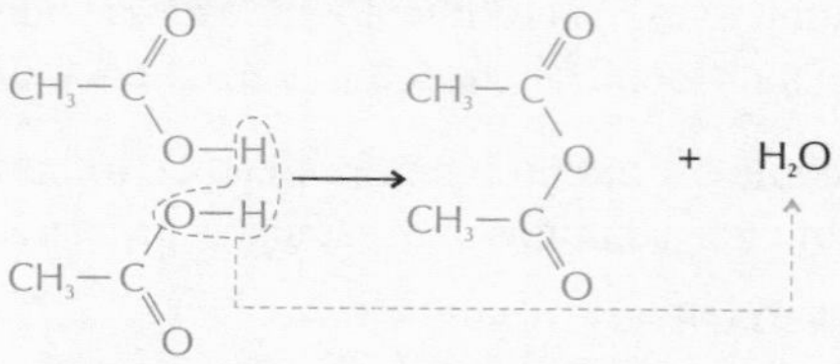

ethanoic acid   ethanoic anhydride

You need to know the reactions of **water, alcohol, ammonia** and **amines** with acid anhydrides. Luckily, they're almost the same as those of acyl chlorides — the reactions are just **less vigorous** and you get a **carboxylic acid** formed instead of HCl.

e.g. $(CH_3CO)_2O_{(l)} + H_2O_{(l)} \rightarrow 2CH_3COOH_{(aq)}$
ethanoic anhydride + water → ethanoic acid

Acyl Chlorides

This page is just for AQA, Edexcel and Nuffield.

Acyl Chloride Reactions are Nucleophilic Addition-Elimination

In acyl chlorides, both the chlorine and the oxygen draw electrons **towards** themselves, so the carbon has a slight **positive** change — meaning it's easily attacked by **nucleophiles**.

Here's the mechanism for a **nucleophilic addition-elimination** reaction between ethanoyl chloride and methanol:

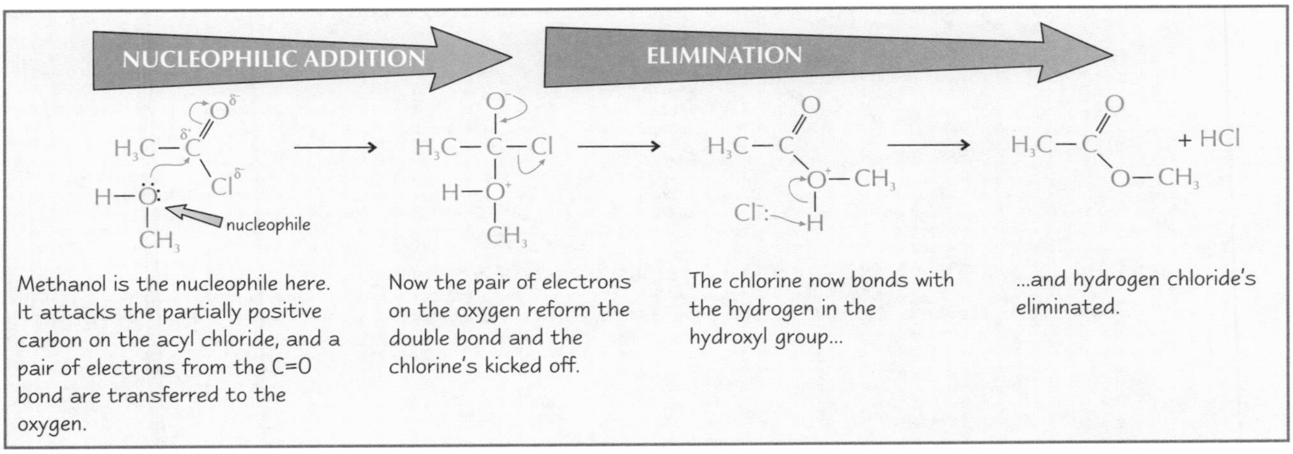

Methanol is the nucleophile here. It attacks the partially positive carbon on the acyl chloride, and a pair of electrons from the C=O bond are transferred to the oxygen.

Now the pair of electrons on the oxygen reform the double bond and the chlorine's kicked off.

The chlorine now bonds with the hydrogen in the hydroxyl group...

...and hydrogen chloride's eliminated.

For the other reactions you need to know, just change the nucleophile to water (H_2O:), ammonia ($\dot{N}H_3$) or an amine (e.g. $CH_3\dot{N}H_2$) — they all work the same way.

Ethanoic Anhydride is Used for the Manufacture of Aspirin   AQA, and Nuffield only

Aspirin is an **ester** — its made by reacting **salicylic acid** (which has an alcohol group) with **ethanoic anhydride** or **ethanoyl chloride**.

Ethanoic anhydride is used in industry because:
- it's **cheaper** than ethanoyl chloride.
- it's **safer** to use than ethanoyl chloride as it's **less corrosive**, reacts **more slowly** with water, and **doesn't** produce dangerous **hydrogen chloride** fumes.

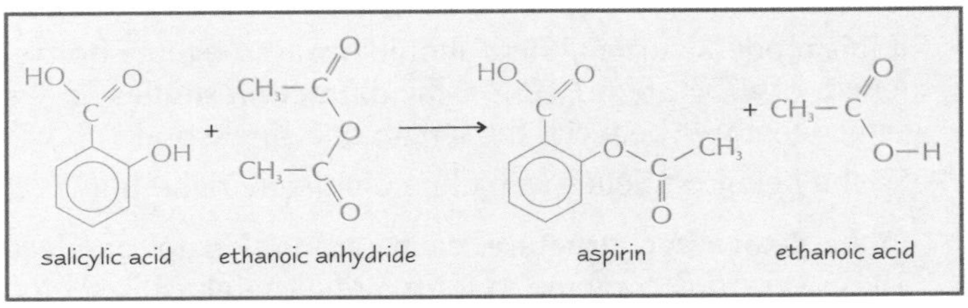

Practice Questions

Q1 Name two things you can react together to make an acyl chloride.

Q2 How can you make an amide from an acyl chloride?

Q3 What by-product is always formed in acid anhydride reactions?

Exam Question

1   Look at the following synthesis route for producing an ester from a carboxylic acid:

$$CH_3CH_2COOH \xrightarrow{Step\ 1} CH_3CH_2COCl \xrightarrow[CH_3OH]{Step\ 2} CH_3CH_2COOCH_3$$

a) Using a suitable reagent, write an equation for Step 1. [2 marks]

b) Name and outline a mechanism for the reaction in Step 2. [5 marks]

c) Suggest an advantage and a disadvantage of using this two-step route rather than a one-step esterification reaction between propanoic acid and methanol. [2 marks]

Anhydrides — aren't they some sort of alien robot...

I could easily lose my mind doing this stuff, let alone a little chlorine particle, and what's worse is I think all those hydrogen chloride fumes are getting to me... I feel kind of dizzy... my head hurts... and I want to lie down and sleep... can't.. keep.. eyes... open... zzzzzzzzzzzzzzzzzzzzzzzzzzzzzzzzzzzz..

Aromatic Compounds

In the old days something just had to be whiffy to qualify as an aromatic compound. Nowadays, they're a bit more fussy about the definition.

Aromatic Compounds are Derived from Benzene

Arenes or **aromatic compounds** contain a **benzene ring**.
Benzene has the formula C_6H_6 — there are two ways of representing it:

> Aliphatic hydrocarbons have no rings of carbon atoms.

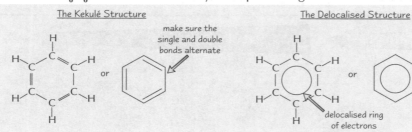

> The Kekulé Structure
> make sure the single and double bonds alternate
> or
> The Delocalised Structure
> or
> delocalised ring of electrons

> Aromatic compounds burn with a smoky flame because such a high proportion of them is carbon.

Arenes are named in two ways. There's no easy rule — you just have to learn these examples:

Some are named as substituted benzene rings...

chloro**benzene**   nitro**benzene**   1, 3-dimethyl**benzene**

...while others are named as compounds with a phenyl group (C_6H_5) attached.

phenol   2-methyl**phenol**   **phenyl**amine

The Bonding in the Benzene Ring's Unusual

The **Kekulé structure** was developed first. This has alternating single and double bonds which can **flip** between the carbons, as shown.

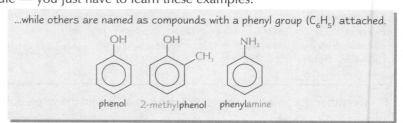

If this model's correct, there should always be three bonds with the length of a C–C bond and three bonds with the length of a C=C bond. But **X-ray diffraction studies** have shown that all the carbon-carbon bonds in benzene are the same length — between the length of a single bond and a double bond.

So the Kekulé structure can't be completely right, but it's still used today as it's useful for drawing reaction mechanisms.

In the **delocalised structure**, each carbon donates an electron from its **p-orbital**. These electrons combine to form a ring of **delocalised electrons**.

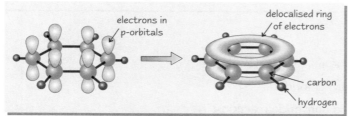

electrons in p-orbitals

delocalised ring of electrons

carbon

hydrogen

> Benzene's a planar (flat) molecule — it's got a ring of carbon atoms with their hydrogens sticking out all on a flat plane.

Delocalisation of Electrons Gives the Molecule Stability

1) Cyclohexene has **one** double bond. When it's hydrogenated, the enthalpy change is **–120 kJmol⁻¹**. If benzene had three double bonds (as in the Kekulé structure), you'd expect it to have an enthalpy of hydrogenation of –360 kJmol⁻¹.

2) But the **experimental** enthalpy of hydrogenation of benzene is **–208 kJmol⁻¹** — far less exothermic than expected.

Energy's put in to break bonds and released when bonds are made. So **more energy** must have been put in to break the bonds in benzene than would be needed to break the bonds in the Kekulé structure.

This difference indicates that benzene is **more stable** than the Kekulé structure would be. This is thought to be due to the **delocalised ring of electrons**.

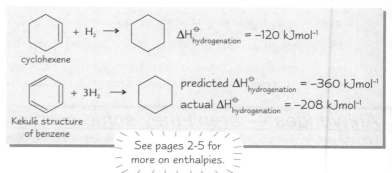

cyclohexene $+ H_2 \rightarrow$ $\Delta H^{\ominus}_{hydrogenation} = -120$ kJmol⁻¹

Kekulé structure of benzene $+ 3H_2 \rightarrow$ predicted $\Delta H^{\ominus}_{hydrogenation} = -360$ kJmol⁻¹
actual $\Delta H^{\ominus}_{hydrogenation} = -208$ kJmol⁻¹

> See pages 2-5 for more on enthalpies.

Aromatic Compounds

Arenes Undergo **Electrophilic Substitution** Reactions...

The benzene ring is a region of **high electron density**, so it attracts **electrophiles**. As the benzene ring's so stable, it tends to undergo **electrophilic substitution** reactions, which preserve the delocalised ring.

...with **Nitronium Ions** as the Electrophile

When you warm **benzene** with **concentrated nitric** and **sulphuric acids**, you get **nitrobenzene**.

Sulphuric acid's a **catalyst** — it helps to make the nitronium ion, NO_2^+, which is the electrophile.

$$HNO_3 + H_2SO_4 \rightarrow H_2NO_3^+ + HSO_4^- \implies H_2NO_3^+ \rightarrow NO_2^+ + H_2O$$

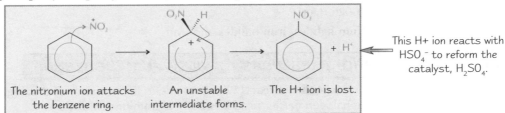

The nitronium ion attacks the benzene ring.

An unstable intermediate forms.

The H+ ion is lost.

This H+ ion reacts with HSO_4^- to reform the catalyst, H_2SO_4.

If you only want one NO_2 group added (**mononitration**), you need to keep the temperature **below 55 °C**. Above this temperature you'll get lots of substitutions.

Nitration reactions are really useful

1) Nitro compounds can be **reduced** to form **aromatic amines** (see page 91). These are used to manufacture **dyes** (see page 94) and **pharmaceuticals**.

2) Nitro compounds **decompose violently** when heated, so they are used as **explosives** — such as 2,4,6-trinitromethylbenzene (**trinitrotoluene** — TNT).

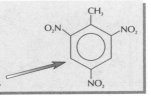

...with **Sulphur Trioxide** as the Electrophile   *Salters* and *Nuffield* only

Sulphur trioxide acts like an **electrophile** — the electronegative oxygens draw the electrons away from the sulphur, making it **slightly positive.** It reacts with arenes in electrophilic substitution reactions, forming a **sulphonic acid**.

You need to either reflux the arene with **concentrated sulphuric acid** for several hours, or warm it with fuming sulphuric acid (which has sulphur trioxide added).

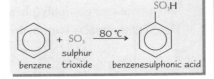

benzene   sulphur trioxide   benzenesulphonic acid

Practice Questions

Q1 Draw Kekulé's proposed structure for benzene.

Q2 What evidence is there for a delocalised electron ring in benzene?

Q3 In the nitration of an arene, what's the role of concentrated sulphuric acid?

Exam Question

1   Two electrophilic substitution reactions of benzene are summarised in the diagram:

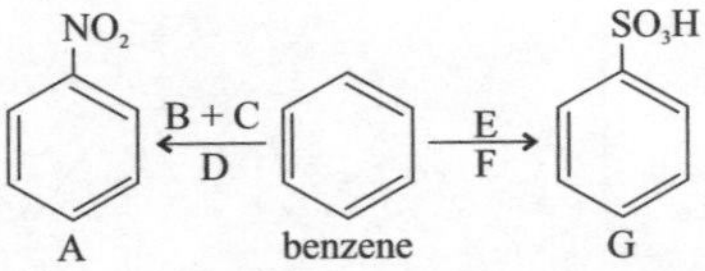

a) i)   Name the product A, and the reagents B and C, and give the conditions D.   [4 marks]
    ii) Outline a mechanism for this reaction.   [3 marks]
    iii) Write equations to show the formation of the electrophile.   [2 marks]

    i)   Name the product G, and the reagent E, and give the conditions F.   [3 marks]
    ii) What is the electrophile in this reaction?   [1 mark]

c)   Why does benzene undergo electrophilic substitution reactions in preference to electrophilic addition reactions?
                                                                                            [2 marks]

Everyone needs a bit of stability in their life...

The structure of benzene is well weird — even top scientists struggled to find out what its molecular structure looked like. If you're asked why benzene reacts the way it does, it's bound to be something to do with the ring of delocalised electrons. Remember there's a hydrogen at every point on the benzene ring — it's easy to forget they're there.

More Reactions of Aromatic Compounds

You're not quite done with benzene yet. Prepare yourself for yet more reactions...

Halogen Carriers Help to Make Good Electrophiles

An electrophile has to have a pretty strong **positive charge** to be able to attack the stable benzene ring. Most compounds just **aren't polarised enough** — but some can be made into **stronger electrophiles** using a catalyst called a **halogen carrier**.

Halogen carriers accept a **lone pair of electrons** from the electrophile. As the lone pair of electrons is pulled away, the **polarisation** in the electrophile **increases** and sometimes a **carbocation** even forms. This makes the electrophile heaps stronger.

Halogen carriers include **aluminium halides**, **iron halides** and **iron**.

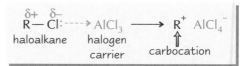

Friedel-Crafts Alkylation Reactions Produce Alkylbenzenes

Friedel and Crafts discovered that a **haloalkane** reacts with benzene in the presence of a **halogen carrier** to produce an **alkylbenzene**. The halogen carrier used is usually **aluminium chloride**.

It's an **electrophilic substitution** reaction, and it needs to be refluxed with a dry ether.

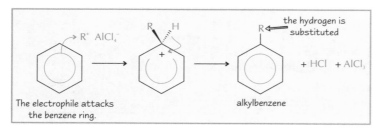

The electrophile attacks the benzene ring.

alkylbenzene

As **alkylbenzenes** are **more reactive** than benzene, it's difficult to stop the reaction at just one substitution — so chances are you'll get **polyalkylation**.

Just for AQA and OCR

Alkylation is used in industry to make **ethylbenzene**, which is used in the manufacture of **poly(phenylethene)** — better known as **polystyrene**.

First, ethene reacts with **hydrogen chloride**:

$$CH_2CH_2 + HCl \rightarrow CH_3CH_2Cl$$

Then a **carbocation** is formed with help from a **halogen carrier**:

$$CH_3CH_2Cl + AlCl_3 \rightarrow CH_3CH_2^+ + AlCl_4^-$$

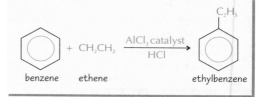

benzene   ethene   ethylbenzene

Friedel-Crafts Acylation Reactions Produce Phenylketones *Not OCR or Nuffield*

Acyl chlorides (see page 82) easily lose their chlorine to form carbocations. They react with benzene in the presence of a **halogen carrier** to produce **phenylketones**.

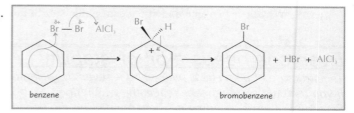

benzene   phenylketone

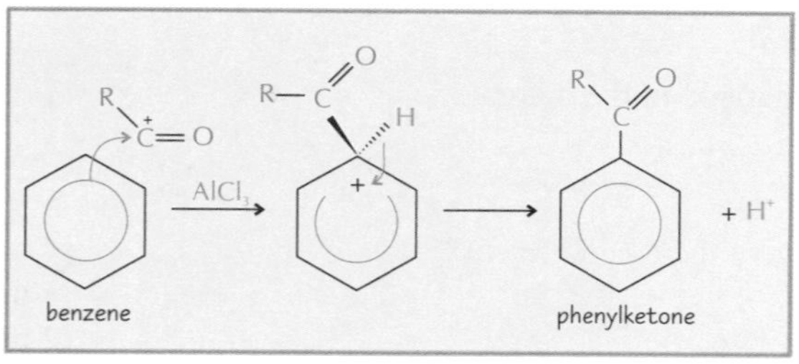

Again, it's an electrophilic substitution reaction, and it needs to be heated under reflux with a dry ether.

Phenylketones are **less reactive** than benzene — so you'll only get one substitution.

Halogen Carriers Help Halogens Substitute the Benzene Ring *Not AQA*

Halogen carriers polarise halogen dimers, such as Br_2 or Cl_2.

The **positively charged** end of the halogen molecule then acts as an **electrophile** and reacts with the benzene ring in the usual **electrophilic substitution** reaction.

benzene   bromobenzene

More Reactions of Aromatic Compounds

If you're doing AQA or Salters you can skip this page — but not the questions.

Benzene Resists **Bromination** More Than **Cyclohexene** Does

Not Edexcel

If you shake **cyclohexene** with **orange bromine water**, it **decolourises** it.
Cyclohexene's double bond is an area of **high electron density** which is easily
attacked by electrophiles. So bromine's **electrophilically ADDED** to cyclohexene.

$$C_6H_{10} + Br_2 \rightarrow C_6H_{10}Br_2$$

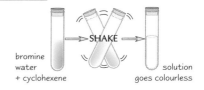

bromine water + cyclohexene — SHAKE → solution goes colourless

Benzene **doesn't** decolourise bromine water — it'd take too much energy to break
the stable delocalised electron ring and add stuff on. (It'll only **substitute** bromine
if there's a **halogen carrier**, as on the bottom of the previous page.)

Benzene Will Do Some **Addition Reactions**

Nuffield only

To add groups to benzene you have to use harsh conditions to break that stable delocalised electron system.

1) Benzene reacts with **hydrogen** in the presence of a **Raney nickel catalyst** at **150 °C** to form **cyclohexane**.

A Raney nickel catalyst has an extremely high surface area.

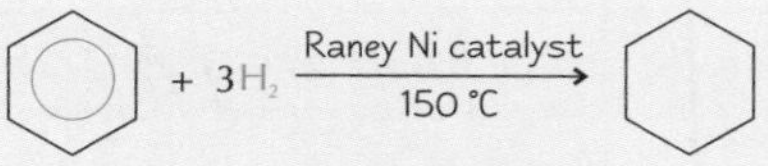

2) Benzene reacts with **chlorine** in **ultraviolet light** to form **1,2,3,4,5,6-hexachlorocyclohexane**.

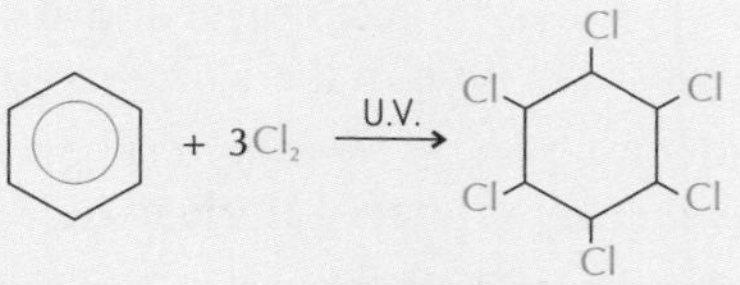

Aromatic Compounds with **Carbon-Containing Side Chains** can be **Oxidised**

Edexcel only

An **alkyl sidechain** on a benzene ring can be **oxidised** by **alkaline potassium manganate(VII) solution** to give a **benzoic acid** (a carboxylic acid with a benzene ring).

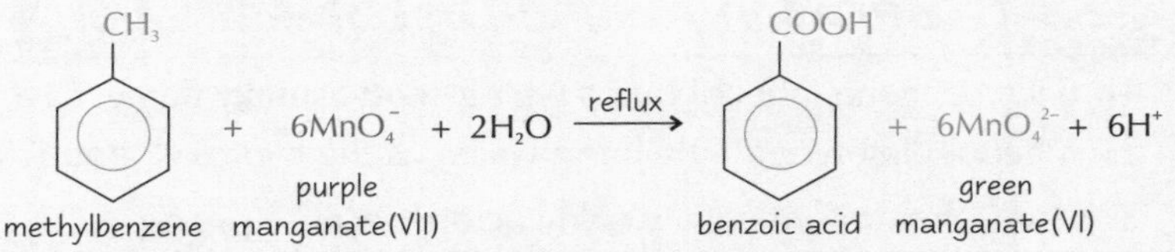

methylbenzene + $6MnO_4^-$ + $2H_2O$ —reflux→ benzoic acid + $6MnO_4^{2-}$ + $6H^+$
purple manganate(VII) — green manganate(VI)

Practice Questions

Q1 How does a halogen carrier work?

Q2 What's produced by: a) alkylation? b) acylation?

Q3 What conditions are needed for the hydrogenation of benzene?

Exam Questions

1 Look at the following synthesis route for manufacturing polystyrene:

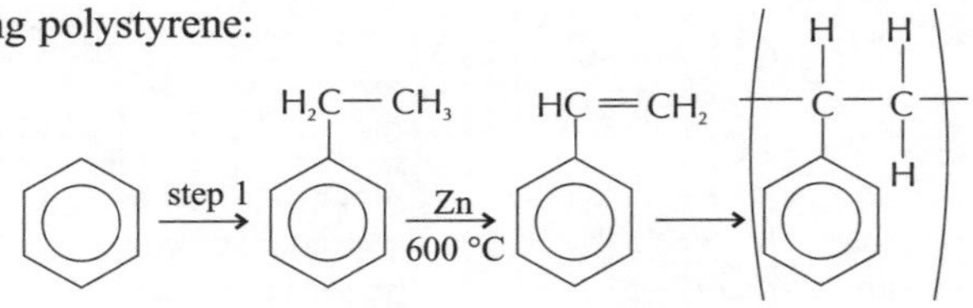

For Step 1,
a) Identify the reagent(s) and catalyst(s) required. [3 marks]
b) Outline a mechanism for the reaction, including an equation to show the formation of the electrophile. [5 marks]
c) Explain the role of the catalyst(s) in the reaction. [3 marks]

2 For the acylation reaction between benzene and ethanoyl chloride:
a) Give suitable catalyst(s) and conditions. [2 marks]
b) Draw the structure of the electrophile. [1 mark]

One ring to rule them all...

Arenes really like their delocalised electron ring — it makes them nice and stable, and they don't want to give it up for anything. They much prefer to be involved in electrophilic substitution reactions, as these let them keep their ring intact. But use harsh enough conditions and you can get some things, such as hydrogen and chlorine, added on.

Phenols

Phenol is the aromatic version of an alcohol. Don't drink them though — they'd get your insides a bit too clean.
Those lucky people doing AQA can skip these two pages

Phenols Have Benzene Rings with –OH Groups Attached

Phenol has the formula **C₆H₅OH**.
Other phenols have various groups attached to the benzene ring:

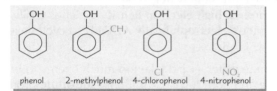

phenol   2-methylphenol   4-chlorophenol   4-nitrophenol

Number the carbons starting from the one with the –OH group.

Test for Phenol Using Iron(III) Chloride Solution   *Salters only*

If you add phenol to **iron(III) chloride solution** and shake, you get a **purple** solution.
Other phenols give other colours.

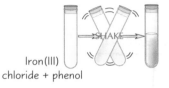

Iron(III) chloride + phenol

Phenols are Used as Antiseptics and Disinfectants   *OCR only*

Phenol's a powerful **antiseptic** and **disinfectant** — it's very corrosive to the skin though.

So it's been replaced by other compounds, including several **phenol derivatives**. ⟹

As well as being less corrosive, the 'new' antiseptics are even better at killing germs.

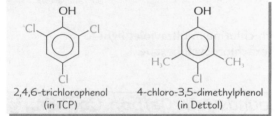

2,4,6-trichlorophenol (in TCP)   4-chloro-3,5-dimethylphenol (in Dettol)

Antiseptics kill bacteria which cause infection in open wounds.

Phenol Burns in Air and Dissolves to Form a Weakly Acidic Solution   *Salters and Nuffield only*

1) Like benzene, phenol burns with a **sooty**, **smoky** flame, as a high proportion of it is **carbon**.

2) Phenol dissolves a little bit in water, as the hydroxyl group's able to form **hydrogen bonds** with water molecules.

3) The solution formed is **weakly acidic** because phenol dissociates in water to form a **phenoxide ion** and an **H⁺ ion**:

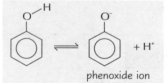

phenoxide ion

Phenols are **stronger** acids than **alcohols**, but **weaker** acids than **carboxylic acids**. Here's why:

The more **stable** the ion formed, the **stronger** the acid...	Greater stability and stronger acids →	*Salters only*

R—O⁻   alkoxide ion   is less stable than   phenoxide ion   is less stable than   R—C(O⁻)(O⁻) carboxylate ion   ● = distribution of negative charge

| Alkoxide ions are formed when alcohols dissociate in water. They don't have double bonds, so there's no delocalisation. The negative charge is concentrated on the oxygen, making it very attractive to H⁺ and very unstable. | Phenoxide ions have a delocalised electron ring, overlapped by the oxygen's p-orbital. The negative charge is spread over the ring, but mostly it's centred on the single oxygen, so the ion is fairly attractive to H⁺ and only moderately stable. | Carboxylate ions have a -COO⁻ functional group. The π bond electron density in the 'double bond' is delocalised over both C–O bonds, mostly near the electronegative oxygens. Because the negative charge is shared between the oxygens, it's less attractive to H⁺ ions and is more stable. |

Acyl Chlorides React with Phenols to Form Esters   *Salters, Edexcel and Nuffield only*

1) The usual way of making an **ester** is to react an alcohol with a carboxylic acid (see page 78). Phenols react **very** slowly with carboxylic acids though, so it's faster to use an **acyl chloride**, such as ethanoyl chloride.

2) **Ethanoyl chloride** reacts slowly with phenol at room temperature, producing the ester **phenol ethanoate** and **hydrogen chloride** gas.

$$H_3C-C(\overset{O}{\underset{Cl}{}}) + HO-C_6H_5 \rightleftharpoons H_3C-C(\overset{O}{\underset{O}{}})-C_6H_5 + HCl$$

ethanoyl chloride   phenyl ethanoate

Phenols

This page is just for OCR, Edexcel and Nuffield.

Phenol reacts with **Bases** and **Sodium** to form **Salts**

1) Phenol reacts with **sodium hydroxide solution** at room temperature to form **sodium phenoxide** and **water**.

2) Phenol **doesn't react** with **sodium carbonate** solution though — sodium carbonate is not a strong enough base.

3) **Sodium phenoxide** is also formed when **sodium** metal is added to liquid phenol. **Hydrogen gas** fizzes off this time.

$$2C_6H_5OH + 2Na \rightarrow 2C_6H_5ONa + H_2$$

Phenol Reacts with **Bromine Water**

If you shake phenol with orange bromine water, it **decolorises**.
A **white precipitate** is formed that smells of **antiseptic** — it's called 2,4,6-tribromophenol.

Remember, benzene doesn't react with bromine water.

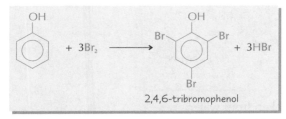

2,4,6-tribromophenol

The electrons in the oxygen's p-orbital are partially **delocalised** onto the benzene ring — this **increases** the **electron density** in the ring, especially at positions 2, 4 and 6.

So bromine's able to attack and substitute itself into the benzene ring.

The hydroxyl group is called an **activating group** because it has made the benzene ring more reactive.

Practice Questions

Q1 What substance can you use to test for phenol? What happens if phenol is present?

Q2 Why are phenols soluble in water?

Q3 How does phenol react with sodium?

Exam Questions

1 Salicylic acid is a precursor of aspirin.
Aspirin is produced when salicylic acid reacts with ethanoyl chloride.

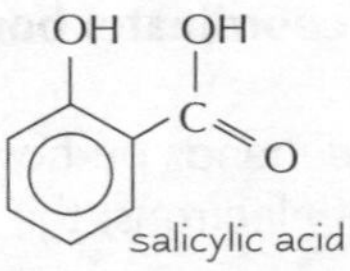

salicylic acid

  a) Draw the structure of aspirin. [1 mark]

  b) What other product is formed in this reaction? [1 mark]

2 Phenol is a stronger acid than ethanol, but a weaker acid than ethanoic acid.
  a) Explain why $C_6H_5O^-$ is more stable than $CH_3CH_2O^-$. [4 marks]

  b) Test results for three solids, D, E and F, are shown here:

Solid	Result with sodium hydroxide solution	Result with sodium carbonate solution
D	solid dissolves	solid dissolves, CO_2 released
E	no reaction	no reaction
F	solid dissolves	no reaction

  Identify which solid is cyclohexanol, which is phenol and which is benzoic acid. [3 marks]

  c) Phenol is shaken with bromine water.
    i) What do you observe? [2 marks]
    ii) Write the equation. [1 mark]
    iii) Benzene doesn't react with bromine water without a catalyst. Explain the difference in reactivity
        between phenol and benzene. [4 marks]

It's the phenol countdown — only 16 pages left of this section...

You might not like this phenol stuff, but if you were a germ, you'd like it even less. If you're ever looking for TCP in the supermarket, try asking for a bottle of 2,4,6-trichlorophenol and see what you get — probably a funny look. Anyway, no time for shopping — you've got to get these pages learned. If you can answer all the questions above, you're well on your way.

Amines

This is a-mean, fishy smelling topic...

Amines are Organic Derivatives of **Ammonia**

If one or more of the **hydrogens** in **ammonia** (NH_3) is replaced with an organic group, you get an **amine**.

methylamine	dimethylamine	trimethylamine	tetramethylamine ion (quaternary ammonium ion)	phenylamine
(primary amine)	(secondary amine)	(tertiary amine)		(primary amine)

aliphatic amines — aromatic amine

Small amines smell similar to **ammonia**, with a **slightly 'fishy'** twist. **Larger amines** smell very '**fishy**'.

Small Amines **Dissolve** in Water to form an **Alkaline** Solution   *Nuffield **only***

1) Small amines are **soluble in water** — the amine group forms **hydrogen bonds** with the water molecules.
The **bigger** the amine, the **greater** the **van der Waals** forces between the amine molecules, and the less soluble it is.

2) If they do dissolve, they form **alkaline** solutions — a hydrogen is taken from water, forming **alkyl ammonium ions** and **hydroxide ions**.

$$CH_3CH_2NH_{2(aq)} + H_2O_{(l)} \rightleftharpoons CH_3CH_2NH_3^+{}_{(aq)} + OH^-{}_{(aq)}$$

Amines have a **Lone Pair of Electrons** that can Form **Dative Covalent Bonds**

AQA and OCR only

Amines act as **bases** because they **accept protons**.
There's a **lone pair of electrons** on the **nitrogen** atom that forms a **dative covalent (coordinate) bond** with an H^+ ion.

The **strength** of the **base** depends on how **available** the nitrogen's lone pair of electrons is. The higher the **electron density** of the lone pair of electrons, the more likely the amine is to accept a proton, and the stronger a base it will be.

Primary aliphatic amines are **stronger** bases than **ammonia**, which is a **weaker** base than **aromatic amines**. Here's why:

The more available the lone pair of electrons, the stronger the base...

Greater availability of lone pair of electrons
Stronger bases

primary aromatic amine (phenylamine)   ammonia   primary aliphatic amine   = distribution of negative charge

The benzene ring draws electrons towards itself and the nitrogen's lone pair gets partially delocalised onto the ring. So the electron density on the nitrogen decreases. This makes the lone pair much less available.

Alkyl groups push electrons onto attached groups. So the electron density on the nitrogen atom increases. This makes the lone pair more available.

Cationic Surfactants are Quaternary Ammonium Salts   *AQA only*

Cationic surfactants are used in things like **fabric conditioners** and **hair products**. They are **quaternary ammonium salts** with at least one long hydrocarbon chain.

They're **positively charged**, so they bind to negatively charged surfaces such as hair and fibre. This gets rid of **static**.

Amines

This page is not for Salters

Aliphatic Amines are made from Haloalkanes or Nitriles

YOU CAN HEAT A HALOALKANE WITH AMMONIA

Amines can be made by heating a **haloalkane** with **ammonia**. You'll get a **mixture** of primary, secondary and tertiary amines, and quaternary ammonium salts, as more than one hydrogen is likely to be substituted. You can separate the products using **fractional distillation**.

E.g.

$$NH_3 + CH_3CH_2Br \longrightarrow CH_3CH_2NH_2 + HBr$$
ammonia → ethylamine

OR YOU CAN REDUCE A NITRILE

You can reduce a nitrile to an amine using $LiAlH_4$ in **dry diethyl ether**, followed by some **dilute acid**. E.g.

$$CH_3C \equiv N \xrightarrow[\text{(2) dilute acid}]{\text{(1) } LiAlH_4, \text{ dry diethyl ether}} CH_3CH_2NH_2$$
ethanenitrile → ethylamine

Another way is to reflux the nitrile with **sodium** metal and **ethanol**.

These are great in the lab, but $LiAlH_4$ and sodium are too **expensive** for industrial use. Industry uses a **metal catalyst** such as platinum or nickel at a high temperature and pressure — it's called **catalytic hydrogenation**.

$$CH_3C \equiv N \xrightarrow[\text{high temp. \& pressure}]{Ni} CH_3CH_2NH_2$$
ethanenitrile → ethylamine

Aromatic Amines are made by Reducing a Nitro Compound *Not Nuffield*

Nitro compounds, such as **nitrobenzene**, are reduced in two steps:

1) Heat a mixture of a **nitro compound**, **tin metal** and **concentrated hydrochloric acid** under **reflux** — this makes a salt.
2) Then to get the **aromatic amine**, you have to add **sodium hydroxide**.

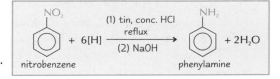

$$nitrobenzene + 6[H] \xrightarrow[\text{(2) NaOH}]{\text{(1) tin, conc. HCl, reflux}} phenylamine + 2H_2O$$

Practice Questions

Q1 What do amines smell like?

Q2 What sort of solution's formed when you mix an amine with water?

Q3 Give a use for quaternary ammonium salts.

Q4 Name two types of molecules that you can make amines from.

Exam Questions

1 a) Explain how methylamine, CH_3NH_2, can act like a base. [1 mark]

b) Methylamine is a stronger base than ammonia, NH_3. However, phenylamine, $C_6H_5NH_2$, is a weaker base than ammonia. Explain these differences in base strength. [4 marks]

2 a) Propylamine can be synthesised from bromopropane. Suggest a disadvantage of this synthesis route. [1 mark]

b) Propylamine can also be synthesised from propanenitrile.
   i) Suggest suitable reagents for its preparation in a laboratory. [2 marks]
   ii) Why is this method not suitable for industrial use? [1 mark]
   iii) What reagents and conditions are used in industry? [2 marks]

You've got to learn it — amine it might come up in your exam...

Rotting fish smells so bad because the flesh releases diamines as it decomposes. Is it fish that smells of amines or amines that smell of fish — it's one of those chicken or egg things that no one can answer. Well, enough philosophical pondering — we all know the answer to the meaning of life. It's 42. Now make sure you know the answers to the questions above.

Amines and Amides

And you thought that was all you needed to know about amines... hee, hee...

Ammonia and Amines are **Nucleophiles**   *This bit's just for AQA and Nuffield*

Ammonia and amines can act as **nucleophiles** because the nitrogen's got a **lone pair of electrons**.
They react with **haloalkanes** in **nucleophilic substitution** reactions. You need to know the mechanism for this:

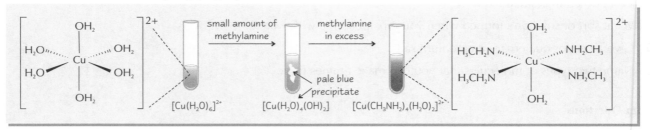

As long as there's some more of the haloalkane, **further substitutions** can take place. They keep happening
until you get a **quaternary ammonium salt**, which can't react any further as it has no lone pair of electrons.

This means you end up with a **mixture** of primary, secondary and tertiary amines, their salts
and a quaternary ammonium salt. You have to separate them out by **fractional distillation**.

Amines will Form a **Complex Ion** With **Copper(II) Ions**   *Just Nuffield*

1) In **copper(II) sulphate** solution, the Cu^{2+} ions form $[Cu(H_2O)_6]^{2+}$ complexes with water. This solution's **blue**.

2) If you add a **small** amount of **methylamine solution** to copper(II) sulphate solution you get a **pale blue precipitate**
    — the amine acts as a **Brønsted-Lowry base** (proton acceptor) and takes two H^+ ions from the complex.
    This leaves copper hydroxide, $[Cu(H_2O)_4(OH)_2]$, which is insoluble.

3) Add more methylamine solution, and the **precipitate dissolves** to form a beautiful **deep blue solution**. Some of the ligands are replaced by methylamine molecules which donate their lone pairs to form dative covalent (coordinate) bonds with the Cu^{2+} ion. This forms soluble $[Cu(CH_3NH_2)_4(H_2O)_2]^{2+}$ complex ions.

See page 39 for more on ligand exchange reactions.

Amines React with **Acids** to Form **Salts**   *Just OCR and Edexcel*

Amines are **neutralised** by **acids** to make an **ammonium salt**.
For example, **ethylamine** reacts with **hydrochloric** acid to form ethylammonium chloride:

$$CH_3CH_2NH_2 + HCl \rightarrow CH_3CH_2NH_3^+Cl^-$$

Amines can be **Acylated** to form **N-substituted Amides**   *Not OCR*

Acylation is when an **acyl group**, $R-C\langle{}^O$, is substituted for an **H atom**.

When amines react with acyl chlorides, a hydrogen atom on the amine is swapped
for the acyl group to produce an N-substituted amide (see page 93). E.g.

The hydrochloric acid formed will react with any excess amine to form
a **salt**, as in the equation above.

Amines and Amides

Amides are Carboxylic Acid Derivatives

Amides contain the functional group **–CONH₂**.

The **carbonyl group** pulls electrons away from the rest of the group, so amides behave differently from amines.

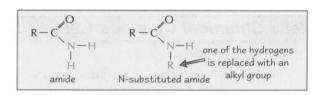

Amides can be Hypothesesed Under Acidic or Basic Conditions   *Salters only*

To **hydrolyse** an amide you can:

Heat it with **dilute acid** to get a **carboxylic acid** and an **ammonium salt**	...or...	...heat it with a **dilute alkali** to get a **carboxylate ion** and **ammonia gas given off**.

$$R-C(=O)NH_2 + H_2O + HCl \longrightarrow R-C(=O)OH + NH_4Cl$$

$$R-C(=O)NH_2 + NaOH \longrightarrow R-C(=O)O^-Na^+ + NH_3$$

Polyamides are polymers in which the monomers are linked by **amide groups**, for example nylon 6,6 or polypeptides — there's lots about these polymers on pages 101-102.

Polyamides are pretty easily **hydrolysed** on heating with **strong acid**, but not with alkali — see page 96 for this reaction.

Nitriles can be Made from Amides   *Just Edexcel*

To make an **nitrile** you need to **dehydrate** the amide. It's done by heating the amide with **phosphorus(V) oxide**.

$$R-C(=O)NH_2 \xrightarrow{P_4O_{10}} R-C\equiv N + H_2O$$
amide　　　　　nitrile

Hofmann Degradation Turns an Amide into a Primary Amine   *Edexcel only*

As well as turning an **amide** into an **amine**, Hofmann degradation also gets rid of a **carbon atom** from the molecule, which is handy for synthesising compounds.

To make an amide react like this, mix it with **bromine** and warm it with a **concentrated** alkali, such as **potassium hydroxide solution**.

This molecule's got one less carbon than before.

$$R-C(=O)NH_2 + Br_2 + 4KOH \longrightarrow R-N H_2 + 2KBr + K_2CO_3 + 2H_2O$$
amide　　　　　　　　　　　　　amine

Practice Questions

Q1 How do you separate the products when ammonia reacts with a haloalkane?

Q2 What's the formula of the precipitate formed when methylamine is added to copper(II) sulphate solution?

Q3 What's Hofmann degradation?

Exam Question

1 Propanoic acid can be made from propanamide directly, or via propanenitrile.

$$CH_3CH_2CONH_2 \xrightarrow{Step\ M} CH_3CH_2COOH$$
$$\xrightarrow{Step\ N} CH_3CH_2CN \xrightarrow{Step\ O}$$

a) i) Which two steps can involve the same reagents? [1 mark]
   ii) Write an equation for one of these two steps. [1 mark]
   iii) What kind of reaction is this? [1 mark]
b) i) Give the reagent(s) and conditions for the third step. [2 marks]
   ii) What kind of reaction is this? [1 mark]

Change the 'n' to a 'd' by swapping an 'H' for an 'O'...

Amines and amides might sound alike, but they are different. The C=O group really changes their reactions. Don't mix them up because this'll mean the examiners have a jolly good giggle at you over their Jammy Dodgers. Make two columns and jot down all the stuff you know about amines on one side, and then all the stuff you know about amides on the other.

Dyes

This page is a killer. *Nuffield **people can skip both these pages.***

Many **Chemical Changes** *Cause Changes in* **Colour**

These chemical changes can be:

- **acid-base or base-acid.** Sometimes colours change when **pH** changes — this is used in **indicators**. (See page 34.)
- **ligand exchange. Complexes** can change colour when their **ligands** are swapped. (See page 66.)
- **oxidation or reduction.** The colour of **transition metal ions** depends on their **oxidation state**. (See page 55.)
- **precipitation.** E.g. adding sodium hydroxide to transition metal ions forms **coloured precipitates**. (See page 39.)
- **polymorphism** — this is where a compound precipitates in different **crystal structures** with different colours, depending on the reaction conditions.

Van Gogh used lead chromate <u>polymorphs</u> to make paint shades ranging from red to orange to bright yellow for his sunflowers.

Azo Dyes *are made in* **Two Steps**   OCR, *Salters* **and** *Edexcel* **only**

First you have to make a **diazonium salt**. Diazonium compounds contain the group –N=N–.

A **Diazonium Salt** *is made by Reacting* **Phenylamine** *with* **Nitrous Acid**

> 'in situ' means in the reaction

Nitrous acid (HNO_2) is **unstable**, so it has to be made *in situ* from sodium nitrite and hydrochloric acid.

$$NaNO_2 + HCl \rightarrow HNO_2 + NaCl$$

Nitrous acid reacts with **phenylamine** and **hydrochloric acid** to form **benzenediazonium chloride**. The temperature **must** be below **10 °C** to prevent a phenol forming instead.

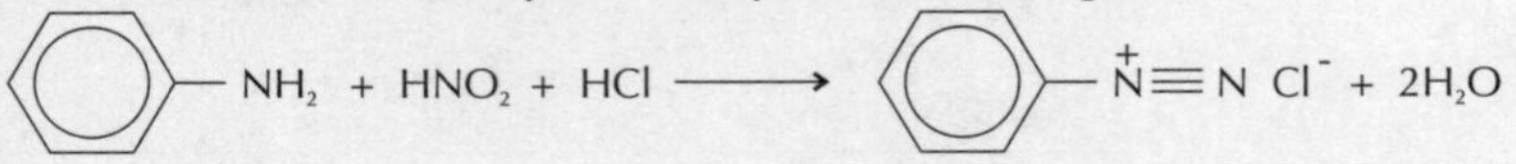

$$\text{C}_6\text{H}_5\text{—NH}_2 + HNO_2 + HCl \longrightarrow \text{C}_6\text{H}_5\text{—}\overset{+}{\text{N}}\equiv\text{N Cl}^- + 2H_2O$$

Now you can make the **azo dye**:

An **Azo Dye** *is made by Coupling a* **Diazonium Salt** *with a* **Coupling Agent**, *Such as* **Phenol**

Phenol first has to be dissolved in **sodium hydroxide** solution to make **sodium phenoxide** solution. It's then stood in **ice**, and chilled **benzenediazonium chloride** is added.

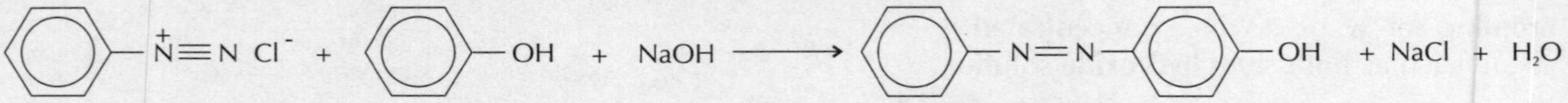

$$\text{C}_6\text{H}_5\text{—}\overset{+}{\text{N}}\equiv\text{N Cl}^- + \text{C}_6\text{H}_5\text{—OH} + NaOH \longrightarrow \text{C}_6\text{H}_5\text{—N}=\text{N—C}_6\text{H}_4\text{—OH} + NaCl + H_2O$$

yellow-orange azo compound

The azo dye **precipitates** out of the solution immediately.

Coupling agents are **aromatic amines** and **phenols**. The lone pairs on their nitrogen or oxygen increase the **electron density** of the benzene ring. This gives the diazonium ion (a **weak electrophile**) something to attack.

Azo compounds made by using different combinations of **diazonium salts** and **coupling agents** are **different colours**. They're mostly shades of red, orange and yellow, but you can get green and blue too.

Aromatic azo compounds are **stable** because the azo functional group –N=N– becomes part of the **delocalised electron system**. Because they're stable, they make great dyes — the molecules don't fall apart, so the colours **don't fade**.

Dyes have to **Attach** *Themselves to Fibres*   *Salters* **only**

A good dye has got to be **colourfast** — it can't **wash out** too easily or **fade** in the light.
It has to be **soluble** though to actually get onto the fibres in the first place. Azo dyes usually have sodium sulphonate ($-SO_3^-Na^+$) groups to increase solubility.

There are quite a few ways dyes can attach themselves to the fibres:

1) Dyes can attach themselves to cotton fibres by **hydrogen bonding** because cotton fibres have **hydroxyl groups**. These dyes aren't very **colourfast** though.

2) **Ionic attractions** can bind dye molecules to **polar** fibres. For example, **amino** groups in nylon or in proteins such as **wool** or **silk** can be protonated ($-NH_3^+$) so that they interact with ionic groups in dyes, such as sulphonates ($-SO_3^-$).

3) **Mordant dyes** need a mordant (a fixing agent) to become attached to fibres. The mordant forms a precipitate on the cloth, which the dye molecules attach themselves to.

4) **Reactive dyes** use strong **covalent bonding** between the dye and the fibre to prevent the dye being washed out.

Dyes

The information on this page is just for OCR (Methods of Analysis and Detection Option) and Salters.

The **Colour** of Organic Molecules Comes from **Chromophores**

The structures in molecules that give them their colour are called **chromophores**.

A chromophore could be a double or triple bond (**C=C, C=O, N≡N, Na=N**), a **delocalised system** or a **lone pair** of electrons.

See page 108 for more on chromophores.

When **light** hits a chromophore, certain wavelengths are **absorbed** as **excited electrons** in the chromophore jump to a **higher energy level**. The range of wavelengths of **visible light** that are not absorbed will be seen as a particular **colour**.

Changing a chromophore structure or attaching **functional groups**, like $-OH$, $-NH_2$ or $-NR_2$, can change the **shade** or **intensity** of a colour.

A **Change** in a **Chromophore** Leads to a Change in Colour

Even a **small change** in a chromophore changes the **wavelength** absorbed, and so changes the colour of the compound. Some chromophores can become **protonated** below a certain pH, and change colour.

Methyl orange is an acid-base indicator and changes from yellow to red below a pH of 3.5.

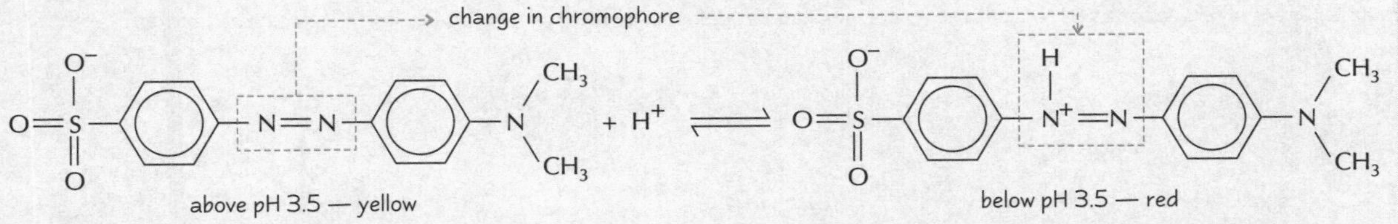

change in chromophore

above pH 3.5 — yellow $\quad + H^+ \rightleftharpoons \quad$ below pH 3.5 — red

Practice Questions

Q1 What's polymorphism?

Q2 Why do diazonium ions attack phenols or phenylamines?

Q3 What's a chromophore? What happens when you change it?

Exam Questions

1 Consider this synthesis pathway:

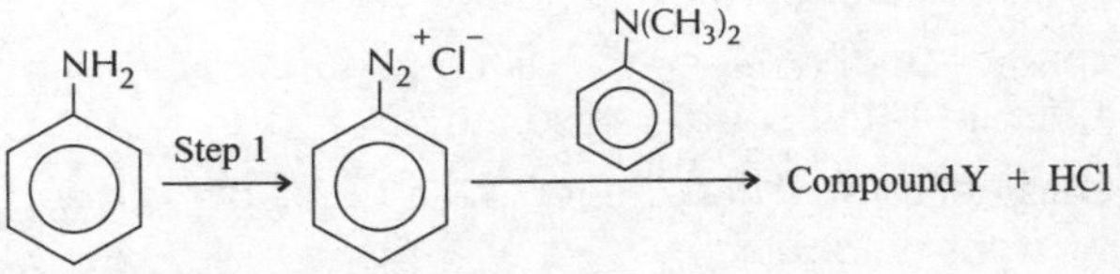

Step 1 Compound Y + HCl

a) i) One reagent for Step 1 is $HCl_{(aq)}$. Give the other reagent for Step 1.
Write an equation to show its generation in situ. [2 marks]
ii) Give the conditions for Step 1. [1 mark]

b) Compound Y is a yellow solid.
i) Draw a possible structure for Compound Y. [1 mark]
ii) What feature(s) are responsible for its yellow colour? [3 marks]
iii) Suggest a use for Compound Y. [1 mark]

2 The compound shown is the dye 'Acid Orange 6'.

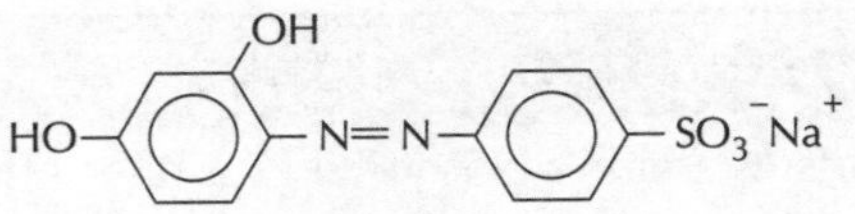

a) Which molecular feature(s) may:

i) be chromophores? [3 marks]
ii) make the dye more soluble? [1 mark]

b) Explain how Acid Orange 6 would bond to silk proteins. [3 marks]

c) Suggest a simple reversible way of changing the colour of Acid Orange 6. [1 mark]

d) Acid Orange 6 is made by a coupling reaction. Draw the structures of the reactants involved. [2 marks]

I asked Van Gogh if he wanted a drink — He said, "Nah, I've got one ear"...

They really need to find a better red dye — like one that doesn't turn everything pink if you accidentally wash a red sock with your white clothes. Lots of things are coming together here — benzene rings, phenylamines, phenols, indicators, jumping electrons, ligand exchange... Check up on anything you can't quite remember and it'll make things heaps easier.

Amino Acids and Proteins

Wouldn't it be nice if you could go to sleep with this book under your pillow and when you woke up you'd know it all.

Amino Acids have an **Amino Group** and a **Carboxyl** Group

An amino acid has a **basic amino group** (NH_2) and an **acidic carboxyl group** (COOH). This makes them **amphoteric** — they've got both acidic and basic properties.

They're **chiral molecules** because the carbon has **four** different groups attached. So a solution of a single amino acid enantiomer will **rotate polarised light** — see pages 70-71.

Glycine's the exception to this as its R group is just a hydrogen.

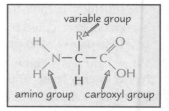
variable group
amino group   carboxyl group

Amino Acids Can Exist As **Zwitterions**

A zwitterion is a **dipolar ion** — it has both a **positive** and a **negative charge** in different parts of the molecule. Zwitterions only exist near an amino acid's **isoelectric point**. This is the **pH** where the **average overall charge** on the amino acid is zero. It's different for different amino acids — it depends on their R-group.

In conditions more **acidic** than the isoelectric point, the $-NH_2$ group is likely to be **protonated**.

At the isoelectric point, both the carboxyl group and the amino group are likely to be ionised — forming an ion called a **zwitterion**.

In conditions more **basic** than the isoelectric point, the –COOH group is likely to **lose** its proton.

low pH   zwitterion   high pH

Paper Chromatography is used to Identify Unknown Amino Acids

Only Salters and Nuffield

Here's how to work out which amino acids are in the mixture:

1) Draw a **pencil line** near the bottom of a piece of chromatography paper and put a **concentrated spot** of the mixture of amino acids on it.

2) Dip the bottom of the paper (not the spot) into a solvent.

3) As the solvent spreads up the paper, the different amino acids move with it, but at **different rates**, so they separate out.

4) When the solvent's **nearly** reached the top, take the paper out and **mark** the **solvent front** with pencil.

5) Amino acids aren't coloured — so you have to spray **ninhydrin solution** on the paper to turn them purple.

6) You can work out the R_f values of the amino acids using this formula:

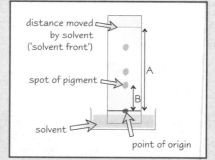
distance moved by solvent ('solvent front')
spot of pigment
A
B
solvent
point of origin

$$R_f \text{ value of amino acid} = \frac{B}{A} = \frac{\text{distance travelled by spot}}{\text{distance travelled by solvent}}$$

Now you can use a **table of known amino acid R_f values** to identify the amino acids in the mixture.

Proteins are Condensation Polymers of Amino Acids   *Not Edexcel*

Proteins are made up of **lots** of amino acids joined together. The chain is put together by **condensation** reactions and broken apart by **hydrolysis** reactions. **Peptide links** are made between the amino acids.

Here's how two amino acids join together to make a **dipeptide**:

Lots of these reactions would happen to make a long chain.

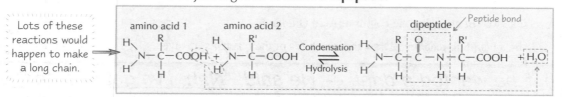
amino acid 1   amino acid 2   Condensation ⇌ Hydrolysis   dipeptide   Peptide bond

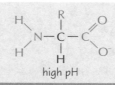

Proteins are really polyamides — the monomers are joined by amide groups. In proteins these are called peptide bonds though.

To break up the protein (**hydrolyse** it) you need to use pretty harsh conditions. **Hot aqueous 6 M hydrochloric acid** is added, and the mixture's heated under reflux for 24 hours. The final mixture's then neutralised.

Amino Acids and Proteins

This page is for AQA, Salters and OCR (Biochemistry Option).

Proteins have **Different** Levels of **Structure**

Proteins are **big, complicated** molecules. They're easier to explain if you describe their structure in four 'levels'. These levels are called the **primary, secondary, tertiary** and **quaternary** structures. You only need to know about the first three (unless you're doing the OCR Biochemistry option — in which case see page 134).

1 The **primary structure** is the **sequence of the amino acids** in the long chain that makes up the protein (the **polypeptide chain**).

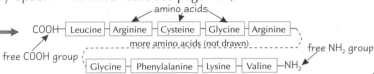

amino acids

COOH—Leucine—Arginine—Cysteine—Glycine—Arginine—

free COOH group   more amino acids (not drawn)   *free NH₂ group*

Glycine—Phenylalanine—Lysine—Valine—NH₂

2 The **peptide links** can form **hydrogen bonds** with each other, meaning the chain isn't a straight line. The shape of the chain is called its **secondary structure**. The most common secondary structure is a **spiral** called an **alpha (α) helix.** ⟶ α helix chain

3 The chain of amino acids is itself often coiled and folded in a characteristic way that identifies the protein. **Extra bonds** can form between different parts of the polypeptide chain, which gives the protein a kind of **three-dimensional shape**. This is its **tertiary structure**.

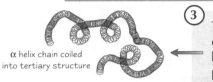

α helix chain coiled into tertiary structure

Different Bonds Hold Proteins Together

The **secondary** structure is held together by **hydrogen bonds** between the peptide links.

The **tertiary** structure is **held together** by quite a few different types of force. These all exist between the **side chains** (R-groups) of the amino acids.

These forces hold the tertiary structure together:

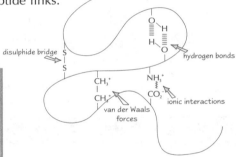

disulphide bridge   *hydrogen bonds*   *van der Waals forces*   *ionic interactions*

1) **Van der Waals** forces — weak attractions between two **non-polar** side groups, e.g. CH_3.
2) **Ionic interactions** between **charged** side groups, like CO_2^- and NH_3^+.
3) **Hydrogen bonding** — between groups such as $-OH$ and $-NH_2$.
4) **Disulphide bridge** — a covalent bond between two sulphur-containing side groups ($-SH$). This type of bond is stronger than the others.

Practice Questions

Q1 What word describes something that can act as an acid and a base?

Q2 What is a zwitterion?

Q3 What's a peptide bond?

Q4 Name the four types of intermolecular forces that are involved in holding together the tertiary structures of proteins.

Exam Question

1 The amino acid serine is otherwise known as 2-amino-3-hydroxypropanoic acid.

a) Draw the structure of serine. [1 mark]

b) When two amino acids react together, a dipeptide is formed.
   i) Explain the meaning of the term dipeptide. [2 marks]
   ii) Draw the structures of the two dipeptides formed when serine and glycine react together. [2 marks]

c) The dipeptides formed can be hydrolysed to give the original amino acids again. Give the reagent(s) and conditions for this reaction. [2 marks]

My top three tides of all time — high tide, low tide and peptide...

The word zwitterion is such a lovely word — it flutters off your tongue like a butterfly. This page isn't too painful — another organic structure, a nice experiment and some stuff on proteins. There's even a two-for-the-price-of-one equation — forwards it's condensation; backwards it's hydrolysis. Remember to learn the conditions for the hydrolysis though.

DNA

OCR (Biochemistry Option) and Salters — get ready to revise your genetic make-up.

DNA is a Polymer of Nucleotides

DNA, DeoxyriboNucleic Acid, contains all the genetic information of an organism.

DNA is made up from lots of **nucleotides**.
Nucleotides are made from the following:

1) A **phosphate group**.

2) A **pentose sugar** — a five-carbon sugar. It's deoxyribose in DNA.

3) A **base** — one of four different bases. In DNA they are **adenine** (A), **cytosine** (C), **guanine** (G) and **thymine** (T).

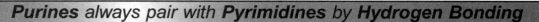

The nucleotides join together to form a **polynucleotide chain**. The bond between each pair of nucleotides forms between the phosphate group of one nucleotide and the sugar of another. This makes what's called a **sugar-phosphate backbone**.

The bases can be in any order, and it's the order of them that holds all the information.

Purines always pair with Pyrimidines by Hydrogen Bonding

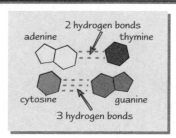

The bases can either be **purines** or **pyrimidines**. The purine bases, adenine and guanine, have two rings of atoms in their structure. Pyrimidine bases, cytosine and thymine, have a single ring of atoms in their structure.

When DNA forms (see below) **purines always** pair with **pyrimidines**. The number of **hydrogen bonds** each can form decides which purine pairs with which pyrimidine. Adenine and thymine can each form **2 hydrogen bonds**, so they pair up. Cytosine and guanine can both form **3 hydrogen bonds**, so they pair up. This is called **complementary base pairing**. When **replicating** genetic information, complementary base pairing makes sure the **order** of bases is **copied** accurately.

DNA is a Double Helix

DNA is made of **two strands** of polynucleotides. These strands spiral around each other to form a **double helix**.

The structure is held together by **hydrogen bonds** between the **base pairs**, rather like rungs of a ladder. As purines always pair with pyrimidines, there are always three rings of atoms between each pair of sugars.

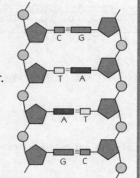

DNA Contains the Basis of the Genetic Code

The **sequence** of bases **determines** the sequence of **amino acids** in a protein. The way that DNA codes for it is called the **genetic code**.

1) DNA codes for specific amino acids with sequences of three bases, called **base triplets**. Different sequences of bases code for different amino acids. For example AGU codes for serine and GUC codes for valine.

2) There are **64** possible **base triplet combinations**, but only about **20** amino acids in human proteins so there are some base triplets to spare. These aren't wasted though:

- some amino acids are coded for by more than one base triplet.
- some base triplets act as 'punctuation' to stop and start production of an amino acid chain.

There's a huge number of possible arrangements of amino acids, allowing for the enormous diversity of proteins.

DNA

Only those of you doing Salters need to learn this page.

Genetic Engineering can be used to Alter the DNA of an Organism

DNA **nucleotides** have the **same** structure in **all organisms**. This means you can put a piece of DNA from one organism into the DNA of another organism. By modifying its DNA, you can get an organism to produce a **new protein** — and proteins control absolutely everything that goes on in your body.

Here's how you can add DNA to bacteria:

1) Cut out the bit of DNA you want using **enzymes**.

2) Insert this new DNA sequence into **plasmids** — these are DNA circles extracted from bacteria. This uses enzymes again.

3) Put the **modified** plasmids into **bacterial host cells**.

4) Multiply the cells in a **fermenter**. When the bacteria reproduce, the DNA is reproduced too.

really useful bit of DNA

plasmid DNA

You've now got **loads** of bacterial cells containing the useful bit of DNA. The bacteria will **make** the **protein** you want — so all you need to do is **isolate** it, then it can be used. Sounds simple, hey!

Genetic Engineering has Important Applications

Here are the applications of genetic engineering that you need to know about:

1) **Producing human proteins**
Sometimes humans can't produce a certain protein that they need — so you can make the protein for them. It doesn't always work, but there are some cases where it does — for example, human growth hormone, Factor 8 for haemophiliacs, and insulin for diabetics.

2) **Producing vaccines**
You can make large amounts of a vaccine to help prevent disease.

3) **Industrial applications**
You can modify bacteria and fungi to treat harmful waste and control pollution.

4) **Genetically modified (GM) plants**
You can use this technique to get higher yields, and increased pest/pesticide/weedkiller resistance in plants — but it's **controversial**, as there are worries over harm to ecosystems and human health.

5) **Gene therapy**
Not very successful at the moment, but scientists are trying to cure some inherited diseases by inserting a functional gene into the body. E.g. this has been tried for cystic fibrosis.

Practice Questions

Q1 What's 'DNA' short for?

Q2 What are the names of the four bases in DNA? Which ones are purines? Which ones are pyrimidines?

Q3 What holds the two strands of DNA together?

Q4 Give three applications of genetic engineering.

Exam Question

1 Children born with a growth hormone deficiency need to have injections of human growth hormone to grow normally. This used to be harvested from the pituitary glands of dead bodies. Now it is produced by genetically modified *E. coli* bacteria.

a) Suggest why harvesting human growth hormone from dead bodies was problematic. [1 mark]

b) Briefly describe:
   i) the structure of DNA, [4 marks]
   ii) how DNA translates into a protein chain. [2 marks]

c) Briefly outline how *E. coli* could be genetically modified to produce human growth hormone. [3 marks]

And then I shall rule the world....

Well, there's plenty of good things all this genetic engineering can do — like making proteins to help people with diabetes or haemophilia. But there's a not-so-good side too. People worry that before long you'll be able to order a blue-eyed, blond-haired baby. But what'll happen if you don't give birth to quite what you ordered — will there be a return policy?

Addition Polymers

Polymers are long molecules made by joining lots of little molecules together.

Polymers can be written with or without the brackets.

Addition Polymers are Formed from Alkenes

The double bonds in alkenes can open up and join together to make long chains called **addition polymers**. It's kind of like they're holding hands in a big line. The individual, small alkenes are called **monomers**.

Poly(phenylethene) is formed from **phenylethene**.

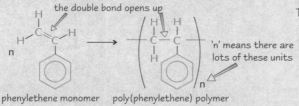

the double bond opens up

'n' means there are lots of these units

This is what a section of the chain would look like.

phenylethene monomer   poly(phenylethene) polymer

section of poly(phenylethene) polymer

Addition polymerisation is a free radical addition reaction.

Poly(chloroethene) is formed from **chloroethene**.

chloroethene monomer   poly(chloroethene) polymer

Polychloroethene's also known as PVC — Poly(vinyl chloride) or Pretty Vulgar Clothing.

Ethene Polymerises by Homolytic Free Radical Addition
Edexcel and Nuffield only

If ethene is heated to **200 °C** at **2000 atm pressure** with a bit of an **oxygen-containing impurity**, **homolytic free radical addition polymerisation** happens and **low density polyethene** forms.

The oxygen-containing impurity acts as an **initiator** — it starts the reaction by generating a **free radical**, RO·.

$$R-O-O-R \longrightarrow 2RO\cdot$$

There are lots of ways the radical could be generated, e.g. a peroxide falling apart.

The free radical then initiates a chain reaction:

INITIATION   The free radical attacks the C=C bond in ethene, making a new free radical.

$$RO\cdot \quad H_2C=CH_2 \longrightarrow ROCH_2CH_2\cdot$$

PROPAGATION   This new free radical attacks more ethene molecules, making longer free radicals. This continues to make a longer and longer chain.

$$ROCH_2CH_2\cdot \quad H_2C=CH_2 \longrightarrow ROCH_2CH_2CH_2CH_2\cdot \text{ etc.}$$

TERMINATION   Two of the free radicals join together, ending the chain reaction.

$$RO(CH_2CH_2)_n\cdot \quad \cdot(CH_2CH_2)_mRO \longrightarrow RO(CH_2CH_2)_n(CH_2CH_2)_mOR$$

The polyethene chains produced have up to 20 000 repeating units, but because of the chain reaction, you get a variety of chain lengths and some are branched.

The Uses of Addition Polymers Depend on their Properties
Edexcel and Salters only

The properties of addition polymers are due to their structures. If they're **regularly** arranged and **unbranched** with **small** side chains, they're able to lie **close** together — this increases the **van der Waals** forces, which makes the polymers **stronger** and **harder** with higher **melting** points.

Name	Structure	Properties	Typical Uses
Polyethene (polythene)		LDPE (low density polyethene) is branched and randomly arranged. HDPE (high density polyethene) is more regularly arranged.	LDPE is weak, flexible and has a low melting point. HDPE is stronger and denser with a higher melting point.	Plastic bags and squeezy bottles. Buckets, food containers, bottles.
Polychloroethene (polyvinyl chloride)		There's an irregular structure, as the big chlorines are randomly orientated. Polarised C-Cl bonds mean dipole-dipole interactions between chains.	Unplasticised (uPVC) is hard and rigid. Plasticisers can be added to make it more flexible	Guttering and window frames. PVC clothes, shoes, electrical insulation.
Polytetrafluoroethene (PTFE)		Big fluorines surround the chain — they've got lots of electrons which increase van der Waals forces, so the chains pack closely together.	Strong, high melting point, very stable and unreactive, and 'non-stick'.	Protecting surfaces from corrosion, reducing friction — non-stick pans, irons, low-friction bearings.
Polyphenylethene (polystyrene)		The benzene rings are huge, but also very electron rich. So there are increased van der Waals forces.	Stiffer than polyethene, but brittle.	Toys and cups. Packaging and insulation.

Addition Polymers

Polymer Properties can be **Modified** to Meet a **Particular Need**   *Salters only*

1) Polymer **properties** can be modified **physically** or **chemically**.

2) When making a polymer, you can include other **monomer** molecules to change the **polymer chain properties** — this is called **co-polymerisation**.
   For example, PVC's brittle at room temperature, so chemists co-polymerise it with an ester (ethyl ethanoate). This molecule has big side chains that stop PVC chains packing as closely together. The intermolecular forces decrease, so the polymer is bendier.

3) **Plasticising** makes the polymer **bendier**. Plasticiser molecules get in between the polymer chains and **reduce** the effect of the intermolecular forces so that the chains can slide around more. The plasticiser molecules must be big enough not to be **volatile** and they must **mix easily** with the polymer — so a polar polymer needs a polar plasticiser.

Some Alkenes form **Atactic**, **Isotactic** or **Syndiotactic** Addition Polymers

Polypropene's a good example: *Not AQA or Nuffield*

Substance	Structure		Properties	Typical Uses
Atactic poly(propene)	methyl groups are randomly orientated	H CH₃H CH₃CH₃H CH₃H H CH₃	Soft and flexible, with a relatively low melting point.	Roofing materials, road paint, glues.
Isotactic poly(propene)	methyl groups all point the same way	H CH₃H CH₃H CH₃ H CH₃ H CH₃	Strong and rigid, with a relatively high melting point.	Rope, carpets and plastic crates.
Syndiotactic poly(propene)	methyl groups are alternately orientated	H CH₃CH₃H H CH₃CH₃H H CH₃	Softer, lower melting point and not as strong as an isotactic isomer	Medical tubing and pouches, shrink-wrapping of food.

It's Hard to **Dispose** of Polymers **Safely**

Disposing of polymers is a big problem:

1) Addition polymers are mostly non-polar so they're difficult to break down by chemicals or hydrolysis.

2) They're **non-biodegradable** (they won't rot) — instead they end up in landfill sites.

3) Burning them produces **toxic** fumes.

But it's not all doom and gloom...

1) Some thermoplastics (plastics that soften when heated) can be **recycled** — they need to be sorted into different types though, which is expensive. Once sorted they can be heated and moulded into new products.

2) You can **crack** plastics (break them by heat) to make **organic feedstock** (raw materials for industry).

3) **Biodegradable** plastics are being developed that break down with the help of **sunlight** or **bacteria**.

Practice Questions

Q1 What's addition polymerisation?

Q2 What conditions is LDPE manufactured under?

Exam Questions

1   Propene, CH_2=CHCH₃, can be polymerised to form poly(propene).
   a) Name this type of polymerisation and draw the repeating unit.   [2 marks]
   b) Describe the structure of isotactic poly(propene). Why is it used to make rope?   [3 marks]
   c) Poly(propene) rope won't rot when it gets wet. Why is this? When is this a disadvantage?   [2 marks]

2   uPVC is a hard, brittle plastic. Adding the plasticiser dioctyl phthalate reduces the melting point.
   a) Draw a section of the polymer before the plasticiser is added. Show three monomer units.   [1 mark]
   b) Why is uPVC hard and brittle? Suggest a use for uPVC.   [3 marks]
   c) Why does adding a plasticiser reduce its melting point? How does this affect the PVC?   [3 marks]

PVC clothes — keep the rain out and the sweat in...
This stuff's very clever. It makes you wonder how anyone ever figured out how to do it. Maybe someone accidentally put some ethene in a very powerful pressure cooker, and ended up with something that was great for making carrier bags from.

Condensation Polymers

Addition polymerisation's not the only sort. There's condensation polymerisation too...

Condensation Polymers Include **Polyesters**, **Polyamides** and **Polypeptides**

1) **Condensation polymerisation** usually involves two different types of monomers.

2) Each monomer has at least **two functional groups**. Each functional group reacts with a group on another monomer to form a link, creating polymer chains.

3) Each time a link is formed, a water molecule is lost — that's why it's called **condensation** polymerisation.

Humans have used **natural** condensation polymers, such as silk (a protein), for thousands of years, but **synthetic** condensation polymers, such as nylon, have only been around since World War II.

Reactions Between **Dicarboxylic Acids** and **Diamines** Make **Polyamides**

Carboxyl groups react with **amino** groups to form **amide links**.

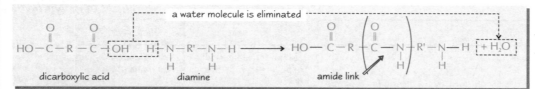

Dicarboxylic acids and diamines have functional groups at each end of the molecule, so long chains can form.

Example **Nylon 6,6** — made from **1,6-diaminohexane** and **hexane-1,6-dicarboxylic acid**.

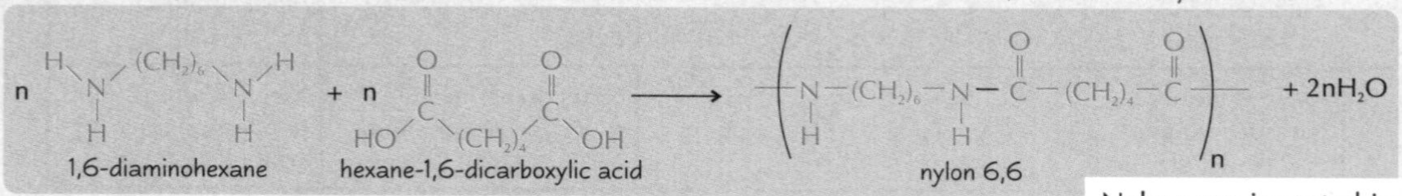

Nylon fibre is very **strong**, **elastic** and quite **abrasion-resistant**. Today it's used to make ropes, tyre cords, carpets and clothes. It's also formed into tough, rigid, solid shapes to make machine parts.

> Nylon was invented in 1935 by chemists who were trying to imitate the peptide links in silk and wool proteins.

Example **Kevlar** — made from **benzene-1,4-diamine** and **benzene-1,4-dicarboxylic acid**.

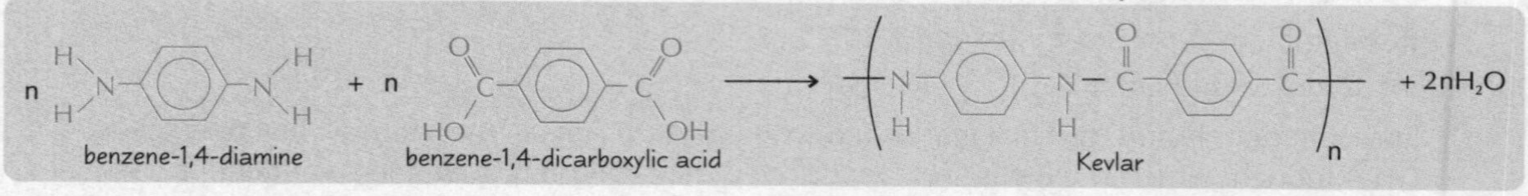

Kevlar is really **strong** and **light** — five times stronger than steel. It's not stretchy, and is quite stiff.

It's most famous in bulletproof vests, but is also used in sports equipment, brake parts and gears, mooring ropes for supertankers, and (because it's **heat resistant**) in firesuits.

Reactions Between **Dicarboxylic Acids** and **Diols** Make **Polyesters**

Carboxyl groups react with **hydroxyl** groups to form **ester links**.

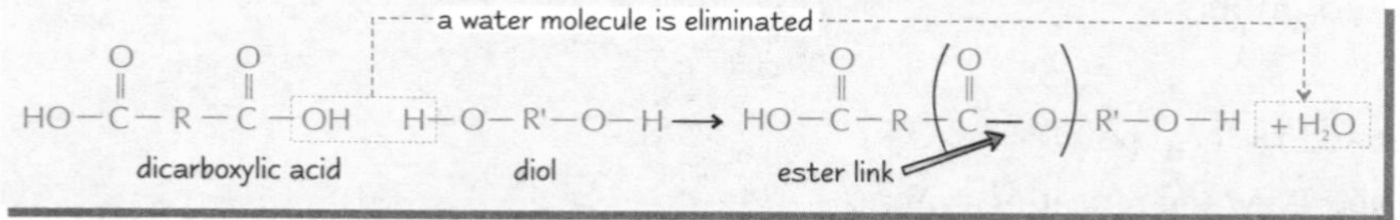

Example **Terylene (PET)** — formed from **benzene-1,4-dicarboxylic acid** and **ethane-1,2-diol**.

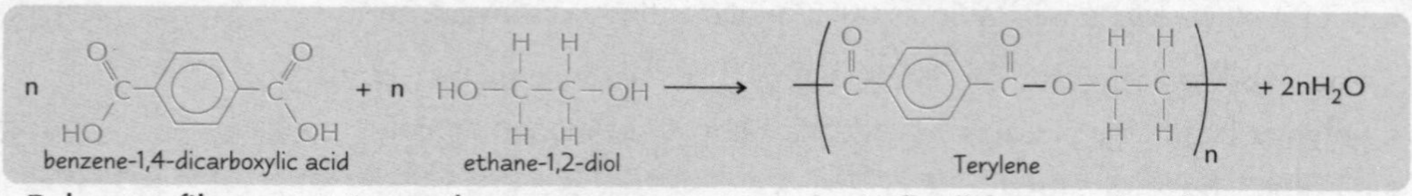

Polyester fibres are **strong** (but not as strong as nylon), **flexible** and **abrasion-resistant**.

Terylene is used in **clothes** to keep them crease-free and make them last longer. Polyesters are also used in **carpets**.

You can treat polyesters (by stretching and heat-treating them) to make them stronger. Treated Terylene's used to make fizzy drink bottles and food containers.

Condensation Polymers

Polypeptides are Condensation Polymers Too   *Not Edexcel*

Polypeptides and proteins are **natural condensation polymers**. They're formed when **amino acids** join up in a long chain.

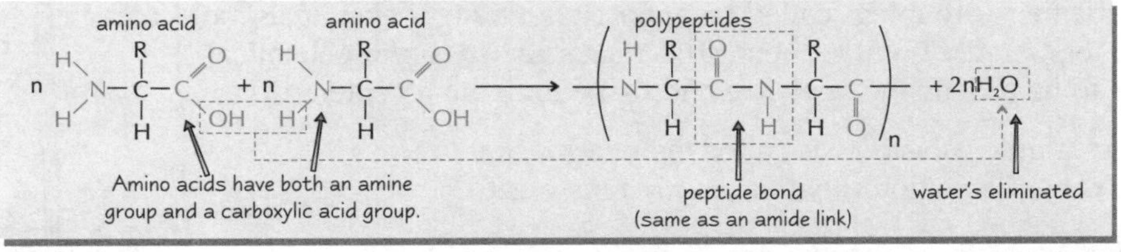

Amino acids have both an amine group and a carboxylic acid group.

peptide bond (same as an amide link)

water's eliminated

See page 96 for amino acids.

Intermolecular Forces Hold Condensation Polymer Chains Together   *Not AQA or Nuffield*

The forces that hold condensation polymer chains together are similar to those found between addition polymers. They include:

1) **Weak van der Waals forces** — these get bigger if there are electron-rich groups like benzene rings.
2) **Permanent dipole-dipole interactions** between polarised C=O and N–H groups on adjacent chains.
3) In polyamides there are strong **hydrogen bonds** between polarised C=O and N–H groups on adjacent chains. The shorter the monomer units, the more hydrogen bonds there are in a chain of a certain length, so the stronger the polymer will be.

If a condensation polymer has permanent dipole-dipole interactions or hydrogen bonding it'll form **very strong linear polymers** that can be spun into **fibres**.

Polyesters and Polyamides are Biodegradable

Amide links in polyamides and **ester links** in polyesters can be easily **hydrolysed** — so they're biodegradable.

These links are found in nature, so there are **fungi** and **bacteria** that are able to degrade them. It's not all hunky-dory though. It takes absolutely ages for synthetic polyamides and polyesters to decompose — e.g. nylon takes 40 years.

Nylon and polyester can be **recycled** — you can buy polyester fleece jackets made from recycled PET bottles.

Practice Questions

Q1 What types of monomer are polyesters formed from?

Q2 What's Kevlar used for?

Q3 Are polyesters and polyamides biodegradable?

Exam Questions

1   a) Nylon 6,6 is the most commonly produced nylon. A section of the polymer is shown below:

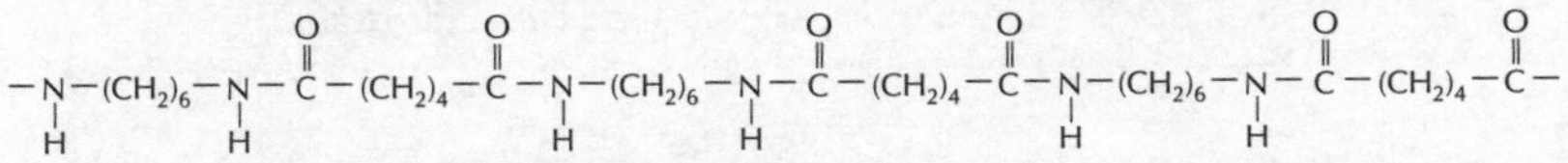

   i) Draw the structural formula of the monomers from which nylon 6,6 is formed. It is not necessary to draw the carbon chains out in full.   [2 marks]
   ii) Suggest why this nylon polymer is called nylon 6,6.   [1 mark]
   iii) Give a name for the linkage between the monomers in this polymer.   [1 mark]

   b) A polyester is formed by the reaction between the monomers hexanedioic acid and 1,6-hexanediol.
   i) Draw the repeating unit for the polyester.   [1 mark]
   ii) Explain why this is an example of condensation polymerisation.   [1 mark]

   c) In terms of intermolecular interactions, explain why nylon 6,6 is stronger than the polyester.   [3 marks]

2   Dissolving stitches used in operations are made from hydroxycarboxylic acid polyesters. Explain why the stitches dissolve in the human body. Why is it not possible to use a polymer like poly(propene) for this purpose?   [4 marks]

Never miss your friends again — form a polymer...

It's a job for Q designing all these polymers — bulletproof vests, nylon parachutes — just think, you could be the next mad inventor, working for the biggest secret agency in the world. And you'd have a really fast car, which obviously would turn into a yacht with the press of a button... and retractable wings so you could fly... just think of the possibilities.

Medicines

Last page of the section. Yippee I hear you say. **AQA, OCR** *and* **Nuffield** **people can skip these two pages.**

Medicines Contain **Pharmacophores**

In virtually every cell of our body there are things called **receptor sites**. Natural chemicals/molecules fit into these receptor sites and temporarily bond to them. This triggers a series of **biochemical reactions**. The chemicals have to be exactly the right **size and shape** to fit the receptor.

The part of a molecule that gives a drug its activity is called the **pharmacophore**. Chemists can design drugs with pharmacophores that fit exactly into target **receptor sites**.

In designing a drug you need to consider if you want the drug to **increase** the response that happens naturally, or **decrease** it.

If you want a drug to increase the natural response you use an **agonist** drug. This binds to the receptor and **triggers** a **response**.

To decrease the response you use an **antagonist** drug. This binds to the receptor and **blocks** it.

drug

signal to the cell

receptor

> The 'fit' of a **pharmacophore** into a **receptor site** depends on:
>
> 1) **Size** and **shape** — it's got to have a particular structure that will fit into the receptor site.
>
> 2) **Bond formation** — functional groups in the pharmacophore form **temporary bonds** with functional groups in the receptor. These are mostly **ionic** interactions or **intermolecular** forces. Covalent bonding is permanent, so would irreversibly block the receptor.
>
> 3) **Orientation** — if the pharmacophore has **optical** or **geometric** isomers, then only one of the isomers will fit (see pages 70-71).

Chemists are Important in the **Design** and **Synthesis** of **New Medicines**

Salters only

Here are some of the things chemists think about and do when they're trying to make a new drug:

> **Is there a natural compound already used to treat the disease?**
> Years ago foxglove leaves were used to treat heart failure — the drug's now been isolated and it's still used today.

> **Is there a compound in the body that's involved in the process?**
> This could give you a good starting point. Noradrenaline is a good example — see page 105.

> **Molecule screening**
> This is where loads of molecules are tested to see if any of them bind to the target receptor.

> **Modifying the Compound**
> Now you've got something which interacts with the receptor site, you can modify its structure to make it fit better. This is where functional groups are added, removed or changed.

> **Testing the Compound**
> Now you've modified it, you've got to test it to see if it works. This often happens again and again to make the best, safest drug possible.

> **Drug**

Computer Modelling Techniques *Speed Up* Molecule Screening and Drug Design

Salters only

Testing for a molecule that will fit a receptor perfectly takes loads of time and is expensive. It's sped up by screening the compounds in batches of 50-100, but this still takes **ages**, and there's no guarantee you'll find something that works.

Computer 3D modelling can speed things up. **Databases** of 3D models of compounds can be searched to find compounds that may fit the **3D model of the target receptor site**, or to find ones containing particular structures and functional groups. This cuts down the number of compounds you need to test in the lab.

Rational drug design is a bit different. It involves building a new compound from scratch using a 3D model of the receptor site.

Classic molecule screening is like looking for a key to fit a lock. Rational drug design is like using the structure of the lock to make your own key.

Medicines

Modifying the Pharmacophore Changes the Pharmacological Activity

You can **change** the **pharmacophore** to make a drug **more effective** or to reduce its side effects.

For example, **noradrenaline**'s naturally found in the body and has been modified to treat different disorders.

Noradrenaline — expands airways and increases your heart rate and blood pressure

Salbutamol — treats asthma symptoms, without the raised heart rate and blood pressure

Isoprenaline — used to increase heart rate for some heart problems

Some Drugs are Fat-Soluble and Some are Water-Soluble

The **solubility** of a drug affects how **long** the drug stays in the body.

If you want the drug to only stay in the body for a **short time**, you can modify it to contain lots of **polar groups** (–OH, C=O, N–H). These form **hydrogen bonds** with water, so are **water-soluble**. They'll be **excreted** quickly in urine.

On the other hand, if you want the drug to stay in the body for a **long time**, you can make it **fat-soluble**. Fat-soluble substances are mainly **covalent** and **non-polar**. They're likely to stay in fatty tissue and can go through cell membranes. Chemists need to take extra care considering the dosage — they could build up to **toxic levels** in the body.

It Can Take 12 Years and Cost £350 Million to get a Drug onto the Market

Salters only

Once you've found a useful-looking compound, there are still a few hurdles to get over:

1) Is it **safe**? **In vitro testing** (in test tubes and Petri dishes) and **animal testing** check this.
   If the compound fails these tests, it goes back to the chemists to have its structure modified, before retesting.

2) How can the drug best be given — in pill form or by injection? What **dose** is most effective?

3) Small- and large-scale **clinical trials** on humans are carried out, watching out for **side effects**.

4) Finally, the drug has to be **legally licensed** before being launched.

Practice Questions

Q1 What's a pharmacophore?

Q2 What's rational drug design?

Q3 What's an agonist drug? What's an antagonist drug?

Exam Question

1   The structures of three non-steroidal anti-inflammatory drugs which share the same pharmacophore are shown below.

Ibuprofen   Naproxen   Fenoprofen

a) Explain the term pharmacophore.   [2 marks]

b) Draw the structure of the pharmacophore found in these three compounds.   [1 mark]

c) Identify a functional group which might make these compounds soluble in water. Explain your answer.   [2 marks]

d) Suggest a way these compounds can bond to receptor sites.   [2 marks]

e) The pharmacophore in these drugs is chiral.
    i) Mark the chiral centre with an asterisk on your diagram for part (b).   [1 mark]
    ii) What implications will this have for the drug's pharmacological activity?   [1 mark]

Why's there no aspirin in the jungle? The paracetamol...

Finding a drug to treat something can take ages, but getting it on the market can take even longer — you've got to be sure it's as safe as possible. Before you scamper away excitedly to the next section, make sure you've got these last few pages firmly in your mind and can answer all the questions without checking back at the page. I'd have a chocolate biscuit too.

Identifying Functional Groups

*If you want to figure out what flavour ice cream you've got, you taste it. If you want to know what chemical you've got, you use one of these tests. Salters **can skip these two pages**.*

Organic Molecules have Functional Groups

A **functional group** is the bit of a molecule that gives it its **characteristic properties** (and part of its name).
As you'll know from the last section, a particular functional group will take part in certain **reactions**.
If you're not sure what compound you've got, these tests will help.

Alcohols can be Primary, Secondary or Tertiary　　*Edexcel only*

Alcohols with just one hydroxyl group (–OH), can be **primary**, **secondary** or **tertiary**.

On the **carbon** with the **hydroxyl** group attached:

1) A **primary** alcohol has **two hydrogen atoms** and just **one alkyl group**.
2) A **secondary** alcohol has **just one hydrogen atom** and **two alkyl groups**.
3) A **tertiary** alcohol has **no hydrogen atoms** and **three alkyl groups**.

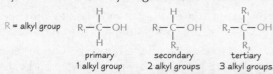

> It's the same naming system for haloalkanes and amines.

Acidified Potassium Dichromate(VI) is Used to Oxidise Alcohols　　*Edexcel only*

If you know you've got some sort of alcohol, you can use the **oxidising** agent **acidified potassium dichromate(VI)** to help you tell which type of alcohol you've got. You need to warm your alcohol with it and watch for a colour change.

RESULTS

PRIMARY – the orange dichromate slowly turns green as an **aldehyde** forms (then eventually a carboxylic acid).

SECONDARY – the orange dichromate slowly turns green as a **ketone** forms.

The test shows the same result for primary and secondary alcohols — you have to use one of the tests on the opposite page to find out if an aldehyde or ketone has formed.

TERTIARY – nothing happens — boring, but easy to remember.

> The colour change is the orange dichromate(VI) ion, $Cr_2O_7^{2-}$, being reduced to the green chromium(III) ion, Cr^{3+}. See page 58.

Phosphorus(V) chloride Identifies an Alcohol　　*Edexcel only*

React an alcohol with phosphorus(V) chloride and you get a **chloroalkane**. It's a good way to test for alcohols.

Add **phosphorus(V) chloride** to the unknown liquid.
If -OH is present, you'll get **misty fumes** of HCl which turn damp **blue litmus red**.

$$ROH_{(l)} + PCl_{5(l)} \rightarrow RCl_{(l)} + HCl_{(g)} + POCl_{3(l)}$$

Use Bromine to Test for Unsaturation　　*Edexcel only*

To see if a compound is unsaturated (contains any double bonds), all you need to do is shake it with **orange bromine solution**.

The solution will quickly **decolourise** if the substance is unsaturated.

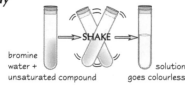

bromine water + unsaturated compound　　SHAKE　　solution goes colourless

Use Silver Nitrate to Identify a Halide Group　　*Edexcel only*

To find out what halide is present in a particular compound, you can do this test:

1) Warm the solution with **sodium hydroxide** (or another alkali). This is called **alkaline hydrolysis**.
2) Then acidify it by adding **dilute nitric acid**.
3) Add **silver nitrate solution** and see what colour precipitate you get.

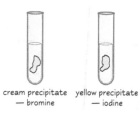

white precipitate — chlorine　　cream precipitate — bromine　　yellow precipitate — iodine

Identifying Functional Groups

Test for a **Carbonyl Group** Using **2,4-dinitrophenylhydrazine**...   *Not AQA or Salters*

You can test to see if an unknown liquid is an aldehyde or ketone (a carbonyl compound) by adding **2,4-dinitrophenylhydrazine (2,4-DNPH)** in methanol and acid (**Brady's reagent**).

 RESULT   If a carbonyl group's present, you'll get a **bright orange precipitate**.

See page 74 for carbonyl compounds.

...or Find out if it's an **Aldehyde** or **Ketone** Using **Fehling's Solution** or **Tollens' Reagent**

You can heat the substance in question with either **Fehling's solution** or **Benedict's solution**.

 RESULTS   An **ALDEHYDE** will give a **brick-red precipitate** of copper(I) oxide.
A **KETONE** won't react.   ⟵ *Not OCR*

OR   Heat the substance with **Tollens' reagent**, which is **silver nitrate dissolved in aqueous ammonia**.

 RESULTS   An **ALDEHYDE** will give a **silver mirror** on the test tube.
A **KETONE** won't react.   ⟵ *Not Salters or Nuffield*

Use **Iodine** to Detect a **Methyl Carbonyl Group**   *Edexcel only*

Heat the unknown substance with either **iodine** in **aqueous sodium hydroxide** or with **potassium iodide** in **sodium chlorate(I) solution**.

RESULT   **Pale yellow crystals** with an antiseptic smell are produced if a methyl carbonyl group is present. The crystals are triiodomethane (iodoform).

Methyl carbonyl groups have a CH_3 group next to a C=O group.

Test for a **Carboxylic Acid** Using a **Hydrogencarbonate**   *Edexcel only*

Add some **sodium (or potassium) hydrogencarbonate** to the unknown compound.

RESULT   If a carboxylic acid is present **carbon dioxide gas bubbles out** — and this'll turn lime water milky.

$$RCOOH_{(aq)} + NaHCO_{3(s)} \rightarrow RCOONa_{(aq)} + H_2O_{(l)} + CO_{2(g)}$$

Practice Questions

Q1   What test could you do to identify a tertiary alcohol?
Q2   How could you distinguish between ethanal and propanone?
Q3   How could you distinguish between ethanal and propanal? (Hint: one of them has a methyl carbonyl group.)
Q4   How could you distinguish between chlorobutane and bromobutane?

Exam Questions

1   Draw as many conclusions as possible about the structure of the organic substances A and B from the following test results.

   a)   Substance A is a neutral substance. It gives no reaction when warmed with Fehling's solution, but it gives an orange precipitate when added to a solution containing 2,4-DNPH. When it's heated with iodine in aqueous sodium hydroxide, pale yellow crystals form.   [5 marks]

   b)   If Substance B is warmed with Fehling's solution, a red precipitate forms.
When heated with iodine in aqueous sodium hydroxide, no crystals form.   [2 marks]

2   3-chloro-3-ethylpentane reacts with aqueous sodium carbonate to give two products. One is an alcohol and the other an alkene. Both substances are liquids at room temperature and can be separated by fractional distillation.

   a)   Describe one test that would identify which of the two substances is the alkene.   [3 marks]

   b)   When the alcohol is warmed with acidified potassium dichromate, no reaction occurs.
What does this tell you about the alcohol?   [1 mark]

Now don't go getting testy with me — I didn't write the syllabus...

Lots of straightforward learning on this page — but make sure you do learn it. Questions on how to test for things are bound to come up somewhere. Once you reckon you know it all, make yourself a table with three columns — what you're trying to identify, the test, and what the positive result is. Then fill it in. If you get stuck, you've got more learning to do.

UV–Visible Spectra

If you're doing AQA you can skip these two pages.
UV–visible spectroscopy is kind of like colorimetry — you can use it for colourless compounds too though.

Electrons can Absorb Energy and Become Excited to Higher Energy Levels

Electrons in a molecule are usually at their **lowest possible energy levels**. But if the molecule **absorbs energy**, the electrons become **excited** and move up to a **higher energy level**. This is an **electronic transition**.

The molecule might absorb wavelengths of light from the **visible region** of the spectrum.
In this case, other wavelengths of visible light will be transmitted and the compound will look **coloured**.

Alternatively, the molecule might absorb **UV light** and transmit **all** the wavelengths of visible light.
In this case, it's look **colourless**, as our eyes can't detect UV light. This is where **spectrometers** come in.

Spectrometers tell you which wavelengths of visible light **and** UV light have been absorbed.
A colorimeter is a special type of spectrometer.

Chromophores Absorb UV or Visible Light

OCR (Methods of Analysis and Detection Option), Salters, Edexcel and Nuffield only.

Chromophores are **structural** features in a molecule that absorb energy in the **UV** or **visible regions** of the spectrum.
If they absorb visible light, they'll be responsible for the compound's **colour** — see page 95.

A chromophore could contain:

1 **A double or triple bond** — e.g. H–C–C≡N (with H above and below the C)

2 **A lone pair of electrons** — e.g. N with H, H, H (ammonia, lone pair shown)

3 **A delocalised electron system** — such as a benzene ring: ← See page 84.

or, a conjugated electron system — these have **alternating double and single bonds**.
The electrons in conjugated systems **spread out** over all the atoms to form a **delocalised system**.

conjugated system = delocalised system

You've got to able to predict if a molecule's able to absorb UV or visible light — so just check for one of these structures.

U.V. and Visible Light Spectroscopy Helps to Identify Organic Compounds

Each chromophore absorbs at a **certain wavelength** — so they have a **characteristic light absorption** pattern.
This information's used to help **identify** organic compounds.

FOR EXAMPLE **Carotene**, the pigment in carrots, has got a conjugated system containing eleven double bonds — so it's a pretty long molecule.

This is the visible spectrum of **carotene**:

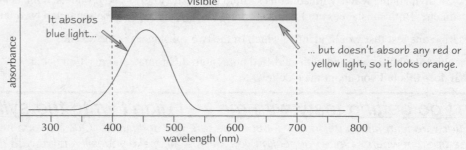

It absorbs blue light...

... but doesn't absorb any red or yellow light, so it looks orange.

UV–Visible Spectra

Increasing Conjugation Decreases the Gap Between Energy Levels

OCR (Methods of Analysis and Detection Option) only

Electrons in conjugated systems are **delocalised** and need **less energy** to excite them than if they were in a single bond.

Delocalisation causes the **upper energy levels** to be **stabilised** and **lowered** in **energy** — so the gaps between the energy levels are **smaller**. The **longer** a conjugated system is, the **smaller** the gaps between the energy levels.

Less energy is needed to excite the electrons to a higher energy level, so **absorption** shifts towards a **longer wavelength**.

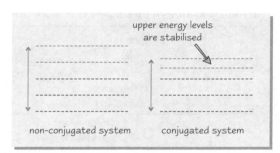

upper energy levels are stabilised

non-conjugated system   conjugated system

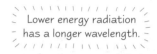

Lower energy radiation has a longer wavelength.

So, to sum it up:

> The **longer** the conjugated system, the **smaller** the gap between energy levels, and the **longer** the wavelength absorbed.

This often **shifts the wavelength** absorbed into the visible range — making the compound coloured.

Visible Spectroscopy is Used to Help Restore Paintings
Just Salters and Nuffield

The **pigments** used in **paintings** can be identified using visible spectroscopy. This helps to date them, and is useful in restoration, because you can make sure you're using exactly the same colour paint.

There are two types of visible spectroscopy used to analyse paints:

1) **REFLECTANCE SPECTROSCOPY**
   White light is shone onto the paint and the spectrum of the **reflected light** is analysed. It shows which wavelengths have been absorbed, so it helps to identify the pigment.

2) **TRANSMISSION SPECTROSCOPY**
   A tiny bit of the pigment is scraped off the painting and dissolved in a solvent. Then white light's shone **through** the solution and the spectrum's analysed.

Practice Questions

Q1 Name three groups that a molecule could contain if it absorbs light from the visible/UV part of the spectrum.

Q2 What is a chromophore?

Q3 Why does the conjugation of an electron system shift absorbance towards a longer wavelength?

Exam Question

1 Look at the diagram of cyanidin, a purple pigment found in flowers.

   a) What is it about its structure that indicates it will be coloured?   [5 marks]

   b) What does the colour of the pigment tell you about the light
       absorbed by the pigment.   [1 mark]

   c) In acid solution the molecule becomes red.
       Which part of the molecule must have changed for this colour change to happen?   [1 mark]

Carotene's the pigment in carrots, so Ovaltine must be the pigment in ovals...

One of the world's unanswerable questions — do we all see the same colours? You'll never know if your best mate sees blue, while you're seeing pink. I wonder what the world would be like if it was all in black and white. I'd half expect a dashing stranger to sweep me off my feet, like in one of those old-fashioned romantic movies. I can only dream...

Atomic Emission Spectroscopy

These pages are just for OCR (Methods of Analysis and Detection Option) and Salters — the rest of you can jump with joy and skip merrily onto page 114.

c = fλ Links **Wavelength** to **Frequency** — OCR (Methods of Analysis and Detection Option) only

Electromagnetic radiation has a **frequency** (number of waves per second) and a **wavelength**.
If you know one of these, you can work out the other by using the formula below:

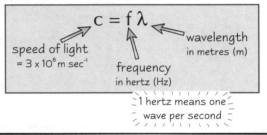

$$c = f\lambda$$

speed of light
= 3 × 10⁸ m sec⁻¹

frequency
in hertz (Hz)

wavelength
in metres (m)

~1 hertz means one wave per second~

Example: Blue light has a wavelength of 450 nm. Calculate its frequency.
~nanometres (nm) are metres × 10⁻⁹~

$$c = f\lambda$$
$$3 \times 10^8 = 450 \times 10^{-9} \times f$$
$$f = \frac{3 \times 10^8}{450 \times 10^{-9}} = 6.67 \times 10^{14} \text{ Hz}$$

Emission Spectra — Excited **Electrons** Releasing **Electromagnetic Energy**

OK...here's a bit of brain-frying stuff coming at you —

1) Atoms in their **ground state** have all their electrons in their **lowest** possible energy levels. Each energy level is given a **quantum number**, n. The value of n for 1s electrons is 1.

2) If the electrons take in **energy** from their surroundings, they can move to **higher energy levels**, which are usually further from the nucleus. At higher energy levels, electrons are said to be **excited**.

3) The energy levels all have **certain fixed energies** — they're **discrete**. Electrons can jump from one energy level to another by **absorbing** or **releasing** a fixed amount of energy.

4) Energy is related to **wavelength**. When **electromagnetic radiation** is passed through a gaseous element, the electrons only absorb **certain wavelengths**, corresponding to **differences between the energy levels**. That means the radiation passing through has certain wavelengths missing. A spectrum of this radiation is called an **atomic absorption spectrum**. The missing wavelengths show up as **dark bands**.

5) When electrons **drop** to lower energy levels, they **give out** energy. This produces lines in the spectrum too — but this time it's called an **emission spectrum**. The wavelengths in an emission spectrum are the **same** as those missing in the corresponding absorption spectrum.

6) Each element has a **different** electron arrangement, so the wavelengths of radiation absorbed and emitted are different. This means the **spectrum** for each element will be unique too.

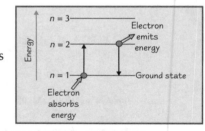

The Atomic Emission Spectrum of **Hydrogen** is made up of **Sets of Lines**

1) Spectra often seem to make as much sense as bar codes. But the emission spectrum of **hydrogen** is fairly simple because hydrogen only has **one** electron that can move.

2) The atomic emission spectrum of hydrogen shows **sets of lines** — you need to know about three of them, the **Lyman series**, the **Balmer series** and the **Paschen series**.

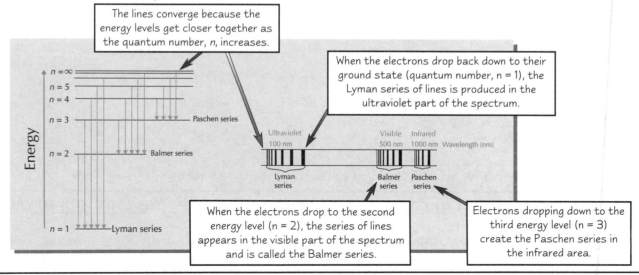

The lines converge because the energy levels get closer together as the quantum number, n, increases.

When the electrons drop back down to their ground state (quantum number, $n = 1$), the Lyman series of lines is produced in the ultraviolet part of the spectrum.

When the electrons drop to the second energy level ($n = 2$), the series of lines appears in the visible part of the spectrum and is called the Balmer series.

Electrons dropping down to the third energy level ($n = 3$) create the Paschen series in the infrared area.

Atomic Emission Spectroscopy

Ionisation Energy can be Calculated from Emission Spectra

OCR (Methods of Analysis and Detection Option) only

The energy released is related to the frequency of the wavelength emitted. This equation is used to convert between the two:

$$E = hf$$ where f = frequency (hertz/Hz) and h = Planck's constant (6.63×10^{-34} Js)

As you can see, the lines in any series on an emission spectrum get closer and closer together — towards a **convergence limit**. This is the point at which an electron breaks free from the atom, and the atom becomes ionised.

convergence limit

The amount of energy needed to cause **one mole of electrons** to break free from one mole of gaseous atoms is the **ionisation energy** for that element. All of these electrons must start off in their **ground state**, so in the case of hydrogen it's the **Lyman series** that you need to consider when calculating an ionisation energy.

> **Example:** The lines in the Lyman series in hydrogen's emission spectrum converge at a frequency of 3.27×10^{15} Hz. Calculate hydrogen's ionisation energy.
>
> $E = hf = 6.63 \times 10^{-34} \times 3.27 \times 10^{15}$
> $= 2.168 \times 10^{-18}$ joules per atom.
>
> So E for 1 mole of atoms $= 2.168 \times 10^{-18} \times 6.02 \times 10^{23}$
> $= 1.305 \times 10^{6}$ Jmol^{-1}
> $= 1305$ kJmol^{-1}

This is Avogadro's number — the number of atoms in a mole.

New Elements have been Discovered in the Emission Spectra of Stars

Emission spectra are produced by **very hot conditions** — like on stars.

Helium was first discovered in the **Sun's atmosphere** in 1868, 27 years before it was found on Earth. During a **total eclipse** of the Sun, the Moon blocked the main body of the Sun, leaving parts of the atmosphere still visible. The emission spectrum showed previously **unseen lines**, which were due to the element helium.

The composition of other stars' atmospheres can be found using their emission spectra.

Emission Spectra Help People Restore Old Paintings | *Salters only*

Like visible spectra, emission spectra are useful for analysing paint samples to see what pigments they contain.

A **high-energy pulse** of laser **light** is focused onto a **tiny sample of paint**. The **pigment** is **vaporised** and then the **vapour** is passed between two electrodes to give it **more energy** and **excite the electrons** even more.

This produces an **emission spectrum**, which is used to **identify the pigment**.

The **sample used** is so **small** (about 10^{-5} g) that **no visible damage** needs to be done to the picture.

Practice Questions

Q1 What's the formula that links frequency and wavelength for light?

Q2 What causes the release of the energy that's detected in an emission spectrum?

Q3 What's the convergence limit of a spectrum?

Q4 Which series of lines in the atomic emission spectrum of hydrogen corresponds to electrons falling back to the second energy level?

Exam Questions

1 The first ionisation energy of lithium is 519 kJmol^{-1}. What is the frequency of the line in the emission spectrum of lithium that enabled this ionisation energy to be calculated? [3 marks]

2 Show that UV light of wavelength 300 nm contains more energy than infrared light of wavelength 800 nm. [5 marks]

I hope you're nearly as excited as the electrons...

I'm reaching my own special convergence limit — the lines on my forehead are getting closer and closer together. Any minute now and I'll be charged and ready to go 'blaaaaa'. But now I've lost all my energy and am about to hit my ground state — my nice comfy bed. Sounds nice, hey — but you've got to learn all this stuff first, I'm afraid.

Chromatography and Electrophoresis

These pages are for OCR (Methods of Analysis and Detection Option) and Salters — the rest of you can either flip over and forget you even read the title, or you can read it for the sheer personal satisfaction of it.

Chromatography is Good for **Separating** and **Identifying** Things

Chromatography is used to **separate** stuff in a mixture — once it's separated out, you can often **identify** the components.

There are quite a few different types of chromatography — but the ones you need to know about are **paper chromatography**, **thin-layer chromatography** (TLC) and **gas-liquid chromatography** (GLC). They all have two phases:

1) A **mobile phase** — where the molecules can move. This is always a liquid or a gas.
2) A **stationary phase** — where the molecules can't move. This must be a solid, or a liquid held in a solid.

The components in the mixture separate out as the mobile phase moves through the stationary phase.

Paper Chromatography Separates Components by **Partition**

In paper chromatography:
1) The **mobile phase** is a **solvent**, such as ethanol, which passes over the stationary phase.
2) The **stationary phase** is **water in the fibres of the paper.**

In **paper chromatography**, the different components **partition** themselves between the two phases. Each component's got a different **partition coefficient** (see page 27) which determines how much it'll dissolve in each phase. This means the components will move up the paper at **different speeds**.

The details of how to do **paper chromatography** are on page 96. Whatever type of mixture you're separating, you'll end up with some **R$_f$ values** (also explained on page 96). You have to compare your R$_f$ values with tables of known values. Watch out though — the R$_f$ values depend on the type of paper and solvent you use.

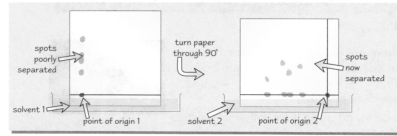

Sometimes the solvent doesn't completely separate out all the chemicals. In this case you need to use **two-way chromatography**, which uses a second solvent to complete the separation.

Thin-Layer Chromatography Separates Components by **Adsorption**

Thin-layer chromatography (TLC) is pretty similar to paper chromatography. The mobile phase is again a solvent, but the stationary phase is a **thin layer of solid**, such as silica gel, on a **glass plate**.

How quickly each component moves depends on how strong its **adsorption** to the stationary phase is. Molecules that have **strong** adsorption will move more **slowly**. The adsorption might be caused by **polar interactions**, for example.

Gas-Liquid Chromatography is a Bit More **High-Tech**

In **gas-liquid chromatography** (GLC) the stationary phase is a **viscous liquid**, such as an oil, which coats the inside of a long tube.

The tube's **coiled** to save space, and built into an oven. The mobile phase is an **unreactive carrier gas** such as nitrogen.

Like with paper chromatography, the components are separated by **partition**. Each component takes a different amount of time from being **injected** into the tube to being **recorded** at the other end. This is the **retention time** — it's what's used to identify the component.

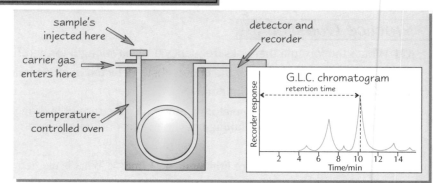

The **area** under each peak (or the height if they're very sharp peaks) tells you the relative **amount** of each component.

GLC is used to:
1) find the **proportions** of various **esters in oils** used in **paints** — you only need a small sample of paint. This lets picture restorers know exactly what paint was originally used.
2) Find the **level** of **alcohol** in **blood or urine** — the results are **accurate** enough to be used as evidence in court.

Chromatography and Electrophoresis

This page is just for OCR (Methods of Analysis and Detection Option) — you lucky people.

Electrophoresis is Chromatography with Electricity

Electrophoresis is the movement of **charged particles** in a fluid or gel through an electric field. It can be used to separate many mixtures.

The sample is placed on a **strip of gel** which contains a buffer solution to keep the pH constant. The strip is then put in an **electric field** and molecules from the sample are **attracted** to the **negative electrode** (if they're positive) or the **positive electrode** (if they're negative).

The speed at which the molecules move depends on:

1) **Overall charge** — the greater a molecule's charge, the faster it moves.
2) **Molecular size** — smaller molecules move faster.

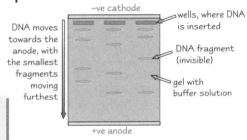

electrophoresis gel soaked in buffer solution

negatively charged particles   positively charged particles

This is a good way to separate **amino acids** —

Amino acids exist as **zwitterions** (**dipolar** ions) at a pH value called their **isoelectric point** (see page 96). At pH values **above** their isoelectric point they tend to become **negative** ions, and at pH values **below** their isoelectric point they tend to become **positive** ions. Because different amino acids have different isoelectric points, they're easily separated by electrophoresis.

Electrophoresis is Used to Make a Genetic Fingerprint

You can **cut up** parts of a person's DNA and **separate** out the pieces using **electrophoresis** — it makes a **genetic fingerprint**.

① Cells are **broken up** by mixing them with **detergents** and **enzymes**. The DNA's then separated from the rest of the cell and cut up into lots of pieces using more enzymes.

② **Electrophoresis** is then used to **separate** the pieces of DNA by **size**. But the DNA fragments **aren't visible** to the eye — you have to do something else to them before you can **see their pattern**.

–ve cathode

wells, where DNA is inserted

DNA moves towards the anode, with the smallest fragments moving furthest

DNA fragment (invisible)

gel with buffer solution

+ve anode

③ A **nylon membrane** is placed over the electrophoresis gel and the DNA fragments **bind** to it. The membrane's then soaked in a solution containing **phosphorus-32**, a **radioactive isotope**, which **binds** to the DNA fragments.

And one day you'll be a fingerprinter too — just like me and Grandad. It's genetic, you see.

④ The nylon membrane is then put on top of unexposed **photographic film**. The film print will show **dark patches** where the radioactive isotope is **present**, which **reveals the position** of the DNA fragments. The pattern is different for every human (except identical twins).

Practice Questions

Q1 How does paper chromatography work?

Q2 What's two-way chromatography?

Q3 What is the retention time?

Exam Question

1 Look at the diagram of a chromatogram produced using TLC on a mixture of substances A and B.

a) Calculate the R$_f$ value of the A spot. [2 marks]

b) Explain why substance A has moved further up the plate than substance B. [3 marks]

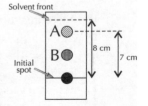

Solvent front

A

B

Initial spot

8 cm

7 cm

A little bit of TLC is what you need...

Electrophoresis might sound complicated, but it's not too bad. In fact, genetic fingerprinting is pretty cool. It's used in paternity testing — seeing if a man's really someone's father, and also in getting evidence to prove criminal cases — for example, if they find a hair at the scene of a murder, they can check if it belongs to any of the suspects.

Mass Spectrometry

A mass spectrometer ionises organic molecules and then breaks lots of the ions up into fragments. Not good news for the molecule, but very handy for scientists. A mass spectrum can tell you the molecular mass, molecular formula, empirical formula, structural formula and your horoscope for the next fortnight.

Mass Spectroscopy can be used to Find Out Relative Molecular Mass (M_r)

Electrons in the spectrometer **bombard** the sample molecules and **break electrons off**, forming ions. A **mass spectrum** is produced, showing the **relative amounts** of ions with different mass-to-charge ratios.

The molecular ion $M^+_{(g)}$ gives the peak in the spectrum with the **second highest mass-to-charge ratio** — so it's the last peak but one on the spectrum. This peak's called the **M peak**. The mass/charge (m/z) value of the M peak is the **molecular mass** of the molecule. ⟵ *This is assuming the peak corresponds to singly charged ions, which is a pretty safe assumption.*

The mass spectrum for ethanol (CH_3CH_2OH) is shown below. Ethanol has the molecular ion $CH_3CH_2OH^+$, and a molecular mass of 46.

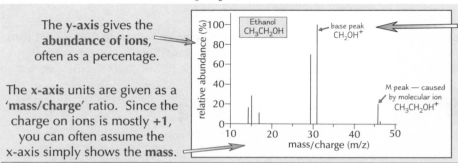

The **y-axis** gives the **abundance of ions,** often as a percentage.

The **x-axis** units are given as a '**mass/charge**' ratio. Since the charge on ions is mostly **+1**, you can often assume the x-axis simply shows the **mass**.

The highest peak is called the **base peak**, and its **relative abundance** is set at **100%.** It's due to a particularly **stable fragment** ion, such as **carbocations** and **acylium** (RCO) ions. All the other peak heights are given as percentages of it.

The Molecular Ion can be Broken into Smaller Fragments

For everyone except OCR people who aren't doing the Methods of Analysis and Detection Option.
The bombarding electrons make some of the molecular ions break up into **fragments**. These all show up on the mass spectrum, making a **fragmentation pattern**. Fragmentation patterns are actually pretty cool because you can use them to identify **molecules** and even their **structure**.

For propane, the molecular ion is $CH_3CH_2CH_3^+$, and the fragments it breaks into include CH_3^+ ($M_r = 15$) and $CH_3CH_2^+$ ($M_r = 29$).

Only the **ions** show up on the mass spectrum — the **free radicals** are 'lost'.

$$CH_3CH_2CH_3^+ \nearrow CH_3CH_2\bullet \;+\; CH_3^+$$
free radical · ion

$$\searrow CH_3CH_2^+ \;+\; \bullet Ch_3$$
ion · free radical

EXAMPLE

Use this mass spectrum to work out the structure of the molecule:

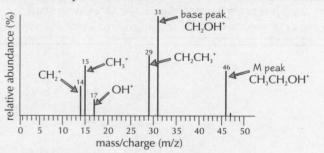

Fragment	Molecular Mass
CH_3	15
C_2H_5	29
C_3H_7	43
OH	17
CHO	29
COOH	45

Some common groups to look for:

The M_r of this molecule is 46.

To work out the structural formula, you've got to **guess** what **ions** could have made the other peaks from their m/z values. The m/z value of a peak matches the **mass** of the ion that made it (assuming it's got a 1+ charge).

For instance, this molecule's got a peak at 15 m/z, so it's likely to have a **CH_3** group. It's also got a peak at 17 m/z, so it's likely to have an **OH group**. Other ions are matched to the peaks on the spectrum above.

Once you think you know what the structure is, draw it out, and work out its **molecular mass**. It should be the same as the m/z value of the M peak.

Make sure you can find a fragment in the molecule that could make **every** peak in the spectrum.

ethanol's molecular formula
H H
| |
H—C—C—O—H
| |
H H

Mass Spectrometry

This page is only for OCR (Methods of Analysis and Detection Option) and Nuffield.

Use the **M+1 Peak** to Work Out the **Number** of **Carbon Atoms** in a Molecule

The peak **1 unit** to the right of the M peak is called the **M+1 peak**. It's mostly due to the **carbon-13 isotope** which exists naturally and makes up about 1.1% of carbon.

You can use the M+1 peak to find out **how many carbon atoms** there are in the molecule.
There's even a handy formula for this (as long as the molecule's not too huge):

$$\frac{\text{height of M+1 peak}}{\text{height of M peak}} \times 100 = \text{number of carbon atoms in compound}$$

EXAMPLE

If a molecule has a molecular peak with a relative abundance of 44.13%, and an M+1 peak with a relative abundance of 1.41%, how many carbon atoms does it contain?

$(1.41 \div 44.13) \times 100 = 3.195$ So the molecule contains 3 carbon atoms.

An **M+2 Peak** Tells Us the Compound Contains **Chlorine** or **Bromine**

If a molecule's got either **chlorine** or **bromine** in it you'll get an **M+2 peak** as well as an M peak and an M+1 peak.
Both chlorine and bromine have natural isotopes with different masses and they all show up on the spectrum.

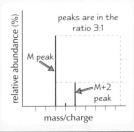

Chlorine's two isotopes, **Cl-35** and **Cl-37**, occur in the ratio 75:25, or **3:1**.

So if a molecule contains chlorine, it will give an **M peak** and an **M+2 peak** with heights in the ratio **3:1**.

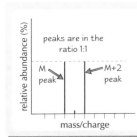

Bromine's got two isotopes, **Br-79** and **Br-81**, that occur in **equal amounts**. So if a molecule contains **bromine**, the M peak and M+2 peak will both have the **same** height.

If you spot an **M+4 peak** it's because the molecule contains **two** atoms of the **halogen**.
(This changes the ratios of the M and M+2 peaks though.)

Practice Questions

Q1 How can you work out the molecular mass from a mass spectrum?

Q2 How can you work out how many carbon atoms a molecule contains?

Q3 How can you spot that a molecule contains bromine?

Exam Question

1   On the right is the mass spectrum of an alkyl halide.
    Use the spectrum to answer this question.

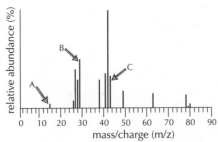

a)   What is the mass of most molecules of this alkyl halide?   [1 mark]

b)   Give the formulae of the fragments labelled A, B and C.   [3 marks]

c)   Does the molecule contain chlorine or bromine? Explain how you know.   [2 marks]

d)   Suggest a molecular formula for this molecule and draw its structure. Explain your suggestion.   [5 marks]

Mass spectrometry — Weight Watchers for molecules...

So mass spectrometry's a bit like weighing yourself, then taking bits off your body, weighing them separately, then trying to work out how they all fit together. Luckily you won't get anything as complicated as a body, and you won't need to cut yourself up either. Good news all round. Learn this page and watch out for the M+1 and M+2 peaks in the exam.

Infrared Spectroscopy

Eeek... more spectroscopy. Infrared (IR to its friends) radiation has less energy than visible light, and a longer wavelength.

Infrared Spectroscopy Lets You Identify Organic Molecules

Infrared spectroscopy produces **scary** looking graphs. Just learn the basics and you'll be fine.

1) In infrared (IR) spectroscopy, a beam of **IR radiation** goes through the sample.

2) The IR energy is absorbed by the **bonds** in the molecules, increasing their **vibrational** energy.

3) **Different bonds** absorb **different wavelengths**. Bonds in different **places** in a molecule absorb different wavelengths too — so the O–H group in an **alcohol** and the O–H in a **carboxylic acid** absorb different wavelengths.

4) This table shows what **frequencies** different groups absorb —

Functional group	Where it's found	Frequency/ Wavenumber (cm⁻¹)	Type of absorption
O–H	alcohols	3200 - 3750	strong, broad
O–H	carboxylic acids	2500 - 3300	medium, very broad
C–O	alcohols, carboxylic acids and esters	1100 - 1310	strong, sharp
C=O	aldehydes, ketones, carboxylic acids and esters	1680 - 1800	strong, sharp

This tells you what the trough on the graph will look like.

O–H groups tend to have broad absorptions — it's because they take part in hydrogen bonding.

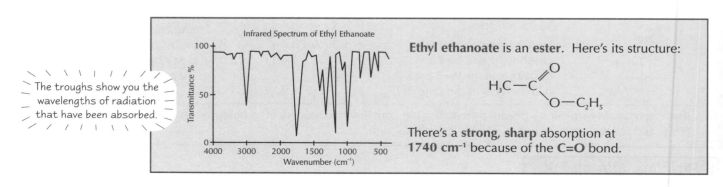

The troughs show you the wavelengths of radiation that have been absorbed.

Ethyl ethanoate is an **ester**. Here's its structure:

There's a **strong**, **sharp** absorption at **1740 cm⁻¹** because of the **C=O** bond.

The Fingerprint Region Identifies a Molecule

There's a region between **1000 cm⁻¹** and **1550 cm⁻¹** on the spectrum called the **fingerprint** region. It's **unique** to a **particular compound**. You can check this region of an unknown compound's IR spectrum against those of known compounds. If it **matches up** with one of them, hey presto — you know what the molecule is.

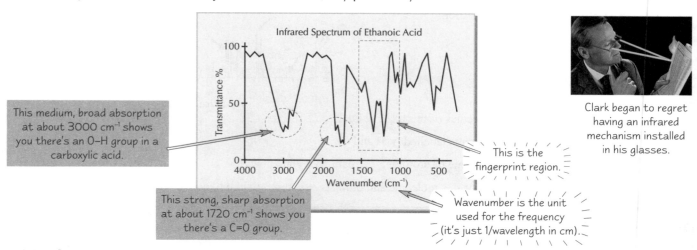

This medium, broad absorption at about 3000 cm⁻¹ shows you there's an O–H group in a carboxylic acid.

This strong, sharp absorption at about 1720 cm⁻¹ shows you there's a C=O group.

This is the fingerprint region.

Wavenumber is the unit used for the frequency (it's just 1/wavelength in cm).

Clark began to regret having an infrared mechanism installed in his glasses.

AQA only: The fingerprint region's got an extra use as well — if you know what molecule you've got, you can tell how pure it is. Any extra peaks in this region will be due to impurities.

Infrared Spectroscopy

This page is for Salters only.

Infrared Spectroscopy Helps Catch **Drunk Drivers**

If a person's suspected to be drink driving, they're **breathalysed**. A very quick test's done by the roadside — if it says the driver's over the limit, they're taken into a police station for a more **accurate test** using **IR spectroscopy**.

The **amount** of alcohol is found by measuring the **intensity** of **one** of the **troughs** in the IR spectrum. The **trough** used corresponds to a **C–H bond** — it's chosen because it's **not affected** by the **water vapour** in the suspect's breath.

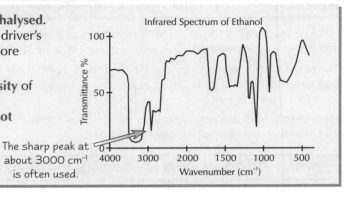

The sharp peak at about 3000 cm⁻¹ is often used.

Aspirin is made from **Salicylic Acid**

The ancient Greeks used white willow bark to reduce pain — the problem was it also caused vomiting. The painkilling ingredient was **salicylic acid**.

This is now used to make **aspirin**, which has fewer side effects.

Salicylic acid

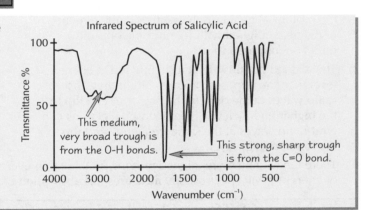

This medium, very broad trough is from the O–H bonds.

This strong, sharp trough is from the C=O bond.

Practice Questions

Q1 Which parts of a molecule absorb infrared radiation?

Q2 What group does a substance contain if its infrared spectrum has a strong, sharp peak at a wavelength of 1700 cm⁻¹?

Q3 What's the fingerprint region?

Exam Question

1 A molecule with a molecular mass of 74 gives the following IR spectrum.

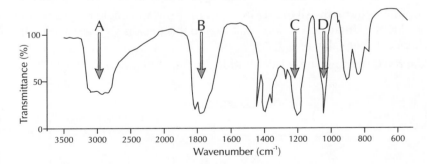

a) Which functional groups produce the troughs labelled A – D? [4 marks]

b) Suggest a molecular formula and name for this molecule. Explain your suggestion. [3 marks]

Ooooh — I'm picking up some good vibrations...

Now, I've warned you — infrared glasses are not for fun. They're highly advanced pieces of technology which if placed in the wrong hands could cause havoc and destruction across the Universe. There's not much to learn on these pages — so make sure you can apply it. You'll be given a data table, so you don't have to bother learning all the wavenumber ranges.

NMR Spectroscopy — The Basics

NMR isn't the easiest of things, so ingest this information one piece at a time — a bit like eating a bar of chocolate (but this isn't so yummy).

NMR Gives You Information About an Organic Molecule's Structure

Nuclear magnetic resonance (NMR) spectroscopy tells you **how many hydrogens** there are in an organic molecule and how they're **arranged**. This lets you work out the **structure** of the organic molecule.

The reason NMR works is that nuclei with **odd** numbers of nucleons (protons and neutrons) have a **spin**. This causes them to have a weak **magnetic field** — a bit like a bar magnet.

Hydrogen nuclei are single **protons**, so they have spin. It's these that are normally looked at in NMR.

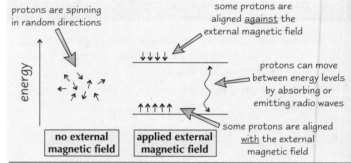

Protons Align in Two Directions in an External Magnetic Field

1) Normally, the protons (hydrogen nuclei) are spinning in **random directions** — so their magnetic fields **cancel out**.

2) But when a strong **external** magnetic field's applied, the protons align themselves either in the direction of the field (**aligned with it**), or in the opposite direction (**opposed to it**).

3) The aligned protons are at a slightly **lower** energy level than the opposed protons. But if they **absorb radio waves** of the right frequency, they can **flip** to the **higher** energy level. The opposed electrons can **emit** radio waves at the same frequency and **flip** to the **lower energy** level.

4) There tends to be more aligned protons, so there's an **overall absorption** of energy. NMR spectroscopy **measures** this **absorption** of energy.

protons are spinning in random directions

some protons are aligned <u>against</u> the external magnetic field

↓↓↓↓

protons can move between energy levels by absorbing or emitting radio waves

↑↑↑↑↑

some protons are aligned <u>with</u> the external magnetic field

no external magnetic field

applied external magnetic field

Protons with Different Environments Absorb Different Amounts of Energy

Protons are **shielded** from the external magnetic field by the surrounding electrons. The groups around a proton affect the amount of electron shielding — for example, the proton might be near something that withdraws electrons. So the **protons** in a molecule feel **different fields** depending on their **environment**. This means they absorb **different** amounts of energy, at **different frequencies**.

Chemical Shift is Measured Relative to Tetramethylsilane

So protons in different environments absorb energy of **different frequencies**. NMR spectroscopy measures these differences relative to a **standard substance** — the difference is called the **chemical shift (δ)**.

The standard substance is usually **tetramethylsilane** (**TMS**), $Si(CH_3)_4$. This molecule has 12 protons all with **identical environments**, so it only produces a **single** absorption peak — and this peak is well away from that of most molecules. Its single peak is given a chemical shift value of 0.

Spectra often show a peak at δ = 0 because some TMS is added to the test compound for calibration purposes.

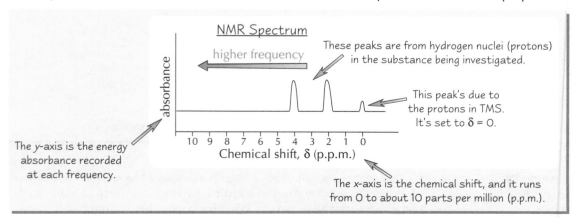

NMR Spectrum

higher frequency

These peaks are from hydrogen nuclei (protons) in the substance being investigated.

This peak's due to the protons in TMS. It's set to δ = 0.

The *y*-axis is the energy absorbance recorded at each frequency.

Chemical shift, δ (p.p.m.)

The *x*-axis is the chemical shift, and it runs from 0 to about 10 parts per million (p.p.m.).

NMR Spectroscopy — The Basics

Each peak on an NMR spectrum is due to one or more protons in a **particular environment**.
The **relative area** under each peak tells you how many protons are in that environment.

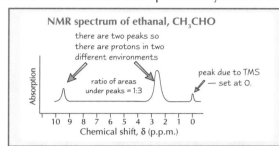

1) There are **two peaks** — so there are **two environments**.

2) The area ratio is **1:3** — so you know there's 1 proton in the environment at δ = 9.5 p.p.m. to every 3 protons in the other environment.

Use a **Table** to Identify the **Proton** Causing the **Chemical Shift**

You use a table like this to **identify** which functional group each peak is due to.

Don't worry — **you don't need to learn it.** You'll be given one in your exam, so use it. The copy you get in your exam may look a little different, and have different values — they depend on the solvent, temperature and concentration.

The chemical shift for R–OH is quite large — this is explained on the next page.

The example above shows that there's 1 hydrogen at δ = 9.5 p.p.m. — this must be due to an **R–CHO** group.
It also shows 3 hydrogens at δ = 1.3 p.p.m. — an **R–CH₃** group.

Chemical shift, δ (p.p.m.)	Type of proton
0.8 – 1.3	R – CH₃
1.2 – 1.4	R – CH₂ – R
2.0 – 2.9	R – COCH₃
3.2 – 3.7	halogen – CH₃
3.6 – 3.8	R – CH₂ – Cl
3.3 – 3.9	R – OCH₂ – R
3.3 – 4.0	R – CH₂OH
3.5 – 5.5	R – OH
7.3	◯ – OH
9.5 – 10	R – CHO
11.0 – 11.7	R – COOH

The protons that cause the shift are highlighted in red. R stands for any alkyl group.

Samples are Dissolved in **Solvents Without Lone Protons**

The sample's got to be dissolved in a solvent that has **no single protons** — or they'll show up as peaks on the spectrum and confuse things. **Deuterated solvents** are often used — their hydrogen atoms have been replaced by **deuterium**. Deuterium's an isotope of hydrogen that's got two nucleons (a proton and a neutron), so there's no overall spin on the nucleus. **CCl₄** can also be used.

Practice Questions

Q1 Which part of the electromagnetic spectrum is absorbed in NMR spectroscopy?
Q2 What happens to the protons when they absorb energy?
Q3 What's a chemical shift?
Q4 What is a deuterated solvent, and why is it used?

Exam Question

1 Look at the NMR spectrum of a primary alcohol.

a) Suggest a solvent that the alcohol might be dissolved in, and explain why this solvent can be used. [3 marks]

b) What are the approximate δ values for the three types of hydrogen atoms? [1 mark]

c) Suggest and explain a possible structure for the alcohol. [4 marks]

Protons are a bit like Kylie — they're spinning around...

The ideas behind NMR are difficult, so don't worry if you have to read these pages quite a few times before they make sense. If you're doing AQA or OCR, you've got to make sure you really understand the stuff on these two pages, cos there's more on the next two — and they aren't any easier. Keep bashing away at it though — you'll eventually go "aaah...I get it".

More NMR Spectroscopy

And now that you know the basics here's the really crunchy bit for you to get your teeth stuck in.
These two pages are only for AQA and OCR.

Spin-Spin Coupling Splits the Peaks in an NMR Spectrum

In high resolution NMR spectra, the peaks are **split** into **smaller peaks**. This is due to the **magnetic fields** of the **neighbouring single protons** interacting with each other, and is called **spin-spin coupling**. Only protons on **adjacent** carbon atoms affect each other.

These **multiple peaks** are called **multiplets**. They always split into the number of neighbouring protons plus one — it's called the **n + 1 rule**. For example, if there are **2 single protons** next door, the peak will be split into 2 + 1 = 3.

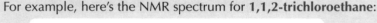

> You can work out the **number** of **neighbouring single protons** by looking at how many the peak splits into:
>
> If a peak's split into **two** (a **doublet**) then there's **one** neighbouring single proton.
> If a peak's split into **three** (a **triplet**) then there are **two** neighbouring single protons.
> If a peak's split into **four** (a **quartet**) then there are **three** neighbouring single protons.

For example, here's the NMR spectrum for **1,1,2-trichloroethane**:

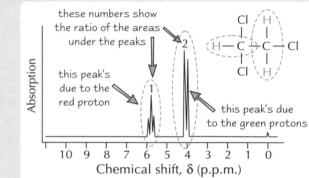

these numbers show the ratio of the areas under the peaks

this peak's due to the red proton

this peak's due to the green protons

The peak due to the green protons is split into **two** because there's **one proton** on the adjacent carbon atom.

The peak due to the red proton is split into **three** because there are **two protons** on the adjacent carbon atom.

When the peaks are split, it's not as easy to see the ratio of the **areas**, so an **integration trace** is often shown. The height increases are proportional to the areas.

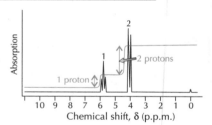

Labile Protons are Identified Using D₂O   *OCR only*

Protons attached to oxygen (OH) or nitrogen (NH) are called **labile protons** because they can be rapidly exchanged with neighbouring molecules. The chemical shift due to them is very variable — they could be involved in **hydrogen bonding** and they're affected lots by the **concentration** and **type** of **solvent**. They make quite a **broad** peak that isn't usually split.

There's a clever little trick chemists use to identify labile protons:

1) Run **two** spectrums of the molecule — one with a little **deuterium oxide**, D_2O, added.

2) If a labile proton is present it'll swap with deuterium and the peak will **disappear**.
   This is because deuterium's got no spin, as it's got an even number of nucleons.

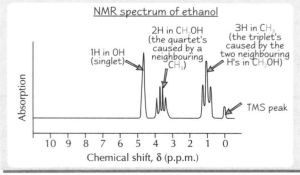

NMR spectrum of ethanol

1H in OH (singlet)

2H in CH₂OH (the quartet's caused by a neighbouring CH₃)

3H in CH₃ (the triplet's caused by the two neighbouring H's in CH₂OH)

TMS peak

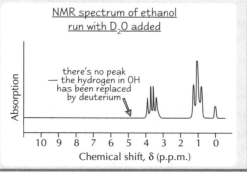

NMR spectrum of ethanol run with D₂O added

there's no peak — the hydrogen in OH has been replaced by deuterium

More NMR Spectroscopy

An NMR Spectrum Gives us a Lot of Information

From an NMR spectrum you can tell these things:

1) The **different proton environments** in the molecule (from the **chemical shifts**).
2) The **relative number** of **protons** in each environment (from the **relative peak area**).
3) The **number of protons adjacent** to a particular proton (from the **splitting pattern**).

Using all this information you can predict **possible structures**, and sometimes the actual structure.

EXAMPLE Using all the NMR spectrum below, predict the structure of the compound.

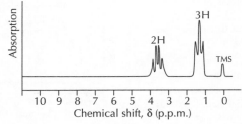

1) The peak at **δ = 2.5 p.p.m.** is likely to be due to an **R–COCH₃** group, and the peak at **δ = 9.5 p.p.m.** is likely to be due to an **R–CHO** group.

2) From the areas, there's one proton in the peak at **δ = 9.5 p.p.m.**, for every three in the peak at **δ = 2.5 p.p.m.**. This fits nicely with the first bit — so far so good.

3) The quartet's got **three** neighbouring protons, and the doublet's got **one** — so it's likely these two groups are next to each other.

Now you know the molecule's got to contain...

...all you need to do is fit them together.

Practice Questions

Q1 What causes the peaks to split?

Q2 What causes a triplet of peaks?

Q3 How can you get rid of a peak caused by an OH group? Explain why this works.

Exam Question

Q1 The NMR spectrum below is that of an alkyl halide. Use the table of chemical shifts on page 119 to answer this question.

a) What is the likely environment of the two protons with a shift of 3.6 p.p.m.?   [1 mark]

b) What is the likely environment of the three protons with a shift of 1.3 p.p.m.?   [1 mark]

c) The molecular mass of the molecule is 64. Suggest a possible structure and explain your suggestion.   [2 marks]

d) Explain the shapes of the two peaks.   [4 marks]

Never mind splitting peaks — this stuff's likely to cause splitting headaches...

Is your head spinning yet? I know mine is. Round and round like a merry-go-round. It's a hard life when you're tied to a desk trying to get NMR spectroscopy firmly fixed in your head. You must be looking quite peaky by now... so go on, learn this stuff, take the dog around the block, then come back and see if you can still remember it all.

Combined Spectral Analysis

Now you know all about the different types of spectroscopy, you've gotta put it all together.

Spectroscopy is all about **Absorbing** or **Emitting** **Electromagnetic Energy**

The spectroscopy techniques in this section all involve **energy** being **absorbed** or **emitted**.
Make sure you know **which region** of the **spectrum** is connected to which **type** of **spectroscopy**.

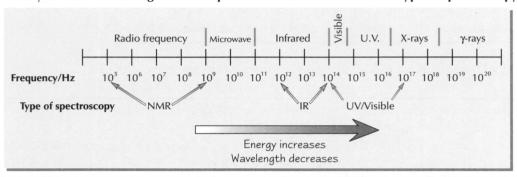

Collect as Much **Information** as You Can Before **Predicting** a **Structure**

It's a good idea to look at as much information as you can before you jump to conclusions and predict a structure.

EXAMPLE The following spectra are all of the same molecule. Deduce the molecule's structure.

The **mass spectrum** tells you the molecule's got a
mass of 44 and it's likely to contain a **CH₃ group**.

The **IR spectrum** strongly suggests a **C=O** bond
in an aldehyde, ketone, ester or carboxylic acid.

The **NMR spectrum** confirms that there's a
COCH₃ group and a **CHO** group.

*If you're doing AQA or OCR you'll
probably get a more complicated
NMR spectrum. See pages 120-121.*

So it looks like the CH₃ group is **right next to** the CHO group.

Putting all this together you get
a molecule with the structure: $H_3C-C{\overset{\displaystyle O}{\diagdown} \atop \diagup}_H$ which is the aldehyde **ethanal**.

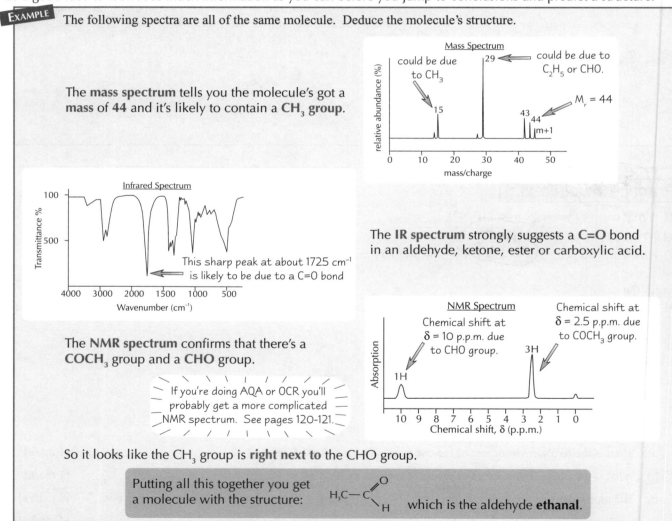

You Often Need **Further Evidence** to **Confirm** the Molecular Structure

You need to be totally sure you've got the **right** molecular structure. Here's what you can do to check:

1) Test its **melting** and **boiling points** — then **compare** them to what they're supposed to be for that molecule.

2) Test to see if it does the **chemical reactions** it's supposed to.

Combined Spectral Analysis

This page is just for Nuffield folk.

The **Empirical Formula** is the **Simplest** Ratio of the **Elements** in a Compound

There are **two ways** to work out an **empirical formula** — which method you use depends on the information you're given.

METHOD 1 — Elemental Percentage Composition

A compound contains 75% carbon and 25% hydrogen. Calculate its empirical formula.
It's easiest if you make a table like this one...

	Carbon	Hydrogen
Percentage	75	25
Divide by the element's atomic mass	75 ÷ 12 = **6.25**	25 ÷ 1 = **25**
Divide by the smallest number	6.25 ÷ 6.25 = **1**	25 ÷ 6.25 = **4**

The ratio of C : H is 6.25 : 25. You divide by the smallest of these numbers to simplify the ratio.

There's one carbon atom to 4 hydrogen atoms so the empirical formula is **CH₄**.

METHOD 2 — Combustion Analysis

Compound X is composed of carbon and hydrogen. 4 g of it completely combusts to give 9 g of water and 11 g of carbon dioxide. Calculate its empirical formula.

The molar mass of CO_2 is 44 g (12 + (16 × 2)).

1) First you have to calculate the **mass of carbon** in 4 g of compound X.
All the carbon in the CO_2 must have come from compound X.
There's 12 g of carbon in every 44 g of CO_2, so in 11 g of CO_2 there's **3 g** of carbon (11 × 12 ÷ 44).
This means there's **3 g** of carbon in 4 g of compound X.

2) Now calculate the **mass of hydrogen** in 4 g of compound X.
All the hydrogen in the water must have come from compound X. There's 2 g of H in every 18 g of H_2O,
so there's **1 g** of H in 9 g of H_2O (9 × 2 ÷ 18 = 1 g). This means there's **1 g** of hydrogen in 4 g of compound X.

Then all you need to do is put these values in the same table as above...

	Carbon	Hydrogen
Mass (g)	3	1
Divide by the element's A_r	3 ÷ 12 = **0.25**	1 ÷ 1 = **1**
Divide by the smallest number	0.25 ÷ 0.25 = **1**	1 ÷ 0.25 = **4**

There's one carbon atom to 4 hydrogen atoms, so the empirical formula is **CH₄**.

To work out the **molecular formula** from the empirical formula you need the compound's **molecular mass** — you could get this from a **mass spectrum**.

The **molecular mass** of substance Y is **30** and its **empirical formula** is **CH₃**.
CH₃ has a mass of 15, so the molecule must be made up of **two CH₃ units**. So the molecular formula is **C₂H₆**.

Practice Questions

Q1 What tests could you do to confirm structures suggested by spectral analysis?

Q2 What's meant by the term empirical formula?

Exam Question

1 Substance Q contains 54.55% C, 9.09% H and 36.36% O by mass. The molecular ion peak in the mass spectrum is at 88. The infrared spectrum has a strong sharp absorptions at 1740 cm⁻¹ and 1240 cm⁻¹.

a) Calculate the empirical and molecular formula of Q. [3 marks]

b) What information does the infrared spectrum data provide? [3 marks]

c) Draw two possible structures consistent with your answers to parts (a) and (b). [2 marks]

d) Explain one further piece of evidence that would help choose the correct structure. [1 mark]

Combined spectral analysis — it's what goes on at ghost hunter conventions...

Use all the information that's given to you. If the examiners included it, then it's likely to contain some information that'll help you figure out the structure. If you think you know what the structure is from a single piece of information, test whether it fits with the other bits. You never know — they might just reveal something really important.

Organic Synthesis

In your exam you may be asked to suggest a pathway for the synthesis of a particular molecule. These pages contain a summary of some of the reactions you should know.

Chemists use **Synthesis Routes** to Get from One Compound to Another

Chemists have got to be able to make one compound from another. It's vital for things like **designing medicines**. It's also good for making imitations of **useful natural substances** when the real things are hard to extract.

If you're asked how to make one compound from another in the exam, make sure you include:

1) any **special procedures**, such as refluxing.

2) the **conditions** needed, e.g. high temperature or pressure, or the presence of a catalyst.

3) any **safety** precautions, e.g. do it in a fume cupboard.

Before learning the vast number of reactions on these pages check which ones you need to know.

If there are things like hydrogen chloride or hydrogen cyanide around, you really don't want to go breathing them in. Stuff like bromine and strong acids and alkalis are corrosive, so you don't want to splash them on your bare skin.

Most of these reactions are covered elsewhere in the book, so look back for extra details.

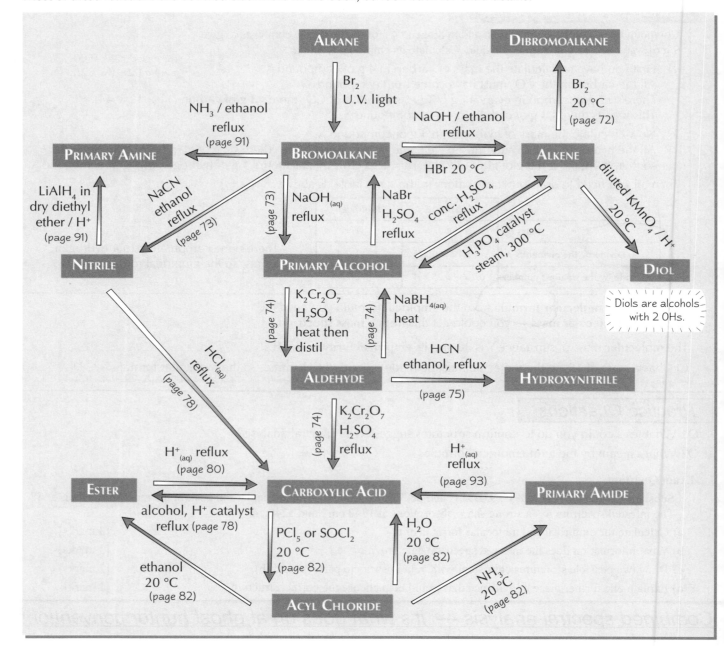

Organic Synthesis

Synthesis Route for Making **Aromatic Compounds**

There are not so many of these reactions to learn — so make sure you know all the nitty-bitty details.
If you can't remember any of the reactions, look back to the relevant pages and take a quick peek over them.

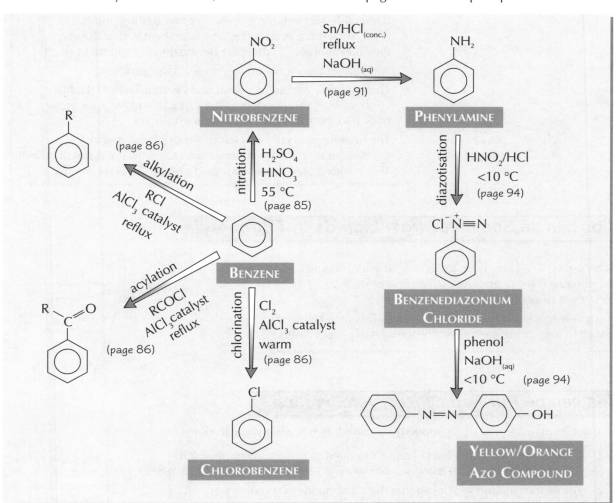

Practice Questions

Q1 How do you convert an ester to a carboxylic acid?

Q2 How do you make an aldehyde from a primary alcohol?

Q3 What do you produce if you reflux a primary amide with an acid?

Q4 How do you make an alkene from a primary alcohol?

Q5 How do you make phenylamine from benzene?

Exam Questions

1   Ethyl methanoate is one of the compounds responsible for the smell of raspberries.
    Outline, with reaction conditions, how it could be synthesised in the laboratory from methanol.   [7 marks]

2   How would you synthesise propanol starting with propane? State the reaction conditions and
    reagents needed for each step and any particular safety considerations.   [8 marks]

I saw a farmer turn a tractor into a field once — now that's impressive...

*There's loads of information here. Tons and tons of it. But you've covered pretty much all of it before, so it shouldn't be
too hard to make sure it's firmly embedded in your head. If it's not, you know what to do — go back over it again.
Then cover the diagrams up, and try to draw them out from memory. Keep going until you can do it perfectly.*

Practical Techniques

You can't call yourself a chemist unless you know these practical techniques. Unless your name's Boots.

Refluxing *Makes Sure You Don't Lose Any* **Volatile** *Organic Substances*

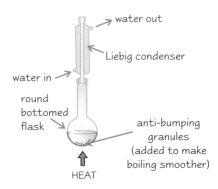

water out

Liebig condenser

water in

round bottomed flask

anti-bumping granules (added to make boiling smoother)

HEAT

Organic reactions are **slow** and the substances are usually **flammable** and **volatile** (they've got **low boiling points**). If you stick them in a beaker and heat them with a Bunsen they'll **evaporate** or **catch fire** before they have **time to react**.

You can **reflux** a reaction to get round this problem.

The mixture's **heated in a flask** fitted with a **vertical 'Liebig' condenser** — this condenses the vapours and **recycles** them back into the flask, giving them **time to react**.

The **heating** is usually **electrical** — hot plates, heating mantles, or electrically controlled water baths are normally used. This **avoids naked flames** that might ignite the compounds.

Solids *can be Separated from* **Liquids** *by* **Filtration**

This bit's easy — all you have to do is **pour** your mixture through some **filter paper**. Any solid bits get stuck, but liquid goes straight through.

It's faster using a Buchner flask and funnel. This uses **reduced pressure** to help bring the molecules through the paper faster.

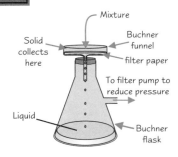

Mixture

Solid collects here

Buchner funnel

filter paper

To filter pump to reduce pressure

Liquid

Buchner flask

Solids *can be* **Purified** *by* **Recrystallisation**

If you've got a small amount of an **impurity** in a **solid**, here's what you can do:

1) **Dissolve** the solid in a hot solvent. You need to use a solvent in which the compound is much more soluble when it's **hot** than when it's cold. ◄
2) Allow the solution to cool so that the compound **recrystallises** from it.
3) **Filter** the mixture — the crystals will collect on the filter paper.
4) Wash the crystals with the **pure cold solvent** and allow them to **dry**.

You might have to filter this solution while it's hot to get rid of any insoluble impurities.

Melting *and* **Boiling** *Points are Good Indicators of* **Purity**

Most **pure substances** have a **specific melting** and **boiling point**. If they're **impure**, the **melting point's lowered** and the **boiling point's raised**. If they're **very impure** melting and boiling will occur across a wide range of temperatures.

Fractional Distillation *Separates Two or More Liquids*

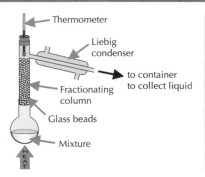

Thermometer

Liebig condenser

to container to collect liquid

Fractionating column

Glass beads

Mixture

Fractional distillation separates liquids with **different boiling points**. The liquid with the lowest boiling point will be distilled first.

The mixture's **heated** in the apparatus shown and the liquid in the flask boils. As the vapour molecules go up the **fractionating column**, they get **cooler**. If the temperature falls below their boiling point, the molecules **condense** and run back down through the glass beads. As the temperature increases, each liquid will reach the top of the column at a different time, in order of their boiling points.

The next page describes exactly how this works.

Practical Techniques

Fractional Distillation **Separates** Components by their **Boiling Points**

This bit's just for Edexcel — the rest of you remember the questions, then you're done.

1) Imagine there's a mixture of **two different liquids**, ethanol and water. Pure ethanol has a boiling point of 78 °C and pure water has a boiling point of 100 °C (at atmospheric pressure). The boiling point of a mixture of ethanol and water will be somewhere between 78 °C and 100 °C, depending on **how much of each** is in the mixture.

If you plot the **boiling points** for all the different **possible compositions**, you get a graph like the one on the right.

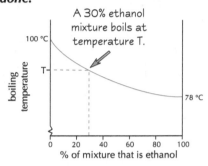

A 30% ethanol mixture boils at temperature T.

2) At the mixture's boiling point there's an **equilibrium** between the liquid and the vapour.

The liquid and the vapour each have a **different composition**. The vapour will contain a greater proportion of ethanol than the liquid, as ethanol's **more volatile**.

You can draw another line on the graph showing the vapour compositions at different boiling temperatures.

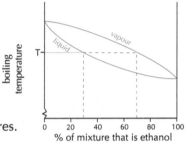

A 30% ethanol mixture at its boiling point has a vapour which is 70% ethanol.

3) The vapour **rises** through the fractionating column and the temperature drops to T_1. This causes the vapour to **condense** back to a liquid.
But as the vapour contains a **greater proportion** of ethanol than the original mixture, it'll condense into a liquid that's also **richer** in ethanol.

At this new temperature (T_1) **another equilibrium** occurs. This time the vapour is even richer in ethanol.

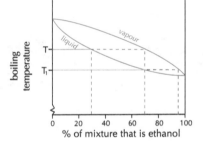

A 70% mixture at its boiling point has a vapour which is 96% ethanol.

4) This keeps happening all the way up the fractionating column. Each new equilibrium is further up the column and contains **purer ethanol**. By the time the vapour reaches the top of the column it will be **pure** ethanol and you can condense it into a different flask.

Practice Questions

Q1 Why is refluxing needed in many organic reactions?

Q2 Why is the melting point helpful in deciding the purity of a substance?

Q3 Draw the apparatus needed for fractional distillation.

Q4 Why is electrical heating often used in organic chemistry?

Exam Question

1 Two samples of stearic acid boil at 69 °C and 64 °C respectively. Stearic acid dissolves in propanone but not in water.

a) Explain which sample is purer. [2 marks]

b) How could the impure sample be purified? [5 marks]

c) How could the sample from b) be tested for purity? [1 mark]

There's just a **fraction** too much information on these pages for me... *boom boom*

And that, my friends, is what chemistry is all about — playing with funny looking pieces of glass and making crystals in pretty colours. Unless you're doing biochemistry, you're definitely on the home straight — only two pages to go. So take a deep breath, turn the page over, and you'll be finished in no time.

The Chemical Industry

These two pages are for Salters only.

Ever wondered what the two main production processes in the chemical industry are? Oh yes you have... read on.

New Products are made on a **Small Scale** First

If you jumped straight in and built a massive chemical plant to make your product and then discovered it didn't work, you'd have wasted a huge amount of money. To avoid expensive mistakes, companies **start small** and then **scale up**.

- Do lots of research.
- Work on a small scale using gram quantities of reactants.

→

- Build a small pilot plant.
- Work on a medium scale using kilogram quantities.

→

- Large scale production in full size plant making tonnes of the stuff.

Construction costs **are a Major Part of the Overall Cost**

Companies try to build their plants using cheap materials like **mild steel** whenever possible. But if the reaction's **corrosive**, they have to use **inert** substances like plastic, stainless steel or ceramic/glass-lined steel. They'll need to think about whether there's going to be **high pressures or temperatures** too.

Pod didn't believe in starting small and constructed an entire universe from Blu-Tack and cheese.

Raw Materials are Converted into **Feedstocks**

Raw materials are the naturally occurring substances which are needed, e.g. crude oil, natural gas, air and limestone. They usually have to be purified first or converted into the **feedstocks** (the actual reactants needed for the process).

For example, to make **ammonia** in the **Haber process**, the **feedstocks** are nitrogen and hydrogen gas. You get these from the **raw materials** air, water and natural gas.

Production Processes are either **Batch** or **Continuous**

	Continuous	**Batch**
What is it?	The reactants continually enter the vessel and the products leave continuously. The reaction doesn't need to be stopped.	The reactants enter the vessel, react and make the product. The product is removed and the vessel cleaned and then used again.
Advantages	Lower labour costs as the process can be easily automated. Large quantities of product can be made non-stop. Less variation in quality.	Small quantities can be made. The reaction vessel can be used to make other products.
Disadvantages	More expensive to build. More expensive to run unless the plant runs at full capacity.	More labour intensive as emptying and cleaning needed. Contamination can occur if cleaning is not thorough.
Examples	Haber process (see page 25); Contact process for making sulphuric acid; blast furnace; making industrial ethanol.	Dye manufacture; aspirin and paracetamol manufacture; steel making.

The Yield and Rate must be **Optimised** by choosing the **Right Conditions**

Time is money in industry, so **as much** product as possible needs to be made as **fast** as possible. Often a **compromise** has to be found between the **amount** of product made and the **time taken** to produce it. **Temperature and pressure** must be chosen carefully, especially for reversible reactions, so that you get a good yield in a short time. A **catalyst** can also be used to speed up the reaction. For example, the **Haber process** uses 200 atm, 450 °C and an iron catalyst to get a 15% yield very quickly (see page 25).

The Chemical Industry

Engineers make use of *Waste Products* to *Reduce Costs*

1) A **co-product** is an expected and unavoidable **second product** — you know you're going to get it because it's in the reaction equation. Where possible, it's **sold** or extracted and used in **another reaction**.

2) A **by-product** is a product of an **unwanted side-reaction** — you usually only get a bit of it. The conditions are normally chosen to try to **minimise** the amount of by-product formed.

3) After the reaction:
   - any co-products and by-products have to be **separated** from the product and removed.
   - any unreacted feedstock is **recycled**, rather than wasted.
   - the waste, known as **waste effluent**, has to be carefully disposed of. There are **legal requirements** about this and heavy fines can be imposed.

4) If the reaction is **exothermic**, there may be waste heat. Engineers try to make use of this heat because it saves money and is better for the environment. Heat exchangers can use excess heat to produce steam or hot water for other reactions.

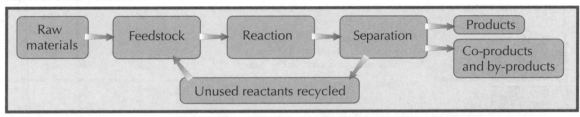

Fixed Costs are Unavoidable, but Variable Costs can be Reduced

Industrial reactions are done to make **money**. But companies start running up bills before they even start making the product. For example, there's research and development to pay for, the cost of having the plant designed and built, the cost of the land to build it on... When they start making the product they'll have to pay for things like raw materials, power, labour costs, and so on.

Some costs are **FIXED** costs — these have to be paid no matter how much product is made. They're things like the cost of designing and building the plant, business rates, and sometimes labour costs.

Some costs can be **VARIABLE** and depend on how much product is made. These include the cost of raw materials and waste effluent treatment.

Companies try to **cut costs** by siting the plant close to the raw materials or near transport links, such as roads or railways. This makes it easier and cheaper to bring the raw materials in and transport the product away. They also need to think about where they're going to get rid of their waste and whether there is a workforce with the right skills available close by.

Practice Questions

Q1 What's a feedstock?

Q2 Give two advantages of a continuous process.

Q3 Give two advantages of a batch process.

Q4 What is the difference between a co-product and a by-product?

Exam Question

1 The petrochemical industry produces large quantities of ethene by catalytic cracking of naphtha with steam in a continuous process.

   a) Give two factors which must be taken into account when siting a new plant. [2 marks]
   b) Give one fixed cost and one variable cost of a cracking plant. [2 marks]
   c) Explain why a 500,000 tonne capacity plant is likely to become uneconomical if it only produces 250,000 tonnes per year. [3 marks]
   d) Give two advantages of using a continuous process over a batch process. [2 marks]

My reaction went wrong and I accidentally produced a bi-plane...

An exam question on the chemical industry is likely to ask something about batch and continuous chemical processes, so they're definitely worth committing to your long-term memory. Remember too that a compromise needs to be found between the yield and the rate of the reaction. The chemical industry isn't there for fun — it's there to make cash. Make sure you understand how waste products can be reused to improve efficiency, and learn a few of the fixed and variable costs that firms incur. I know that's a lot of learnin', but it's the only way to be sure of exam success.

Carbohydrates

Carbohydrates are sugars. They're really important as they're the main energy source in living organisms.
More importantly — they sweeten up your cup of tea. **This section's just for people doing the OCR Biochemistry Option**

Monosaccharides *are the Simplest Type of* Carbohydrate

Carbohydrates contain carbon, hydrogen and oxygen, and have the general formula $C_x(H_2O)_y$. ← *For example, glucose is $C_6H_{12}O_6$.*
They're divided into three types — **monosaccharides**, **disaccharides** and **polysaccharides**.
Monosaccharides can form **straight chain** or **ring** structures:

Straight Chain Structures

A **straight chain monosaccharide** is just a **carbon chain** with a **hydroxyl group** attached to each carbon except for one. This carbon has a **carbonyl group** (C=O) attached to it instead.

Ribose has five carbon atoms, so it's called a **pentose**.
Glucose and **fructose** both have six carbons, so they're **hexoses** — the difference between these two is the position of the carbonyl group.

There are two optical isomers of monosaccharides, labelled D and L. Most natural sugars exist in the D form.

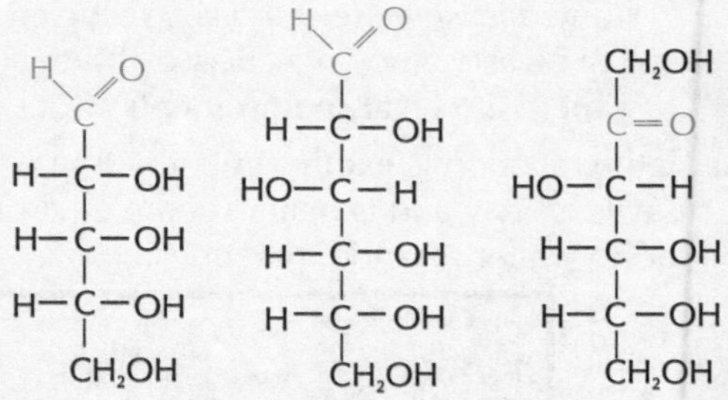

D-ribose
(a pentose)

D-glucose
(a hexose)

D-fructose
(a hexose)

Ring Structures

Glucose forms a ring structure called a **pyranose**.

There are two isomers of glucose, and you've got to know which is which. In **α-glucose**, the –OH group on carbon 1 is drawn **beneath** the ring. In **β-glucose**, the –OH group is drawn **above** the ring.

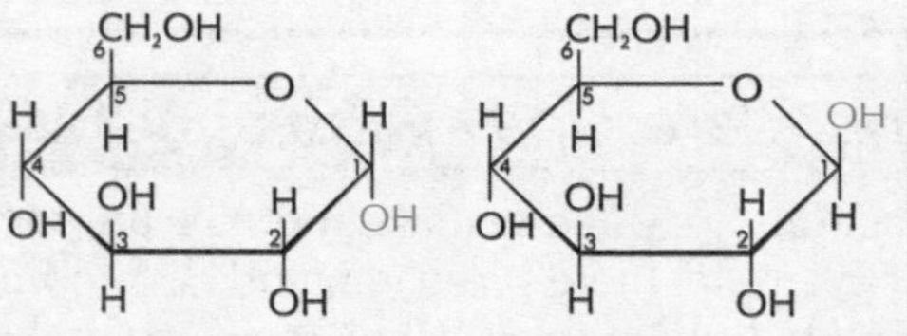

α-D-glucose

β-D-glucose

Disaccharides *Form when* Two *Monosaccharides React Together*

When a disaccharide forms, a water molecule's **lost** — so it's a **condensation** reaction.

The disaccharide **maltose** is made when two molecules of **α-D-glucose** combine. The bond's called an **α-1,4-glycosidic link**, because it forms between carbon **1** of one molecule and carbon **4** of another.

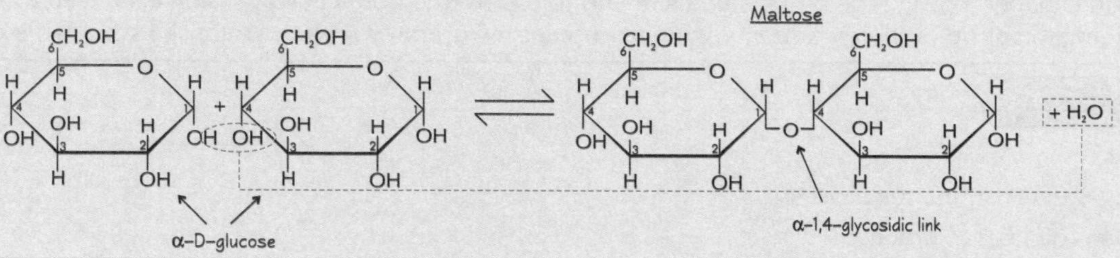

Maltose

α–D-glucose

α–1,4–glycosidic link

+ H_2O

Cellobiose is formed when two molecules of **β-D-glucose** combine. The bond is called a **β-1,4-glycosidic link**.

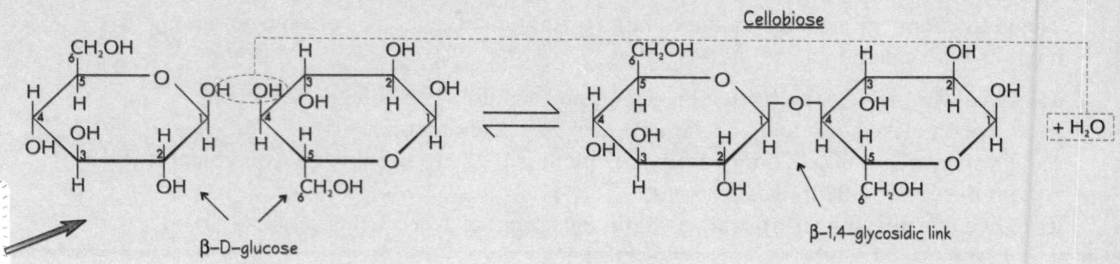

Cellobiose

One of the glucose molecules is turned upside down.

β–D-glucose

β–1,4–glycosidic link

+ H_2O

You can reverse these condensation reactions by **hydrolysis**. You need to use either an **enzyme** and water or **a hot aqueous acid**.

Monosaccharides and many disaccharides **dissolve easily** in water due to loads of **hydrogen bonding** between their –OH groups and those in water. But they won't dissolve in non-polar solvents — no way José.

Carbohydrates

Starch and *Glycogen* are *Polysaccharides...*

Polysaccharides are lots of sugar molecules joined together — they're **condensation polymers** of monosaccharides. **Starch** and **glycogen** are both condensation polymers of **α-D-glucose.**

Starch is a mixture of two different polymers, **amylose** and **amylopectin.**

Amylose is a long unbranched chain in which the monomers are joined by **α-1,4-glycosidic links**. It's like a lot of maltose units joined together.

Amylopectin has lots of amylose-like chains in a branched structure. The branches are every 29-30 glucose units and they're formed by **α-1,6-glycosidic links.**

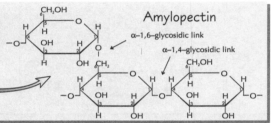

Glycogen has a similar structure to amylopectin, but is even more **branched** — the branches are every 8-12 glucose units.

Starch **stores energy** in **plants**, and glycogen **stores energy** in **animals**. They're great for this because:

1) They're **insoluble** due to their huge molecular mass, so they stay intact in the cell.
   Their insolubility also means that they don't draw water into the cell by **osmosis.**

2) They're **easily hydrolysed** into soluble monosaccharides by **enzymes.**
   The monosaccharides can be **transported** to cells where they **release energy.**

Lucy was once a typical, happy child. But her mother always said that sweets would be the ruin of her.

...and so is Cellulose

Cellulose is a condensation polymer of β–D–glucose. It's like lots of **cellobiose molecules** joined together.

It's **not branched**, so it has a **linear** structure. The chains lie **parallel** to each other, held together by **hydrogen bonds**. This makes it **strong** and **stable**, meaning it's ideal for its use as a **structural material** in plant cell walls.

Cellulose is **insoluble**, like starch and glycogen, but water can get between the bundles of chains into the cell.

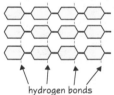

hydrogen bonds

Practice Questions

Q1 Draw diagrams of α–D–glucose and β–D–glucose.

Q2 Name the polysaccharides that are condensation polymers of α–D–glucose.

Q3 Explain how the structure of cellulose is related to its function in the plant cell.

Exam Questions

1 The disaccharide maltose contains two α–D–glucose units.
   a) Deduce the molecular formula of maltose. [1 mark]
   b) What is the name of the link between the α–D–glucose units? [2 marks]
   c) Explain how you would convert maltose back into α–D–glucose. [2 marks]

2 Starch, glycogen and cellulose are all polysaccharides.
   a) Name the two polymers found in starch and briefly outline the structural difference between them. [4 marks]
   b) State the roles of the following;
      i) starch in plants,
      ii) cellulose in plants,
      iii) glycogen in animals. [3 marks]

A little something to sweeten up the page...

Mary Poppins knew the importance of these molecules, and after these pages I reckon you probably will too. Think "beat her up" to remember which way the –OH points on the carbon 1 atom in β-glucose. Another way to look at it is that the β form has –OH groups alternately pointing up and down. Now, where did I put that cup of tea...

Lipids and Membrane Structures

Wow... a whole two pages all about fat. It's not about how to lose those extra pounds though, or how to turn that love handle into a rippling muscle, but all about the molecules and what they do. Read on...

Fatty Acids are Carboxylic Acids

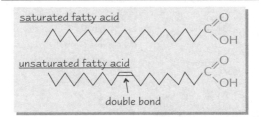

saturated fatty acid

unsaturated fatty acid

double bond

Fatty acids have a long hydrocarbon chain with a **carboxylic acid** group at the end. If the hydrocarbon chain contains **no double bonds** then the fatty acid is **saturated**, but if it contains one or more double bonds then it's **unsaturated**.

Fatty acids can also be written like this ⟶ where 'R' is a hydrocarbon chain.

See page 80 for more on esters.

Triglycerides are Esters of Glycerol with Fatty Acids

Three fatty acids react with glycerol to form **triglyceryl esters** (or triglycerides).

This is called **esterification** — because **ester linkages** are formed. Water's eliminated, so it's a **condensation** reaction too.

Triglycerides are **lipids**.

Glycerol   3 x Fatty Acid   Triglyceride

esterification / condensation

$+ 3H_2O$

When triglycerides are **hydrolysed** they release fatty acids.

If they're fully hydrolysed, glycerol and three fatty acids are formed. But sometimes not all the ester bonds are broken, so you also get **monoglycerides** (with one fatty acid attached) and **diglycerides** (with two fatty acids attached).

Triglyceride   Monoglyceride   Fatty Acids

Hydrolysis (e.g. using lipase)

not all the ester bonds have broken

Lipases (fat-digesting enzymes), certain micro-organisms or hot aqueous acid or alkali can be used to increase the rate of hydrolysis.

Soap is a sodium or potassium salt of a fatty acid. It's made in a process called **saponification.**

You need to heat a triglyceride with an alkali to get a salt of a fatty acid.

Triglyceride

$+ 3KOH$ heat

A potassium salt of a fatty acid. This is soap.

Glycerol

Triglycerides are non-polar, so they're only **soluble** in **non-polar** solvents. See page 7 for why.

Lipids are a Better Energy Store than Carbohydrates

Lipids can release more than **twice** as much energy per gram as carbohydrates. Here's why:

1) To release energy, carbohydrates and lipids need to be **oxidised**. This involves adding **oxygen** to form strong C=O and O–H bonds. Energy is released when these bonds form.

2) There are hardly any oxygen atoms in lipids — in fact they can pretty much be represented as $(CH_2)_n$. So most of a **lipid's mass** is due to **carbon**. Lots of C=O bonds can be formed and lots of **energy released**.

3) Carbohydrates $(CH_2O)_n$ are already **partially oxidised** — they contain O–H, C=O and C–O bonds. This makes them good for **instant access** to energy, but as they're already halfway to being completely oxidised, they don't release that much energy.

Not only are lipids a great energy store, but they're insoluble in water — which streans they don't draw water into body cells by osmosis. Also, when lipids form fat stores, they **protect** internal organs and are useful **heat insulators**. They're basically handy substances to have around.

Lipids and Membrane Structures

Phosphoglycerides are a Major Component of **Cell Membranes**

Phosphoglycerides (phospholipids) look similar to triglycerides, but one of the fatty acids is replaced by a **phosphate** group. The phosphate group is **polar**, so phosphoglycerides have different chemical properties to triglycerides.

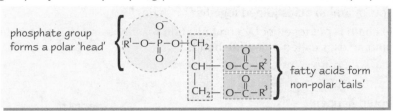

phosphate group forms a polar 'head'

fatty acids form non-polar 'tails'

Phosphoglycerides form **bimolecular layers** (or bilayers) in cells.

1) The polar heads are **hydrophilic**, but the non-polar tails are **hydrophobic**. So when they're put in water they end up with their heads in the water and their tails sticking out.

— Hydrophilic means attracted to water.
— Hydrophobic means repelled by water.

hydrophobic 'tails' in air

hydrophilic 'heads' in water

2) **Bimolecular layers** form when thousands of phosphoglycerides are in a watery environment. They arrange themselves to form a **double layer**, so that all of their tails are shielded from water. **Cell membranes** are bimolecular layers.

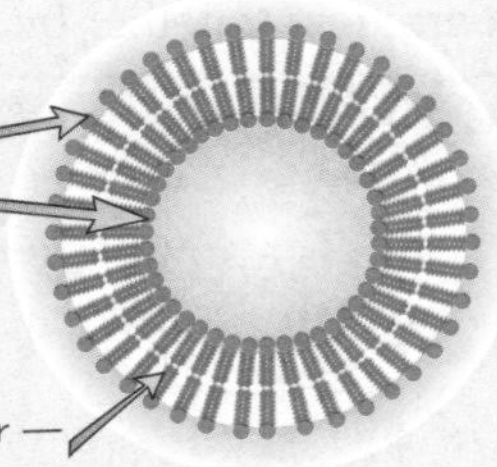

The hydrophilic heads are attracted to water.

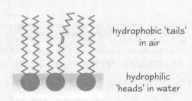

The hydrophobic tails are shielded from water — they're held together by van der Waals forces.

Practice Questions

Q1 Give **three** biological functions of lipids.

Q2 Explain why triglycerides are soluble in non-polar solvents.

Q3 What type of reaction is it when a triglyceride forms from glycerol and fatty acids?

Exam Questions

1 Triglycerides and phosphoglycerides are common in biological systems.

   a) Describe **one** similarity and **one** difference in the structure of triglycerides and phosphoglycerides.   [2 marks]

   b) Explain why the structure of a phosphoglyceride is suited to its role in the formation of bilayers in cells. Include a diagram.   [5 marks]

2 Tristearin is a triglyceride. It has a molecular formula of $C_{57}H_{110}O_6$.

   a) Tristearin can be hydrolysed.
     i) Suggest a way to **increase** the rate of hydrolysis of tristearin.   [1 mark]
     ii) Name the two different types of molecules that form when tristearin is fully hydrolysed.   [2 marks]

   b) Calculate the percentage of oxygen by mass in tristearin.   [2 marks]

   c) Explain whether triglycerides or carbohydrates release more energy per gram on oxidation.   [3 marks]

A little bit of grease to get those brain cogs moving...

The names of these lipids can be tricky, and what's worse, chemists and biologists use different names. Remember phosphoglycerides are often called phospholipids, and it's this type of fat that's found in cell membranes. Now go get yourself a nice, big, cream cake — with extra cream, and chocolate, and more cream. Yummy.

Proteins

You've already covered a lot about amino acids and proteins on pages 96–97.
Here's a bit of extra stuff you need to know for the OCR Biochemistry Option.

Proteins are made from Long Chains of Amino Acids

Proteins are formed from lots of **amino acids** joined together in chains.

Just like carbohydrates, the chain is put together by **condensation** reactions and broken apart by **hydrolysis** reactions. The bond between the amino acids is called a **peptide bond**.

Here's a recap of how two amino acids join together.

Loads of these reactions would happen to join more and more amino acids together to form a protein.

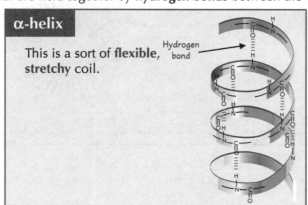

α-helices and β-pleated Sheets make up the Secondary Structure of Proteins

Proteins have a **primary**, **secondary**, **tertiary** and **quaternary** structure. The first three are covered on page 97.

There's a little more you need to know about the **secondary** structure, and this time you also need to know about the **quaternary** structure. Make sure you also know the types of **bonds** holding the protein's structure together.

Proteins have two main **secondary structures** — the **α-helix** and the **β-pleated sheet**.
Both are held together by **hydrogen bonds** between the C=O in one **peptide link** and the N–H in another.

α-helix	β-pleated sheet
This is a sort of **flexible**, **stretchy** coil.	This is where different parts of the same polypeptide chain, or separate polypeptide chains, **line up** side by side. It's flexible, but unlike the α-helix it's not **stretchy**.

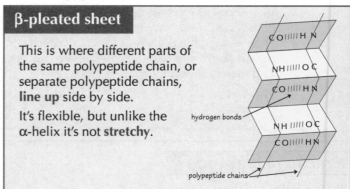

The Quaternary Structure is the Arrangement of Different Polypeptides

Many proteins are made up of two or more polypeptide chains. The **quaternary** structure is how they are arranged.
1) The quaternary structure is held together by the same forces of attraction as the tertiary structure
   — **van der Waals forces**, **ionic interactions**, **hydrogen bonds** and **disulphide bridges**.
2) Any **non-protein molecules** that join to the polypeptides are also part of the quaternary structure.

Haemoglobin Transports Oxygen around the Body

Haemoglobin's a good example of the **quaternary structure** of a protein —

Haemoglobin is the protein in red blood cells that carries oxygen. It has **four** polypeptide chains, two each of two different types. Each polypeptide chain has one **heam** group — a non-protein molecule containing iron (Fe^{2+}).

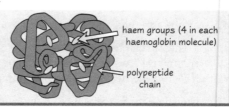

It's the **iron** that allows haemoglobin to carry oxygen. What happens goes like this:
1) Oxygen from your lungs moves into your blood where it combines **reversibly** with Fe^{2+}, in the haem groups. The O_2 molecule uses its lone pair to form a **weak**, **coordinate** bond to the Fe^{2+} (O_2 is acting as a **ligand** — see page 62). Haemoglobin with O_2 attached is called **oxyhaemoglobin** and is red. Each haemoglobin molecule contains 4 haem units and so 4 Fe^{2+} ions. So each haemoglobin molecule can carry four O_2 molecules.
2) When the O_2 is released to the cells, it's replaced by a water ligand to form **deoxyhaemoglobin**. This complex is blue.

Proteins

Proteins can be *Denatured* by *Heavy Metal Ions*, *Heat* or *pH Changes*

The 3D structure of a protein is only held together by weak forces of attraction, like hydrogen bonds or van der Waals forces. These are easily **broken**, and this causes a change in the 3D shape — this is called **denaturation**.
Heavy metal ions, **high temperatures** and **pH changes** all denature proteins:

HEAVY METAL IONS, e.g. Hg²⁺ and Ag⁺

These ions combine with –COO⁻ side groups and disrupt the **ionic bonds** in the tertiary structure. The –COO⁻ group uses its lone pair to form a coordinate bond with the metal ion. Heavy metal ions also affect –SH groups.

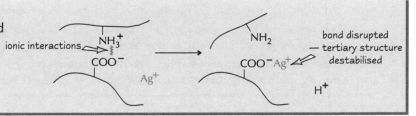

HIGH TEMPERATURE

High temperatures cause the atoms of the protein to vibrate more. **Hydrogen bonds**, **ionic bonds** and **van der Waals** interactions are all overcome.

When heating egg white, the colour changes from transparent to white — it's the protein being denatured.

CHANGES IN pH

Low pH — lots of H⁺ ions about

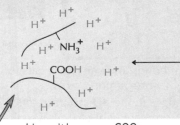

H⁺ ions combine with some –COO⁻ groups to form –COOH. This breaks ionic bonds, so the protein is denatured.

Normal conditions

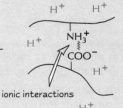

ionic interactions

High pH — hardly any H⁺ ions about

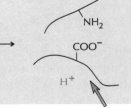

–NH₃⁺ groups lose H⁺ ions into the solution to form –NH₂. This breaks ionic bonds and the protein is denatured.

Practice Questions

Q1 What are the two main secondary structures of proteins?

Q2 How many polypeptide chains make up haemoglobin?

Q3 Give three ways that a protein can be denatured.

Exam Questions

1 There are two main types of secondary protein structure, the α-helix and the β-pleated sheet.

   a) Name the bond/force that holds the secondary protein structure in the correct shape.   [1 mark]

   b) Draw a diagram to show a β-pleated sheet. On your diagram show the bonds that hold the structure together.   [2 marks]

2 Haemoglobin is responsible for the transport of oxygen around the body

   a) Explain the term quaternary structure, using haemoglobin as an example.   [4 marks]

   b) Describe the role of Fe²⁺ in the transport of oxygen.   [3 marks]

Metallica is enough to make your eggs go rotten...

Now if you haven't done so already, go back to pages 96 and 97 and read through the rest of the stuff on proteins. Make sure you know that the tertiary structure is stabilised by bonds between the side groups of amino acids, whilst the secondary structure is stabilised by hydrogen bonding between the residues in the C=O and N–H groups in the peptide bonds.

Enzymes

Enzymes are very important proteins. The basics about them were covered on page 13 — these two pages top you up to what you need to know for the OCR Biochemistry Option. Now repeat after me... enzymes are proteins...

Enzymes *are* Biological Catalysts

Enzymes speed up chemical reactions by acting as biological catalysts.

1) They catalyse every **metabolic reaction** in the bodies of living organisms.

2) Enzymes are **proteins**. Some also have **non-protein components**.

3) Every enzyme has an area called its **active site**. This is the part that the substrate fits into so that it can interact with the enzyme.

Substrates are the molecules that enzymes act on to speed up reactions.

4) Enzymes have an extremely **high activity** — often a million or more reactions per second — wow. They're loads **more efficient** than inorganic catalysts, so they're used to speed up reactions in industry.

Enzymes *have* High Specificity

Enzymes are a bit picky. They only work with **specific substrates** — usually only one. This is because, for the enzyme to work, the substrate has to **fit** into the **active site**. If the substrate's shape doesn't match the active site's shape, then the reaction won't be catalysed. This is called the '**lock and key**' model.

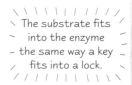
The substrate fits into the enzyme the same way a key fits into a lock.

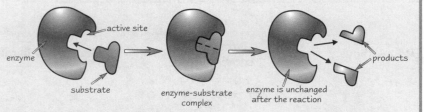

enzyme — *active site* — *substrate* — *enzyme-substrate complex* — *enzyme is unchanged after the reaction* — *products*

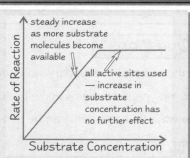
Kevin wondered if he'd ever find the right key.

The substrate is held in the active site by **temporary** bonds such as hydrogen bonds and van der Waals forces. These temporary bonds form between the substrate and "R" groups of the enzyme's amino acids.

Enzymes are **denatured** by the same things as proteins (see page 135). Changes in the **3D structure** affect the **shape** of the **active site**, so the substrate no longer fits. This prevents the enzyme from being able to catalyse the reaction.

Enzyme Activity is Affected by *Substrate Concentration* up to a Point

Substrate concentration affects the rate of reaction up to a certain point. The higher the substrate concentration, the faster the reaction, but only up until a **'saturation' point**. After that, there's so many substrate molecules that the enzymes have about as much as they can cope with, and adding more **makes no difference**.

Rate of Reaction
steady increase as more substrate molecules become available
all active sites used — increase in substrate concentration has no further effect
Substrate Concentration

Enzyme Concentration *Affects the Rate of Reaction*

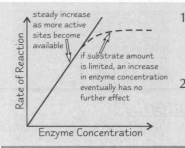
Rate of Reaction
steady increase as more active sites become available
if substrate amount is limited, an increase in enzyme concentration eventually has no further effect
Enzyme Concentration

1) The **more enzyme molecules** there are in a solution, the more likely a substrate molecule is to **collide** with one. So increasing the concentration of enzyme **increases** the rate of reaction.

2) But if the amount of substrate is limited, there comes a point when there's more than enough enzyme to deal with all the available substrate, so adding more enzyme has **no further effect**.

Enzymes

Immobilised Enzymes have many Advantages

Immobilised enzymes are **trapped** in a non-reactive, insoluble material, such as a **fibrous polymer mesh.**

This has the **advantages** that:

1) The enzymes aren't mixed in with the product, so it's easy to recover them and **reuse** them. This keeps **production costs** down.

2) The product isn't **contaminated** by the enzyme, and there tends to be fewer unwanted side reactions.

3) Immobilised enzymes are more **pH-** and **heat-stable**.

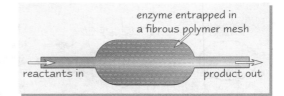

enzyme entrapped in a fibrous polymer mesh

reactants in   product out

The **confectionery industry** uses **immobilised glucose isomerase** to convert glucose to sweeter-tasting fructose.

Inhibitors Slow Down the Rate of Reaction

There are two types of enzyme inhibition — make sure you can tell the difference between the two.

COMPETITIVE INHIBITORS

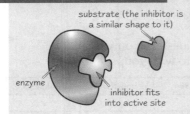

substrate (the inhibitor is a similar shape to it)

enzyme

inhibitor fits into active site

Competitive inhibitors have a **similar shape to the substrate**. They compete with the substrate to bond to the active site, but no reaction follows. Instead they **block** the active site, so **no substrate** can **fit** in it.

How much inhibition happens depends on the **relative concentrations** of inhibitor and substrate — if there's a lot of the inhibitor, it'll take up most of the active sites and very little substrate will be able to get to the enzyme. The amount of inhibition is also affected by how **strongly** the inhibitor bonds to the active site.

Competitive inhibition is **reversible**.

NON-COMPETITIVE INHIBITORS

Non-competitive inhibitors bond to the enzyme **away from its active site**, but this causes the active site to **change shape**. They don't 'compete' with the substrate because even if there's a substrate in the active site, the inhibitor can still fit on.

Non-competitive inhibition may be **reversible** or **irreversible**.

Many **heavy metal ions** act as non-competitive inhibitors, such as Hg^{2+}.

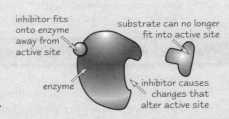

inhibitor fits onto enzyme away from active site

substrate can no longer fit into active site

enzyme

inhibitor causes changes that alter active site

Practice Questions

Q1 Describe the lock and key mechanism of enzyme activity.

Q2 Give four advantages of using immobilised enzymes in industrial manufacturing processes.

Exam Questions

1   In an experiment, the respiratory enzyme succinate dehydrogenase was mixed with the substrate succinate and the inhibitor malonate. The molecules succinate and malonate have a similar molecular structure.

   a)   Suggest the type of inhibition that may occur.   [1 mark]

   b)   Explain how the malonate can act as an inhibitor.   [2 marks]

   c)   Explain what would happen to the reaction rate if the concentration of succinate was increased.   [1 mark]

2   The enzyme alcohol dehydrogenase catalyses the breakdown of the substrate ethanol.
In an experiment, the rate of reaction was measured.

   a)   Explain why the rate eventually reaches a constant value as the concentration of ethanol is increased.   [1 mark]

   b)   Suggest the likely effect on the rate of adding dilute hydrochloric acid. Justify your answer.   [3 marks]

Immobilised enzymes — like lobsters in a lobster pot...

I bet you're feeling completely immobilised. Trapped in a network of study and revision, school and home. Don't forget to check back to page 13 and make sure you know what else upsets enzymes besides inhibitors. It's a hard life for enzymes — the whole world seems to be against them. Wrong temperature, wrong pH, inhibitors — they're all out to get them.

DNA and RNA Structure

DNA — the molecule of life. And unfortunately, just like life, it's complicated.

DNA and RNA are Very Similar Molecules

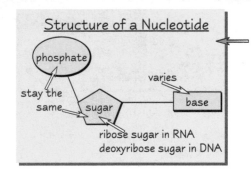

Structure of a Nucleotide

phosphate

stay the same

sugar

varies

base

ribose sugar in RNA
deoxyribose sugar in DNA

DNA and RNA are **nucleic acids** — made up of lots of **nucleotides** joined together. Nucleotides are units made from a **pentose sugar** (with 5 carbon atoms), a **phosphate** group and a **base** (containing nitrogen and carbon).

The sugar in **DNA** nucleotides is a **deoxyribose** sugar — in **RNA** nucleotides it's a **ribose** sugar.

$HOCH_2$ O   OH

H

OH OH

Ribose

$HOCH_2$ O   OH

H

OH H

Deoxyribose

Although DNA is called deoxyribonucleic acid, it still contains oxygen.

Within DNA and RNA, the sugar and the phosphate are the same for all the nucleotides. The only bit that's different between them is the **base**. There are five possible bases and they're split into two groups:

When Jenny started talking about pyrimidine rings, Matt knew it was the right time.

	NAME	BASIC STRUCTURE	BASE	found in DNA	found in RNA
1)	Purine bases	2 rings of atoms	adenine	✓	✓
			guanine	✓	✓
2)	Pyrimidine bases	single ring of atoms	cytosine	✓	✓
			thymine	✓	✗
			uracil	✗	✓

DNA and RNA are Polymers of Mononucleotides

Mononucleotides (single nucleotides) join together by a **condensation reaction** between the **phosphate** of one group and the **sugar** molecule of another. As in all condensation reactions, **water** is a by-product.

DNA is made of **two strands of nucleotides**. RNA has just the one strand. In DNA, the strands spiral together to form a **double helix**. The strands are held together by **hydrogen bonds** between the bases.

DNA

polynucleotide strands

sugar-phosphate backbone

hydrogen bonds between bases, keeping the strands coiled together

bases

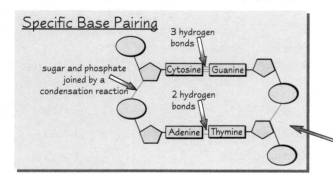

Specific Base Pairing

sugar and phosphate joined by a condensation reaction

3 hydrogen bonds

Cytosine — Guanine

2 hydrogen bonds

Adenine — Thymine

Each base can only join with one particular partner — this is called **specific base pairing**.

1) In DNA **adenine** always pairs with **thymine** (**A - T**) and **guanine** always pairs with **cytosine** (**G - C**).

2) It's the same in RNA, but **thymine**'s replaced by **uracil** (so it's **A - U** and **G - C**).

2 hydrogen bonds form between adenine and thymine.
3 hydrogen bonds form between guanine and cytosine.

DNA's Structure Makes it Good at its Job

1) The job of DNA is to carry **genetic information**. A DNA molecule is very, very **long** — much longer than RNA. It's **coiled** up very tightly, so a lot of genetic information can fit into a **small space** in the cell nucleus.

2) Its **paired structure** means it can **copy itself** — this is called **self-replication**. It's important for cell division and for passing on genetic information to the next generation.

DNA and RNA Structure

mRNA, tRNA and rRNA are Different Types of RNA

There are **three types** of RNA and they're all involved in **making proteins**.

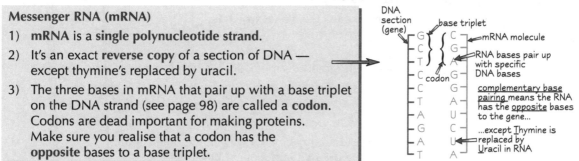

Messenger RNA (mRNA)

1) mRNA is a **single polynucleotide strand**.

2) It's an exact **reverse copy** of a section of DNA — except thymine's replaced by uracil.

3) The three bases in mRNA that pair up with a base triplet on the DNA strand (see page 98) are called a **codon**. Codons are dead important for making proteins. Make sure you realise that a codon has the **opposite** bases to a base triplet.

DNA section (gene) — base triplet — mRNA molecule — RNA bases pair up with specific DNA bases — complementary base pairing means the RNA has the opposite bases to the gene... ...except Thymine is replaced by Uracil in RNA — codon

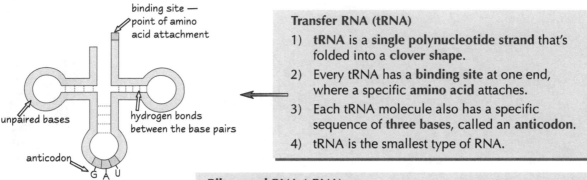

binding site — point of amino acid attachment

unpaired bases

hydrogen bonds between the base pairs

anticodon G A U

Transfer RNA (tRNA)

1) tRNA is a **single polynucleotide strand** that's folded into a **clover shape**.

2) Every tRNA has a **binding site** at one end, where a specific **amino acid** attaches.

3) Each tRNA molecule also has a specific sequence of **three bases**, called an **anticodon**.

4) tRNA is the smallest type of RNA.

Ribosomal RNA (rRNA)

rRNA is made up of polynucleotide strands that are attached to proteins to make things called **ribosomes** (see page 141). It's the largest type of RNA.

Practice Questions

Q1 Name each of the five bases and say which are present in DNA and RNA.

Q2 Draw the molecular structure for ribose and deoxyribose.

Q3 Name the three types of RNA.

Exam Questions

1 The structure of a pentose sugar present in nucleic acids is shown here.

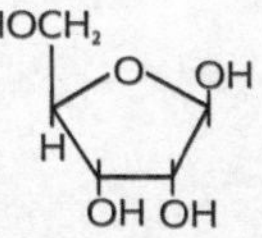

   HOCH₂ O OH
   H
   OH OH

   a) Name the pentose sugar. [1 mark]

   b) A pentose sugar, a base and a phosphate group react together to form a nucleotide.

      i) Give **one** difference between a nucleotide and a nucleic acid. [1 mark]

      ii) Draw the basic structure of a single nucleotide. [2 marks]

2 On analysis, a nucleic acid was found to contain the bases A, G, C and T only. 26 % of the bases present were guanine, G.

   a) Suggest, with a reason, whether the nucleic acid was DNA or RNA. [2 marks]

   b) Work out the percentages of A, C and T present in the nucleic acid. [2 marks]

   c) The sequence of bases on a small section of the nucleic acid was –AGCC–. What would the base sequence be in the complementary strand? [2 marks]

Give me a D, give me an N, give me an A! What do you get? — very confused...

Aagghhhh...so many letters and abbreviations. Why can't life just be simple. Notice that cytosine and thymine both have the letter 'y' in them, just like pyrimidine does. It's a good way to remember which bases are pyrimidines. I'm afraid there's nowt else you can do except buckle down, pull your socks up and get all them facts learnt.

DNA Replication and Protein Synthesis

Now you know the structures, here's what they actually do. It's completely mind-boggling.

DNA can Copy Itself — Self-Replication

DNA has to be able to **copy itself** before **cell division** can take place, which is essential for growth and reproduction — pretty important stuff. This is how it's done:

1) The DNA double helix starts to split into two single strands as the hydrogen bonds break — a bit like a zip.

2) Bases on **free-floating nucleotides** in the cytoplasm now pair up with the complementary bases on the nucleotides in the DNA. **Complementary base pairing** makes sure the correct nucleotide joins in the correct place.

3) An enzyme called **DNA polymerase** joins the new nucleotides together to form a polynucleotide chain.

4) This happens on each of the strands to make an **exact copy** of what was on the other strand. The result's **two molecules** of DNA **identical** to the **original molecule** of DNA.

5) This type of copying is called **semi-conservative replication** — as **half** of the new strands of DNA are from the **original** piece of DNA.

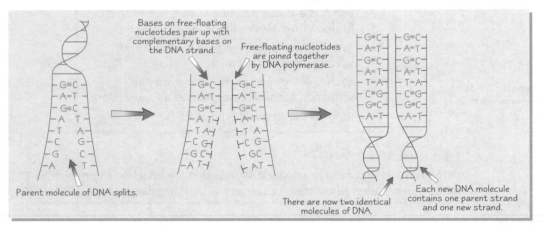

Bases on free-floating nucleotides pair up with complementary bases on the DNA strand.

Free-floating nucleotides are joined together by DNA polymerase.

Parent molecule of DNA splits.

There are now two identical molecules of DNA.

Each new DNA molecule contains one parent strand and one new strand.

The hospital offered a complementary baby pairing facility.

mRNA is Made During Transcription

Messenger RNA (mRNA) is made using DNA as a **template**, in a similar way to DNA replication. The process is called **transcription**.

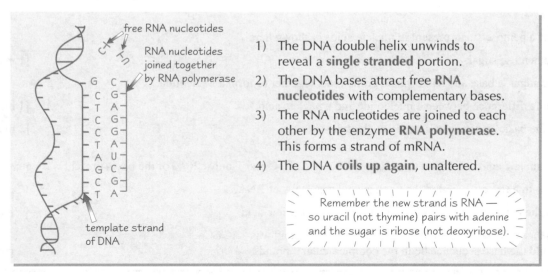

free RNA nucleotides

RNA nucleotides joined together by RNA polymerase

template strand of DNA

1) The DNA double helix unwinds to reveal a **single stranded** portion.

2) The DNA bases attract free **RNA nucleotides** with complementary bases.

3) The RNA nucleotides are joined to each other by the enzyme **RNA polymerase**. This forms a strand of mRNA.

4) The DNA **coils up again**, unaltered.

Remember the new strand is RNA — so uracil (not thymine) pairs with adenine and the sugar is ribose (not deoxyribose).

The newly made mRNA strand does not wind up with the DNA — it's **released** and is free to **move around** the cell. It's small enough to move outside the cell's nucleus into the cytoplasm, where it's used in the next process you've got to learn about — **translation**.

DNA Replication and Protein Synthesis

Proteins are Made During Translation

In **translation**, amino acids are joined together to make a **polypeptide** chain.

1) **Ribosomes** are large complexes made from rRNA and proteins. A **ribosome** attaches to the **mRNA**, and starts to move along it, looking for a **start codon**.

   Start codons have the base sequence **AUG** and indicate that the code for a new polypeptide chain is beginning. They code for the **first** amino acid in the chain (so this is always the same).

2) Once it's found a start codon, the ribosome temporarily pauses, until a **tRNA** with the correct **anticodon bases** pairs with the AUG codon inside the ribosome. The tRNA has an amino acid attached to it.

3) The ribosome then moves three bases forward, and waits for a different tRNA to bring another amino acid into the ribosome. Now there are two amino acids inside the ribosome and the ribosome joins them together with a **peptide bond**.

4) The ribosome moves forwards again. The first tRNA now **leaves** the ribosome and breaks away from its amino acid. A new tRNA brings in the third amino acid of the chain.

5) The process continues in this way until a **stop codon** is reached. The stop codon **doesn't** code for an amino acid. The ribosome **releases** the polypeptide chain at this point.

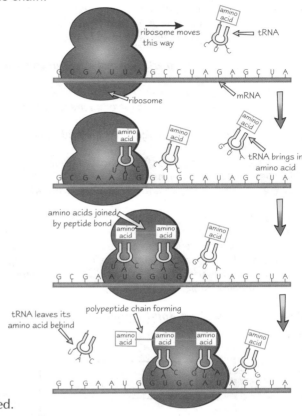

Practice Questions

Q1 What's made during transcription?

Q2 What base does RNA have that DNA doesn't?

Exam Questions

1 The sequence of bases in a small portion of mRNA is as follows: –AAGGUGCAUCGA–

   a) Why couldn't the sequence be from a portion of DNA? [1 mark]

   b) How many amino acids does this sequence code for? Explain your answer. [2 marks]

   c) Write the sequence of bases for the DNA strand from which the portion of mRNA was transcribed. [2 marks]

2 A codon on a section of mRNA has the sequence of bases –GGU–, which corresponds to the amino acid glycine. The mRNA codes for a polypeptide which contains 73 amino acids.

   a) What is the name given to the process of polypeptide synthesis from mRNA? [1 mark]

   b) Explain how this glycine molecule is inserted into the polypeptide. You should use the term <u>anticodon</u> in your answer. [4 marks]

   c) Explain why the mRNA that coded for this polypeptide contains at least 222 bases. [2 marks]

Help...I need a translation...

When you first go through protein synthesis it might make approximately no sense, but I promise its bark is worse than its bite. All those strange words disguise what is really quite a straightforward process — and the diagrams are dead handy for getting to grips with it. Keep drawing them yourself, 'til you can reproduce them perfectly.

Answers

Section One — Energetics

Page 3 — Enthalpy Change

1 a) *The first ionisation enthalpy is the enthalpy change when one mole [1 mark] of gaseous 1+ ions [1 mark] is formed from one mole of gaseous atoms [1 mark].*

   b) $Al^+_{(g)} \rightarrow Al^{2+}_{(g)} + e^-$
   [Correct equation 1 mark. Correct state symbols 1 mark.]

2 a) *Magnesium ions are small [1 mark]. This gives them a high charge density which means strong attraction/bonding [1 mark].*
   Magnesium ions and Oxygen ions are both doubly charged [1 mark].
   This gives a strong attraction/bonding [1 mark].

   b) *High melting point [1 mark].*

   c) *E.g. as a (refractory) lining for furnaces [1 mark].*

3 a) $Ca(NO_3)_2$ $Sr(NO_3)_2$ $Ba(NO_3)_2$ *[1 mark]*

   b) *Ca^{2+} is the smallest cation; Ba^{2+} is the largest [1 mark]. The smaller the cation, the greater the charge density [1 mark]. Ca^{2+} polarises the nitrate ion most, making it very unstable [1 mark] (or Ba^{2+} polarises the least, so the nitrate ion remains stable).*

Page 5 — Lattice and Bond Enthalpies

1 a)

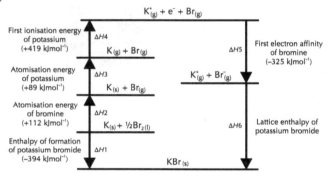

   [1 mark for left of cycle. 1 mark for right of cycle. 1 mark for formulas/state symbols. 1 mark for correct directions of arrows.]

   b) *Lattice enthalpy, $\Delta H6 = -\Delta H5 - \Delta H4 - \Delta H3 - \Delta H2 + \Delta H1$*
   $= -(-325) - (+419) - (+89) - (+112) + (-394)$ [1 mark]
   $= -689$ [1 mark] kJmol⁻¹ [1 mark]
   Award marks if calculation method matches cycle in part (a).

2 a) Bonds broken   Bonds formed
   $4 \times C–H = 4 \times 412 = 1648$ $2 \times C=O = 2 \times 743 = 1486$
   $2 \times O=O = 2 \times 496 = 992$ $4 \times O–H = 4 \times 463 = 1852$
                                              [1 mark]
   $(1648 + 992) - (1486 + 1852) = -698$ kJmol⁻¹ **[1 mark]**
   [1 mark for correct (negative) sign on final answer.]

   b) *(Some of) the bond enthalpies used are averages [1 mark]. These may be different from the actual ones in the substances [1 mark]*
   OR enthalpy cycles use data for the actual compounds shown [1 mark]. Maximum 2 marks.

Page 7 — Enthalpies and Dissolving

1 *Sodium chloride has an entropy increase when it dissolves [1 mark]. Calcium carbonate has an entropy decrease [1 mark].*

2 a)

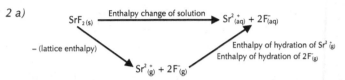

   [1 mark for each of the 4 enthalpy changes labelled, 1 mark for a complete, correct cycle.]
   Don't forget — you have to double the enthalpy of hydration for F^- because there are two in SrF_2.

   b) $-(-2492) + (-1480) + (2 \times -506)$ *[1 mark]* $= 0$ kJmol⁻¹
   [1 mark]

   c) *The entropy change [1 mark]*
   If there's an increase in entropy, SrF_2 is likely to be soluble, but if there's a decrease, then SrF_2 is likely to be insoluble.

3 a) *Insoluble [1 mark]*

   b) *Potassium iodide is ionic [1 mark]; cyclohexane is non-polar [1 mark]. Dissolving would involve the breaking of strong bonds in the ionic lattice [1 mark], but only weak bonds would form between the ions and the cyclohexane [1 mark].*

Page 9 — Entropy and Free Energy Change

1 a) *Reaction is not likely to be spontaneous [1 mark] because there is a decrease in entropy [1 mark].*
   Remember — more particles means more entropy.
   There's 1½ moles of reactants and only 1 mole of products.

   b) $\Delta S_{system} = 26.9 - (32.7 + 102.5)$ *[1 mark]*
   $= -108.3$ *[1 mark]* JK⁻¹mol⁻¹ *[1 mark]*

   c) *Reaction is not likely to be spontaneous [1 mark] because ΔS_{system} is negative/there is a decrease in entropy [1 mark].*

2 a) (i) $\Delta S_{system} = 48 - 70 = -22$ JK⁻¹mol⁻¹ *[1 mark]*
   $\Delta S_{surroundings} = -(-6000)/250 = +24$ JK⁻¹mol⁻¹ *[1 mark]*
   $\Delta S_{total} = \Delta S_{system} + \Delta S_{surroundings} = -22 + 24 = +2$ JK⁻¹mol⁻¹
   [1 mark]

   (ii) $\Delta S_{surroundings} = -(-6000)/300 = +20$ JK⁻¹mol⁻¹ *[1 mark]*
   $\Delta S_{total} = \Delta S_{system} + \Delta S_{surroundings} = -22 + 20 = -2$ JK⁻¹mol⁻¹
   [1 mark]

   b) *It will be spontaneous at 250 K, but not at 300 K [1 mark], because ΔS_{total} is positive at 250 K but negative at 300 K [1 mark].*

Section Two — Kinetics

Page 11 — Reaction Rates

1 a) *E.g. Volume of $CO_{2(g)}$ [1 mark] using a gas syringe [1 mark] / Colour of $Br_{2(aq)}$ [1 mark] using a colorimeter [1 mark] / pH (since $H^+_{(aq)}$ produced) [1 mark] using a pH meter [1 mark].*
   You can follow the rate by monitoring any property that changes.

   b) *Plot a graph of $[Br_{2(aq)}]$ against time [1 mark]. Draw a tangent to the curve at a particular time [1 mark]. Calculate the gradient of the tangent [1 mark].*

   c) *The rate will decrease [1 mark]. There will be less Br_2 molecules per unit volume [1 mark], so there will be less collisions and therefore less successful collisions [1 mark].*

Answers

2 a) Gas volume of $O_{2(g)}$ [1 mark] using, e.g. a gas syringe [1 mark].

b)

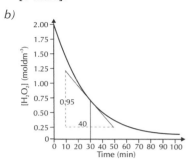

Rate after 30 minutes = 0.95/40
  = 0.02375 [1 mark] moldm⁻³min⁻¹ [1 mark]
Accept rate within range 0.02375 ± 0.005.
[1 mark for [H₂O₂₍aq₎] on y-axis and time on x-axis. 1 mark for points accurately plotted. 1 mark for best-fit smooth curve. 1 mark for tangent to curve at 30 minutes.]

Page 13 — Activation Energy and Catalysts

1 a) It lowers the activation energy [1 mark] by providing an alternative reaction route [1 mark]. This means more molecules of H_2O_2 have sufficient energy to react on collision [1 mark].

b) Heterogeneous catalyst [1 mark].

c) Enzymes will only act on substrates that have the correct shape to fit in the active site [1 mark]. So enzymes will usually only catalyse one reaction for one substrate molecule [1 mark].

Page 15 — Rate Equations

1 a) Rate = $k[NO_{(g)}]^2[H_{2(g)}]$ **[1 mark for correct orders, 1 mark for the rest]**
Sum of individual orders = 2 + 1 = 3rd order overall [1 mark].

b) (i) $0.00267 = k \times (0.004)^2 \times 0.002$ [1 mark]
$k = 8.34 \times 10^4$ dm⁶mol⁻²s⁻¹ **[1 mark for answer, 1 mark for units]**.
Units: k = moldm⁻³s⁻¹/[(moldm⁻³)² × (moldm⁻³)]
= dm⁶mol⁻²s⁻¹.

(ii) It would decrease [1 mark].
If the temperature decreases, the rate decreases too.
A lower rate means a lower rate constant.

Page 17 — Orders of Reactions and Half-Life

1 a) Experiments 1 and 2: [D] doubles and the initial rate quadruples (with [E] remaining constant) [1 mark]. So it's 2nd order with respect to [D] [1 mark].
Experiments 1 and 3: [E] doubles and the initial rate doubles (with [D] remaining constant) [1 mark]. So it's 1st order with respect to [E] [1 mark].
Always explain your reasoning carefully — state which concentrations are constant and which are changing.

b) rate = $k[D]^2[E]$ [1 mark]

c) Any row of experimental data can be picked in this sort of question.
In Experiment 1: $1.30 \times 10^{-3} = k \times 0.2^2 \times 0.2$ [1 mark]
$k = 0.163$ dm⁶mol⁻²s⁻¹ **[1 mark for answer, 1 mark for units]**.
Units: k = moldm⁻³s⁻¹/[(moldm⁻³)² × (moldm⁻³)]
= dm⁶mol⁻²s⁻¹.
Watch out — check the table of rate data carefully. In this case, the initial rate needs to be multiplied by 10⁻³.

Page 19 — Rates and Reaction Mechanisms

1 a) rate = $k[H_{2(g)}][ICl_{(g)}]$ [1 mark]

b) (i) One molecule of H_2 and one molecule of ICl [1 mark]. If the molecule is in the rate equation, it must be in the rate-determining step [1 mark]. The orders of the reaction tell you how many molecules of each reactant are in the rate-determining step [1 mark].

(ii) Incorrect [1 mark]. H_2 and ICl are both in the rate equation, so they must both be in the rate-determining step OR the order of the reaction with respect to ICl is 1, so there must be only one molecule of ICl in the rate-determining step [1 mark].

Section Three — Equilibria

Page 21 — Equilibria

1 $K_c = \dfrac{[H_2][I_2]}{[HI]^2}$ [1 mark] $\Rightarrow [HI]^2 = \dfrac{[H_2][I_2]}{K_c} = \dfrac{2.0 \times 0.3}{0.0167} = 35.93$ [1 mark]

$\Rightarrow [HI] = \sqrt{35.93} = 6.0$ moldm⁻³ [1 mark]

2 a) (i) mass/M_r = 42.5/46 = 0.92 [1 mark]
(ii) moles of O_2 = mass/M_r = 14.1/32 = 0.44 [1 mark]
moles of NO = 2 × moles of O_2 = 0.88 [1 mark]
moles of NO_2 = 0.92 − 0.88 = 0.04 [1 mark]

b) Concentration of O_2 = 0.44 ÷ 22.8 = 0.019 moldm⁻³
Concentration of NO = 0.88 ÷ 22.8 = 0.039 moldm⁻³
Concentration of NO_2 = 0.04 ÷ 22.8 = 1.75 × 10⁻³ moldm⁻³ [1 mark]
$K_c = \dfrac{[NO]^2[O_2]}{[NO_2]^2}$ [1 mark]

$\Rightarrow K_c = \dfrac{(0.039)^2 \times (0.019)}{(1.75 \times 10^{-3})^2}$ [1 mark] = 9.4 [1 mark] moldm⁻³ [1 mark]

(Units = (moldm⁻³)² × (moldm⁻³)/(moldm⁻³)² = moldm⁻³)

Page 23 — Gas Equilibria

1 a) $K_p = \dfrac{p(SO_2)p(Cl_2)}{p(SO_2Cl_2)}$ [1 mark]

b) Cl_2 and SO_2 are produced in equal amounts [1 mark].
$p(Cl_2) = p(SO_2) = 60.2$ kPa [1 mark]
Total pressure = $p(SO_2Cl_2) + p(Cl_2) + p(SO_2)$ [1 mark]
$p(SO_2Cl_2) = 141 − 60.2 − 60.2 = 20.6$ kPa [1 mark]

c) $K_p = \dfrac{(60.2)(60.2)}{(20.6)}$ [1 mark] = 176 [1 mark] kPa [1 mark]

(Units = (kPa × kPa)/ kPa = kPa)

2 a) $p(O_2)$ = ½ × 36 [1 mark] = 18 kPa [1 mark]
b) $p(NO_2)$ = total pressure − p(NO) − p(O₂)
= 99 − 36 − 18 [1 mark] = 45 kPa [1 mark]

c) $K_p = \dfrac{p(NO_2)^2}{p(NO)^2 p(O_2)}$ [1 mark] = $\dfrac{(45)^2}{(36)^2(18)}$ [1 mark]

= 0.0868 [1 mark] KPa⁻¹ [1 mark]

(Units = kPa²/(kPa² × kPa) = kPa⁻¹)

Answers

Page 25 — More on Equilibrium Constants

1 a) T2 is lower than T1 *[1 mark]*.
A decrease in temperature shifts the position of equilibrium in the exothermic direction, producing more product *[1 mark]*.
More product means K_p increases *[1 mark]*.
A negative ΔH means the forward direction's exothermic — it gives out heat.

b) The yield of SO_3 increases *[1 mark]*. (A decrease in volume means an increase in pressure. This shifts the equilibrium position to the right.)
K_p is unchanged *[1 mark]*.

Page 27 — More on Equilibrium Constants

1 a) $K_{sp} = [Pb^{2+}][SO_4^{2-}]$ *[1 mark]*
b) $[Pb^{2+}] = [SO_4^{2-}]$ *[1 mark]* $[Pb^{2+}]^2 = 2.5 \times 10^{-8}$ *[1 mark]*
$[Pb^{2+}] = 1.58 \times 10^{-4}$ $moldm^{-3}$ *[1 mark]*
Mr of $PbSO_4 = 207 + 32 + (16 \times 4) = 303$
solubility of $PbSO_4 = 1.58 \times 10^{-4} \times 303 = 0.0479$ gdm^{-3}
[1 mark]
c) $[SO_4^{2-}] = 0.15$ (because H_2SO_4 is a strong acid) *[1 mark]*
$[Pb^{2+}] = (2.5 \times 10^{-8}) \div 0.15 = 1.67 \times 10^{-7}$ $moldm^{-3}$ *[1 mark]*
solubility of $PbSO_4 = 1.67 \times 10^{-7} \times 303 = 5.05 \times 10^{-5}$ gdm^{-3}
[1 mark]

2 $[Ca^{2+}] = 0.01 \div 2 = 0.005$ $moldm^{-3}$
$[SO_4^{2-}] = 0.006 \div 2 = 0.003$ $moldm^{-3}$ *[1 mark]*
$[Ca^{2+}][SO_4^{2-}] = 0.005 \times 0.003 = 1.5 \times 10^{-5}$ mol^2dm^{-6} *[1 mark]*
1.5×10^{-5} is less than K_{sp}, so $CaSO_4$ won't precipitate
[1 mark].

3 Let amount of iodine in water = x
$85.5 = \dfrac{1 - x / 50}{x / 50} = \dfrac{1 - x}{x}$ *[1 mark]*
$85.5x = 1 - x$ *[1 mark]* $x = 1 \div 86.5 = 0.0116$ g *[1 mark]*

Page 29 — Acids and Bases

1 a) (i) A proton donor *[1 mark]*
(ii) A proton acceptor *[1 mark]*
b) (i) $HSO_4^- \rightarrow H^+ + SO_4^{2-}$ *[1 mark]*
(ii) $HSO_4^- + H^+ \rightarrow H_2SO_4$ *[1 mark]*

2 Weak acids dissociate (or ionise) a small amount *[1 mark]* to produce hydrogen ions (or protons) *[1 mark]*.
$HCN \rightleftharpoons H^+ + CN^-$ or $HCN + H_2O \rightleftharpoons H_3O^+ + CN^-$
[1 mark for correct formulas, 1 mark for equilibrium sign.]

Page 31 — pH Calculations

1 a) $K_a = \dfrac{[H^+][A^-]}{[HA]}$ or $K_a = \dfrac{[H^+]^2}{[HA]}$ *[1 mark]*

b) $K_a = \dfrac{[H^+]^2}{[HA]}$ $\Rightarrow$ $[HA]$ is 0.280 because very few HA will dissociate *[1 mark]*.
$[H^+] = \sqrt{(5.60 \times 10^{-4}) \times 0.280} = 0.0125$ $moldm^{-3}$ *[1 mark]*
$pH = -log_{10}[H^+] = -log_{10}(0.0125) = 1.90$ *[1 mark]*

2 $[H^+] = 10^{-2.65} = 2.24 \times 10^{-3}$ $moldm^{-3}$ *[1 mark]*
$K_a = \dfrac{[H^+]^2}{[HX]}$ *[1 mark]* $= \dfrac{[2.24 \times 10^{-3}]^2}{[0.15]}$
$= 3.34 \times 10^{-5}$ *[1 mark]* $moldm^{-3}$ *[1 mark]*

3 a) $K_w = [H^+][OH^-]$ *[1 mark]*, 1.0×10^{-14} mol^2dm^{-6} *[1 mark]*
b) $[OH^-] = 0.0370$ *[1 mark]*
$[H^+] = K_w \div [OH^-] = (1 \times 10^{-14}) \div 0.0370 = 2.70 \times 10^{-13}$
[1 mark]
$pH = -log_{10}[H^+] = -log_{10}(2.70 \times 10^{-13}) = 12.57$ *[1 mark]*

Page 33 — Titrations and pH Curves

1 - pH at equivalence for nitric acid is 7, whereas pH at equivalence for ethanoic acid is greater than 7.
- pH at start for nitric acid is lower than for ethanoic acid.
- near vertical bit is bigger/slightly steeper for nitric acid than for ethanoic acid.
[1 mark for each difference, up to a maximum of 2.]

2 a) 25 cm^3 *[1 mark]*
b) (i) 3-5 *[1 mark]* (ii) 1-2 *[1 mark]*

Page 35 — Indicators, pH Curves and Calculations

1 a) moles of ethanoic acid = $(25.0 \times 0.350) \div 1000$
$= 0.00875$ *[1 mark]*
Volume of KOH = (number of moles $\times$ 1000) $\div$ molar concentration $= (0.00875 \times 1000) \div 0.285$ *[1 mark]*
$= 30.7$ cm^3 *[1 mark]*
b) Thymol blue *[1 mark]*. It's a weak acid/strong base titration so the equivalence point is above pH 8 *[1 mark]*.

2 a) $H_2C_2O_4 + OH^- \rightarrow HC_2O_4^- + H_2O$ *[1 mark]*
b) Overall the equation is: $H_2C_2O_4 + 2OH^- \rightarrow C_2O_4^{2-} + 2H_2O$
moles of NaOH = $(38.4 \times 0.25) \div 1000$
$= 0.0096$ moles *[1 mark]*
moles ethanedioic acid = $\frac{1}{2} \times 0.0096 = 0.0048$ moles
[1 mark]
concentration of ethanedioic acid = $(0.0048 \times 1000) \div 25.0$
$= 0.192$ $moldm^{-3}$ *[1 mark]*
Make sure you read the question carefully — in fact, read it twice.
38.4 cm^3 is the <u>total</u> volume of NaOH added. So you have to work out the complete equation involving the loss of <u>both protons</u> on ethanedioic acid.

Page 37 — Buffers

1 a) $K_a = \dfrac{[C_6H_5COO^-][H^+]}{[C_6H_5COOH]}$ *[1 mark]*

$\Rightarrow [H^+] = 6.4 \times 10^{-5} \times \dfrac{0.40}{0.20} = 1.28 \times 10^{-4}$ $moldm^{-3}$ *[1 mark]*
$pH = -log_{10}[1.28 \times 10^{-4}] = 3.9$ *[1 mark]*
b) $C_6H_5COOH \rightleftharpoons H^+ + C_6H_5COO^-$ *[1 mark]*
Adding H_2SO_4 increases the concentration of H^+ *[1 mark]*.
The equilibrium shifts left to reduce concentration of H^+, so the pH will only change very slightly *[1 mark]*.

2 a) $CH_3(CH_2)_2COOH \rightleftharpoons H^+ + CH_3(CH_2)_2COO^-$ *[1 mark]*
b) $[CH_3(CH_2)_2COOH] = [CH_3(CH_2)_2COO^-]$,
so $[CH_3(CH_2)_2COOH] \div [CH_3(CH_2)_2COO^-] = 1$ *[1 mark]*
and $K_a = [H^+]$. $pH = -log_{10}[1.5 \times 10^{-5}] = 4.8$
[1 mark]
If the concentrations of the weak acid and the salt of the weak acid are equal, they cancel from the K_a expression and the buffer pH = pK_a.

Answers

Page 39 — Metal-Aqua Ions

1 Fe^{3+} has a higher charge density than Fe^{2+} *[1 mark]*. This means Fe^{3+} polarises water molecules more *[1 mark]*, weakening the O-H bond more *[1 mark]* and making it more likely that H^+ ions are released into the solution *[1 mark]*.

2 A blue precipitate *[1 mark]* of copper hydroxide forms in the blue solution *[1 mark]* of copper sulphate. On addition of excess ammonia, the precipitate dissolves *[1 mark]* to give a deep blue solution *[1 mark]*.
$Cu(H_2O)_6^{2+} + 2OH^- \rightleftharpoons Cu(H_2O)_4(OH)_2$ *[1 mark]* $+ 2H_2O$
$Cu(H_2O)_4(OH)_2 + 4NH_3 \rightleftharpoons$
$\qquad Cu(NH_3)_4(H_2O)_2^{2+}$ *[1 mark]* $+ 2H_2O + 2OH^-$
[1 additional mark for each balanced equation. 2 marks in total.]

Section Four — Elements of the Periodic Table

Page 41 — Period 3 Elements and Oxides

1 a) $4Na_{(s)} + O_{2(g)} \rightarrow 2Na_2O_{(s)}$
$4Al_{(s)} + 3O_{2(g)} \rightarrow 2Al_2O_{3(s)}$
$Si_{(s)} + O_{2(g)} \rightarrow SiO_{2(s)}$
$4P_{(s)} + 5O_{2(g)} \rightarrow P_4O_{10(s)}$
[1 mark for the correct formula of each oxide, 1 mark for each correctly balanced equation.]
Examiners love to test out your ability to write balanced equations — so take care with both the formulas and the balancing.

b) Na_2O — giant *[1 mark]*, ionic *[1 mark]*
Al_2O_3 — giant *[1 mark]*, ionic,
$\qquad$ with some covalent character *[1 mark]*
SiO_2 — giant *[1 mark]*, covalent *[1 mark]*
P_4O_{10} — simple molecular *[1 mark]*, covalent *[1 mark]*

c) Na_2O — pH 12–14 accepted *[1 mark]*
Al_2O_3 — insoluble *[1 mark]*
SiO_2 — insoluble *[1 mark]*
P_4O_{10} — pH 2–4 accepted *[1 mark]*

d) Phosphorus(V) oxide is covalent *[1 mark]*, so it has molecules held together by weak intermolecular bonds/van der Waals forces *[1 mark]*. Sodium oxide is ionic *[1 mark]*, so it has strong electrostatic forces holding its ions together *[1 mark]*. Van der Waals forces need less energy *[1 mark]* to overcome.
You need to mention both compounds to get full marks.

Page 43 — Period 3 Chlorides and Group 4

1 a) $NaCl_{(s)} \rightarrow Na^+_{(aq)} + Cl^-_{(aq)}$
$AlCl_{3(s)} + 3H_2O_{(l)} \rightarrow Al(OH)_{3(s)} + 3HCl_{(aq)}$
$PCl_{5(s)} + 4H_2O_{(l)} \rightarrow H_3PO_{4(aq)} + 5HCl_{(aq)}$
[1 mark for each correct formula of a chloride, 1 mark for each correctly balanced equation, total 6 marks.]

b) Sodium chloride — pH 7 *[1 mark]*; it's ionic, so it breaks up into ions. These ions don't affect the amount of H^+ or OH^- ions in the solution *[1 mark]*.
Aluminium chloride — pH 3 *[1 mark]*; it hydrolyses with water to form hydrochloric acid *[1 mark]*.
Phosphorus(V) chloride — pH 0–1 *[1 mark]*; it hydrolyses with water to form hydrochloric acid and phosphoric acid $(H_3PO_{4(aq)})$ *[1 mark]*.

Page 45 — The Nitrogen Cycle and Water

1 a) The boiling point of water is higher than the other hydrides of Group 6 *[1 mark]*. Water has strong intermolecular forces between molecules/hydrogen bonds *[1 mark]*, unlike the other hydrides. These take lots of energy to overcome/break *[1 mark]*.

b) There are hydrogen bonds between water molecules *[1 mark]*. When water freezes these hydrogen bonds cause the molecules to arrange themselves in a more open, regular structure *[1 mark]*. There's more space between the molecules, so it's less dense *[1 mark]* and floats. There's no hydrogen bonding in hydrogen sulphide *[1 mark]*, so it becomes more dense *[1 mark]* as it freezes.
Make sure you compare water to the other hydrides in parts a) and b).

c) Water's high enthalpy of vaporisation allows it to act as a climate control *[1 mark]*. In the hotter areas of the globe, such as at the equator, water in the sea absorbs lots of energy *[1 mark]* and some evaporates *[1 mark]*. Air currents move the water vapour towards the colder areas of the globe, such as the poles. It releases energy there *[1 mark]* as it condenses *[1 mark]*.

Section Five — Electrochemistry and Transition Elements

Page 47 — Oxidation and Reduction

1 a) $Ti + (4 \times -1) = 0$, $Ti = +4$ *[1 mark]*
b) $(2 \times V) + (5 \times -2) = 0$, $V = +5$ *[1 mark]*
c) $Cr + (4 \times -2) = -2$, $Cr = +6$ *[1 mark]*
d) $(2 \times Cr) + (2 \times -7) = -2$, $Cr = +6$ *[1 mark]*

2 a) $2MnO_4^-_{(aq)} + 16H^+_{(aq)} + 10I^-_{(aq)} \rightarrow 2Mn^{2+}_{(aq)} + 8H_2O_{(l)} + 5I_{2(aq)}$
[1 mark for correct reactants and products, 1 mark for correct balancing]
You have to balance the number of electrons before you can combine the half-equations. And always double-check that your equation definitely balances. It's easy to slip up and throw away marks.

b) Mn has been reduced *[1 mark]* from +7 to +2 *[1 mark]*
I^- has been oxidised *[1 mark]* from −1 to 0 *[1 mark]*

c) Reactive metals have a tendency to lose electrons, so are good reducing agents *[1 mark]*. I^- is already in its reduced form *[1 mark]*.

Page 49 — Electrode Potentials

1 a)

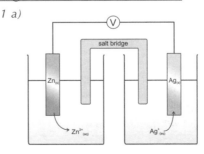

[1 mark for correctly labelled electrodes, 1 mark for salt bridge, 1 mark for external circuit.]

b) $+0.80 V - (-0.76 V) = 1.56 V$ *[1 mark]*

c) The concentration of Zn^{2+} ions or Ag^+ ions was not 1.00 $moldm^{-3}$ *[1 mark]*. The pressure wasn't 101 kPa *[1 mark]*.

d) $Zn_{(s)} + 2Ag^+_{(aq)} \rightarrow Zn^{2+}_{(aq)} + 2Ag_{(s)}$ *[1 mark]*

e) The zinc half-cell. It has a more negative standard electrode potential/it's less electronegative *[1 mark]*.

Answers

Page 51 — The Electrochemical Series

1 a) $Zn_{(s)} + Ni^{2+}_{(aq)} \rightleftharpoons Zn^{2+}_{(aq)} + Ni_{(s)}$ **[1 mark]**
$E^\circ = (-0.25) - (-0.76) = +0.51$ V **[1 mark]**

b) $2MnO^-_{4(aq)} + 16H^+_{(aq)} + 5Sn^{2+}_{(aq)} \rightleftharpoons$
$\quad 2Mn^{2+}_{(aq)} + 8H_2O_{(l)} + 5Sn^{4+}_{(aq)}$ **[1 mark]**
$E^\circ = (+1.52) - (+0.15) = +1.37$ V **[1 mark]**

c) No reaction **[1 mark]**. Both reactants are in their oxidised form **[1 mark]**.

d) $Ag^+_{(aq)} + Fe^{2+}_{(aq)} \rightleftharpoons Ag_{(s)} + Fe^{3+}_{(aq)}$ **[1 mark]**
$E^\circ = (+0.80) - (+0.77) = +0.03$ V **[1 mark]**

Page 53 — Applying Electrochemistry

1 a) (i) Magnesium reacts in preference to iron **[1 mark]**, because it has a more negative E° value **[1 mark]**. All the magnesium must corrode away before the iron will rust **[1 mark]**.

(ii) The coating forms a barrier **[1 mark]**. Zinc is more reactive than iron, so if iron is exposed zinc it reacts in preference to it because it has a more negative E° value **[1 mark]**. All the zinc must corrode away before the iron will rust **[1 marks]**.

(iii) Tin forms a barrier **[1 mark]** which prevents water/oxygen coming in contact with the iron **[1 mark]**.
[1 mark for each point. Maximum 8 marks]

A sacrificial metal must have a more negative electrode potential than the metal it's protecting. Tin's electrode potential is less negative than iron's, so it's just a barrier to keep out air and water.

b) (i) e.g. ships' hulls / underground pipes
(ii) e.g. buckets
(iii) e.g. food cans
[All three objects = 2 marks, any two objects = 1 mark]

Page 55 — Transition Metals — The Basics

1 a) $1s^2 2s^2 2p^6 3s^2 3p^6 3d^{10}$ or $[Ar]3d^{10}$ **[1 mark]**

b) No, it doesn't **[1 mark]**. Cu^+ ions have a full 3d subshell **[1 mark]**.

c) copper(II) sulphate $(CuSO_{4(aq)})$ **[1 mark]**

d) The outer s and d energy levels are very close together **[1 mark]**, so different numbers of electrons can be gained or lost using fairly similar amounts of energy **[1 mark]**.

Page 57 — Transition Metals — Vanadium and Cobalt

1 a) The solution would turn from yellow to blue **[1 mark]**, to green **[1 mark]** to violet **[1 mark]**.

b) The yellow to blue colour change shows a reduction of +5 (VO_2^+) **[1 mark]** to +4 (VO^{2+}) **[1 mark]**.
$Zn_{(s)} + 4H^+ + 2VO_2^+_{(aq)} \rightleftharpoons 2VO^{2+}_{(aq)} + 2H_2O_{(l)} + Zn^{2+}_{(aq)}$
[1 mark]
The blue to green colour change shows a reduction of +4 (VO^{2+}) to +3 (V^{3+}) **[1 mark]**.
$Zn_{(s)} + 4H^+ + 2VO^{2+}_{(aq)} \rightleftharpoons 2V^{3+}_{(aq)} + 2H_2O_{(l)} + Zn^{2+}_{(aq)}$
[1 mark]
The green to violet colour change shows a reduction of +3 (V^{3+}) to +2 (V^{2+}) **[1 mark]**.
$Zn_{(s)} + 2V^{3+}_{(aq)} \rightleftharpoons 2V^{2+}_{(aq)} + Zn^{2+}_{(aq)}$ **[1 mark]**

Page 59 — Transition Metals — Chromium and Copper

1 a) $Cr_2O_7^{2-}_{(aq)} + 14H^+_{(aq)} + 6Fe^{2+}_{(aq)} \rightleftharpoons 2Cr^{3+}_{(aq)} + 7H_2O_{(l)} + 6Fe^{3+}_{(aq)}$
[1 mark for correct reactants and products, 1 mark for correct balancing.]

b) Chromium has been reduced **[1 mark]** from +6 to +3 **[both states for 1 mark]**. Iron has been oxidised **[1 mark]** from +2 to +3 **[both states for 1 mark]**.

2 a) A is copper(II) sulphate (allow Cu^{2+} ions) **[1 mark]**.
B is copper **[1 mark]**.

b) $Cu_2SO_{4(s)} \rightleftharpoons CuSO_{4(aq)} + Cu_{(s)}$ **[1 mark]**
(allow $2Cu^+_{(aq)} \rightleftharpoons Cu^{2+}_{(aq)} + Cu_{(s)}$)

c) Disproportionation **[1 mark]**

Remember — in redox reactions, one species is reduced and a different species is oxidised. But in disproportionation reactions, one species is oxidised and reduced.

Page 61 — Transition Metals — Titrations and Calculations

1 a) $(0.200 \times 30.00) \div 1000 = 0.00600$ moles of $S_2O_3^{2-}$
[1 mark]

b) number of moles of $S_2O_3^{2-} : I_2 = 2 : 1$
So the number of moles of iodine = $0.00600 \div 2 = 0.00300$
[1 mark]

c) number of moles of $I_{2(aq)} : Cu^{2+}_{(aq)} = 1 : 2$
So the number of moles of copper(II) ions = $0.00300 \times 2 = 0.00600$ **[1 mark]**

d) Mass of Cu present = moles $\times M_r = 0.00600 \times 63.5 = 0.381$ g
[1 mark]
Percentage of Cu = $(0.381 \div 0.500) \times 100 = 76.2\%$ **[1 mark]**
You've got to look at the equations to find out how many moles of one thing reacts with another. This question's nice as it gives you the balanced equations — you mightn't always be so lucky though.

Page 63 — Complex Ions — The Basics

1 a) $A = [Cu(H_2O)_6]^{2+}$ **[1 mark]**, $B = [CuCl_4]^{2-}$ **[1 mark]**,
$C = [Cu(NH_3)_4(H_2O)_2]^{2+}$ **[1 mark]**

b) $[Cu(H_2O)_6]^{2+}_{(aq)} + 4Cl^-_{(aq)} \rightleftharpoons [CuCl_4]^{2-}_{(aq)} + 6H_2O_{(l)}$ **[1 mark]**

c)
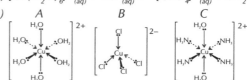

A = octahedral, B = tetrahedral, C = octahedral.
[1 mark for each correct diagram, one mark for each shape name.]

Page 65 — Complex Ions — Ligand Exchange

1 a) $[Co(H_2O)_6]^{2+}$ **[1 mark]**

b) $[CoCl_4]^{2-}$ **[1 mark]**

c) $[Co(H_2O)_6]^{2+}_{(aq)} + 4Cl^-_{(aq)}$ **[1 mark]** $\rightleftharpoons [CoCl_4]^{2-}_{(aq)} + 6H_2O_{(l)}$
[1 mark]

d) Increase the concentration of $Cl^-_{(aq)}$ ions by adding concentrated HCl **[1 mark]**. Or heat the mixture so the equilibrium shifts to the right-hand side **[1 mark]**.

2 $[Ag(H_2O)_2]^+_{(aq)} + 2NH_{3(aq)} \rightleftharpoons [Ag(NH_3)_2]^+_{(aq)} + 2H_2O_{(l)}$
[1 mark]

$K_{stab} = \dfrac{[Ag(NH_3)_2]^+}{[[Ag(H_2O)_2]^+][NH_3]^2}$ **[1 mark]**

Units = $dm^6 mol^{-2}$ **[1 mark]**

Answers

Page 67 — Transition Metal Ion Colour

1 a) The 3d orbitals split into two different energy levels *[1 mark]*.

b) The two complexes have different ligands *[1 mark]* and different coordination numbers *[1 mark]*, so the energy gap between their d orbitals will be different *[1 mark]*. This means different wavelengths/frequencies/energies of light will be absorbed *[1 mark]*.

c)

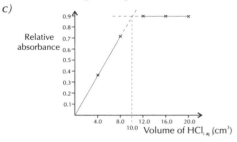

[1 mark for axes, 1 mark for correctly plotted points, 1 mark for extrapolated lines to form intersect.]

d) $(0.1 \times 5.0) \div 1000 = 5.0 \times 10^{-4}$ moles *[1 mark]*

e) 10.0 cm^3 of 0.2 moldm^{-3} HCl contains $(10.0 \times 0.2) \div 1000 = 2.0 \times 10^{-3}$ moles of Cl^- *[1 mark]*
Ratio $Cu^{2+} : Cl^- = 5.0 \times 10^{-4} : 2.0 \times 10^{-3} = 1 : 4$ *[1 mark]*
So formula = $[CuCl_4]^{2-}$ (Charge = $(1 \times 2) + (4 \times -1) = -2$)
[1 mark for CuCl$_4$, 1 mark for charge of 2–]

Page 69 — Uses of Transition Metals

1 a) $S_2O_8^{2-}{}_{(aq)} + 2I^-{}_{(aq)} \rightarrow I_{2(aq)} + 2SO_4^{2-}{}_{(aq)}$ *[1 mark]*

b) The thiosulphate ions reacts with the iodine formed to produce iodide ions *[1 mark]*. Only when all the thiosulphate ions have been used up will any iodine formed react with the starch *[1 mark]*.

c) Fe^{3+} ions oxidise I^- ions to I_2 *[1 mark]*
$2Fe^{3+}{}_{(aq)} + 2I^-{}_{(aq)} \rightarrow I_{2(aq)} + 2Fe^{2+}{}_{(aq)}$ *[1 mark]*
Fe^{2+} ions can now reduce peroxodisulphate ions to sulphate ions *[1 mark]*.
$S_2O_8^{2-}{}_{(aq)} + 2Fe^{2+}{}_{(aq)} \rightarrow 2Fe^{3+}{}_{(aq)} + 2SO_4^{2-}{}_{(aq)}$ *[1 mark]*

d) Cu^+ ions *[1 mark]*
Copper(II) sulphate would work exactly the same way as the iron(III) chloride. The copper(II) ions are reduced to copper(I) ions, and then oxidised back to copper(II). It's a catalyst, so it has to be regenerated at the end.

2 Haemoglobin is a molecule found in blood with a central Fe^{2+} ion coordinately bonded with 5 nitrogen atoms *[1 mark]*. Water occupies the sixth ligand position *[1 mark]*. Ligand substitution occurs and oxygen molecules occupy the site of the water ligands, allowing oxygen to be transported around the body *[1 mark]*. Some toxins such as carbon monoxide form stronger bonds and will preferentially occupy the sixth site, preventing oxygen being transported *[1 mark]*.

Section Six — Organic Chemistry

Page 71 — Isomerism

1 a) The property of having stereoisomers, which are molecules with the same molecular formula and with their atoms arranged in the same way *[1 mark]*, but with a different orientation of the bonds in space *[1 mark]*.

b) For example:

[1 mark for each correct structure, 1 mark for correct cis/trans labels.]

c) (i) For example:

[1 mark for each correctly drawn structure — they don't have to be orientated in the same way as in the diagram above, as long as the molecules are mirror images of each other.]

(ii) An asymmetric carbon/a chiral carbon/a carbon with four different groups attached *[1 mark]*.

(iii) Shine (monochromatic) plane-polarised light through a solution of the molecule *[1 mark]*. The enantiomers will rotate the light in opposite directions *[1 mark]*.

2 a)

[1 mark]

The chiral carbon is the one with 4 different groups attached.

b) (i) A mixture of equal quantities of each enantiomer of an optically active compound *[1 mark]*.

(ii) Smaller doses needed *[1 mark]*, less side-effects (because there's no D-DOPA enantiomer) *[1 mark]*.

Page 73 — Reaction Mechanisms

1 a) Excess methane *[1 mark]* and UV light/sunlight *[1 mark]*.

b) Homolytic/free radical *[1 mark]* substitution *[1 mark]*.

2 a) (i) Attracted to areas of high electron density/negative charge *[1 mark]*

(ii) $H_3C-\underset{\underset{Br}{|}}{\overset{\overset{CH_3}{|}}{C}}-CH_3$

[1 mark]

Its carbocation intermediate is more stable *[1 mark]*, as it has more alkyl groups pusing electrons towards the positive charge to stabilise it *[1 mark]*.

b) The S_N1 mechanism from page 73 is needed here — remember to change :CN$^-$ to :OH$^-$.

[1 mark for choosing the S_N1 mechanism, 1 mark for each of the two steps correctly drawn]

Take care when drawing mechanisms — the curly arrows have got to come from the lone pair of electrons, not from the negative charge.

Answers

Page 75 — Aldehydes and Ketones

1 a) Propanal *[1 mark]*   *[1 mark]*

Propanone *[1 mark]*   *[1 mark]*

b) (i) Nucleophilic addition *[1 mark]*

(ii) From propanal,   *[1 mark]*

This product *[1 mark]* will be produced in a racemic mixture, because it's asymmetrical/contains a chiral carbon *[1 mark]*.

From propanone,   *[1 mark]*

c) (i) $CH_3CH_2CHO + 2[H] \rightarrow CH_3CH_2CH_2OH$ *[1 mark]*
(ii) $NaBH_4$ *[1 mark]* in aqueous methanoic conditions *[1 mark]*, or, $LiAlH_4$ *[1 mark]* in dry diethyl ether *[1 mark]*. *[2 marks maximum]*

Page 77 — Aldehydes, Ketones and Grignard Reagents

1 a) They're both carbonyl compounds *[1 mark]*.
b) (i) It's an aldehyde *[1 mark]*.
(ii) It's a ketone *[1 mark]*.
c) (i) Ethanal *[1 mark]*

(ii)   *[1 mark]*

The only aldehyde with a methyl carbonyl group is ethanal.
d) Recrystallise it and measure its melting point *[1 mark]*, then compare it with known melting points of the derivatives *[1 mark]*.

2 a) Reflux bromomethane *[1 mark]* with magnesium *[1 mark]* in dry diethyl ether *[1 mark]*. $CH_3Br + Mg \rightarrow CH_3MgBr$ *[1 mark]*

b) (i) 2-methylpropan-2-ol *[1 mark]*   *[1 mark]*

(ii) Acid must be added to hydrolyse the intermediate *[1 mark]*.

Page 79 — Carboxylic Acids

1 a)   *[1 mark]*

b) Butan-1,2,3,4-tetraol *[1 mark]*   *[1 mark]*

$LiAlH_4$ *[1 mark]*, in dry diethyl ether *[1 mark]*.
c) Potassium hydroxide/potassium carbonate/potassium hydrogencarbonate *[1 mark]*
You might never have heard of 2,3-dihydroxybutanedioic acid before, but that's OK. You just have to apply what you know about carboxylic acids to it. Potassium bitartrate ends in -ate, so it's a salt.

2 a) Propanoic acid partially dissociates in water *[1 mark]* to release H^+ ions *[1 mark]*. Propanol doesn't do this *[1 mark]*.
b) Add a carbonate/hydrogencarbonate *[1 mark]*. Propanol will show no reaction *[1 mark]*. Propanoic acid will produce bubbles of carbon dioxide *[1 mark]*.
c) (i) $CH_3CH_2COOH + CH_3CH_2CH_2OH \rightleftharpoons$ $CH_3CH_2COOCH_2CH_2CH_3 + H_2O$
[1 mark for correct equation, 1 mark for reversible arrow.]
(ii) Distil off the ester as it's formed *[1 mark]*.
Removing the ester shifts the equilibrium to the right to replace it.

Page 81 — Esters

1 a) 2-methylpropyl ethanoate *[1 mark]*
b) Food flavouring/perfume *[1 mark]*
c) Ethanoic acid *[1 mark]*   *[1 mark]*

2-methylpropan-1-ol *[1 mark]*   *[1 mark]*

This is acid hydrolysis *[1 mark]*
d) With sodium hydroxide, sodium ethanoate is produced, but in the reaction in part (c), ethanoic acid is produced *[1 mark]*.

2 a) Containing no double bonds in the carbon chain *[1 mark]*.
b) The higher the proportion of saturated fatty acids, the higher the melting point of the butter *[1 mark]*.

Page 83 — Acyl Chlorides

1 a) $CH_3CH_2COOH + PCl_5 \rightarrow CH_3CH_2COCl + POCl_3 + HCl$
or $CH_3CH_2COOH + SOCl_2 \rightarrow CH_3CH_2COCl + SO_2 + HCl$
[1 mark for suitable reagent, 1 mark for correct equation]
b) It's a nucleophilic addition-elimination *[1 mark]*. See page 83 for the mechanism *[1 mark for each correct step, maximum 4 marks. Marks are deducted for curly arrows in the wrong place or forgetting that HCl is also made.]*
c) Advantage: faster reaction/no need to heat/reaction goes to completion/no equilibrium so better yield (of end product) *[any one for 1 mark]*
Disadvantage: toxic/corrosive hydrogen chloride gas produced (in both steps) *[1 mark]*

Answers

Page 85 — Aromatic Compounds

1 a) (i) A: nitrobenzene *[1 mark]*
    B + C: concentrated nitric acid *[1 mark]* and concentrated sulphuric acid *[1 mark]*
    D: warm, not more than 55 °C *[1 mark]*
    When you're asked to name a compound, write the name, not the formula.
  (ii) See page 85 for the mechanism. *[1 mark for each of the three steps.]*
  (iii) $HNO_3 + H_2SO_4 \rightarrow H_2NO_3^+ + HSO_4^-$ *[1 mark]*
      $H_2NO_3^+ \rightarrow NO_2^+ + H_2O$ *[1 mark]*
  b) (i) G: benzenesulphonic acid *[1 mark]*
    E + F: <u>concentrated</u> sulphuric acid, heat under reflux OR <u>fuming</u> sulphuric acid, at room temperature *[1 mark for reagent, 1 mark for appropriate conditions for that reagent]*
  (ii) Sulphur trioxide/SO_3 *[1 mark]*
  c) Benzene is stabilised due to its delocalised electrons *[1 mark]*. Addition would disrupt the stability/substitution preserves the delocalisation *[1 mark]*.

Page 87 — More Reactions of Aromatic Compounds

1 a) Reagent: ethene / CH_2CH_2 *[1 mark]* and HCl *[1 mark]*
    Catalysts: a halogen carrier e.g. $AlCl_3$ *[1 mark]*
  b) The electrophile is formed as follows:
    $CH_2CH_2 + HCl \rightarrow CH_3CH_2Cl$ *[1 mark]*
    then $CH_3CH_2Cl + AlCl_3 \rightarrow CH_3CH_2^+ + AlCl_4^-$ *[1 mark]*
    For the mechanism see page 86. *[1 mark for correct structures of intermediates, 1 mark for the correct structure of products, 1 mark for correct positions of curly arrows.]*
  c) The catalyst polarises *[1 mark]* the chloroethane molecule to produce a carbocation/positive charge/positively charged ion *[1 mark]*. This can then act as an electrophile/attack/react with the benzene ring *[1 mark]*.
2 a) Catalyst: any halogen carrier e.g. $AlCl_3$ *[1 mark]*.
    Conditions: dry ether, reflux *[1 mark]*
  b) $H_3C-C^+_{\setminus O}$ *[1 mark]*

Page 89 — Phenols

1 a)   *[1 mark]*

  b) Hydrogen chloride gas/HCl *[1 mark]*
    Just ignore the carboxyl group and pretend it's a phenol reacting with the acyl chloride.

2 a) In $C_6H_5O^-$, the oxygen's lone pairs of electrons are partially delocalised onto the benzene ring/the oxygen's p-orbital overlaps with the benzene ring's π bond system *[1 mark]*, which spreads the negative charge of the phenoxide ion out over the ring *[1 mark]*.
    $CH_3CH_2O^-$ has no delocalisation *[1 mark]*, so the negative charge is concentrated on the oxygen *[1 mark]*.
  b) D: benzoic acid *[1 mark]*
    E: cyclohexanol *[1 mark]*
    F: phenol *[1 mark]*
    Carboxylic acids are stronger acids than phenols, and phenols are stronger acids than alcohols without benzene rings.

c) (i) The bromine water decolourises *[1 mark]* and a white precipitate's formed that smells of antiseptic *[1 mark]*.

  (ii)   *[1 mark]*

  (iii) Bromine isn't a strong enough electrophile *[1 mark]* to react with the stable benzene ring *[1 mark]*. In a phenoxide ion the oxygen lone pairs are partially delocalised onto the ring, increasing the electron density *[1 mark]* and making the benzene ring more susceptible to electrophilic attack *[1 mark]*.

Page 91 — Amines

1 a) It can accept protons/H^+ ions *[1 mark]*.
  b) Methylamine is stronger as the nitrogen lone pair is more available *[1 mark]* — the methyl group/CH_3 pushes electrons onto/increases electron density on the nitrogen *[1 mark]*. Phenylamine is weaker as the nitrogen lone pair is less available *[1 mark]* — nitrogen's electron density is decreased as it's partially delocalised around the benzene ring *[1 mark]*.
2 a) You get a mixture of primary, secondary and tertiary amines, and quaternary ammonium salts *[1 mark]*.
  b) (i) $LiAlH_4$ and dry diethyl ether *[1 mark]*, followed by dilute acid *[1 mark]* OR reflux *[1 mark]* with sodium metal and ethanol *[1 mark]*.
  (ii) It's too expensive *[1 mark]*.
  (iii) Metal catalyst such as platinum or nickel *[1 mark]* and high temperature and pressure *[1 mark]*.

Page 93 — Amines and Amides

1 a) (i) M and O *[1 mark]*
    These two reactions both use a dilute acid. For reaction O check back to page 78.
  (ii) $CH_3CH_2CONH_2 + H_2O + HCl \rightarrow CH_3CH_2COOH + NH_4Cl$
    or $CH_3CH_2CN + 2H_2O + HCl \rightarrow CH_3CH_2COOH + NH_4Cl$ *[1 mark]*
  (iii) Hydrolysis *[1 mark]*
  b) (i) Phosphorus(V) oxide/P_4O_{10} *[1 mark]* and heat *[1 mark]*
  (ii) Dehydration *[1 mark]*

Page 95 — Dyes

1 a) (i) Nitrous acid/HNO_2 *[1 mark]*
      $NaNO_{2(aq)} + HCl_{(aq)} \rightarrow HNO_{2(aq)} + NaCl_{(aq)}$ *[1 mark]*
  (ii) Temperature below 10 °C *[1 mark]*
  b) (i)   *[1 mark]*

  (ii) Benzene rings *[1 mark]*, N=N/azo group *[1 mark]*, nitrogen lone pair *[1 mark]*.
  (iii) Dye *[1 mark]*
    This is actually Butter Yellow, which was once used to colour margarine — until they found out it was carcinogenic (cancer-inducing) and banned it. Eeek...

2 a) (i) Benzene rings *[1 mark]*, N=N/azo group *[1 mark]*, oxygen lone pairs *[1 mark]*.
  (ii) $-SO_3^-Na^+$/sodium sulphonate group *[1 mark]*
  b) Ionic attraction *[1 mark]* between $-SO_3^-$ (in dye) *[1 mark]* and $-NH_3^+$ (in protein) *[1 mark]*.
  c) Increase the pH (of the solution) *[1 mark]*.
    (This removes the proton from −OH.)

  d)   *[1 mark for each]*

Answers

Page 97 — Amino Acids and Proteins

1 a)

HOCH₂–C–COOH with NH₂ on top and H on bottom **[1 mark]**

You'll need a cool head for this one. Start with a 3-carbon chain and then add the groups — you start numbering the carbons from the one with the carboxyl group.

b) (i) Two amino acids joined together **[1 mark]** by a peptide/amide/–CONH– link **[1 mark]**.

ii)

H₂N–C–C–N–C–COOH (with CH₂/OH, O, H groups) **[1 mark]**

H₂N–C–C–N–C–COOH (with O, CH₂/OH, H groups) **[1 mark]**

The amino acids can join together in either order — that's why there are two dipeptides.

c) Hot aqueous 6 M hydrochloric acid/HCl **[1 mark]**, reflux for 24 hours **[1 mark]**.

Page 99 — DNA

1 a) E.g. not enough bodies/risk of catching disease from bodies/public distaste **[1 mark]**

b) (i) Two sugar-phosphate polymers/backbones **[1 mark]** form a double helix **[1 mark]** held together by hydrogen bonding **[1 mark]** between complementary pairs of bases **[1 mark]**.

(ii) Each three base pairs (base triplet) codes for one amino acid **[1 mark]**. The sequence of base triplets codes for the sequence of amino acids in the protein **[1 mark]**.

c) Genes that code for human growth hormone **[1 mark]** are inserted into a plasmid/bacterial DNA **[1 mark]**, and placed into a bacterial host cell **[1 mark]**.

Page 101 — Addition Polymers

1 a) Addition polymerisation **[1 mark]**

–C–C– polymer repeat unit (with H, H, H, CH₃) **[1 mark]**

b) Methyl groups are pointing in the same direction **[1 mark]**, so chains can lie close together, increasing van der Waals/temporary dipole-induced dipole interactions **[1 mark]**. This makes the fibres very strong/have a high tensile strength **[1 mark]**.

c) Poly(propene) is chemically inert/resistant to chemical attack/can't be hydrolysed **[1 mark]**. It's a disadvantage when you need to dispose of the rope **[1 mark]**.

2 a)

–C–C–C–C–C–C– repeat unit (with H, H, H, H, H, H / H, Cl, H, Cl, H, Cl) **[1 mark]**

uPVC is just unplasticised PVC.

b) Polarised C-Cl bonds **[1 mark]** create strong permanent dipole-dipole interactions between the chains **[1 mark]**. Guttering/window frames or a similar use for a strong hard plastic **[1 mark]**.

c) Plasticiser molecules get in between the polymer chains, and reduce the effect of intermolecular forces **[1 mark]**, so chains can move around more **[1 mark]**. This makes PVC more flexible **[1 mark]**.

Page 103 — Condensation Polymers

1 a) (i)

H₂N–(CH₂)₆–NH₂   HO₂C–(CH₂)₄–CO₂H **[1 mark each]**

(ii) There are six carbon atoms in each monomer/reagent **[1 mark]**.

Don't forget to count the carbons in the carboxyl groups too.

(iii) Amide/peptide link **[1 mark]**

b) (i)

⎛ O   O   ⎞
⎝ C–(CH₂)₄–C–O–(CH₂)₆–O ⎠

or

⎛ O   O   ⎞
⎝ O–(CH₂)₆–O–C–(CH₂)₄–C ⎠ **[1 mark]**

(ii) For each link formed, one water molecule is eliminated **[1 mark]**.

c) Nylon-6,6 has hydrogen bonding **[1 mark]** between C=O and N-H groups **[1 mark]**. The polyester doesn't have hydrogen bonding **[1 mark]**.

2 Water in the body **[1 mark]** slowly hydrolyses **[1 mark]** the ester links **[1 mark]** in the polyester. Poly(propene) does not hydrolyse/is chemically unreactive/inert **[1 mark]**.

Page 105 — Medicines

1 a) The part of a drug that fits into a receptor **[1 mark]** and gives the drug its activity **[1 mark]**.

b) 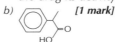 **[1 mark]**

This is a skeletal structure so the H's aren't shown.

c) –COOH/carboxyl group **[1 mark]**. This group can hydrogen bond with water **[1 mark]**.

d) Hydrogen bonding **[1 mark]** between the carboxyl group and polar groups in the receptor site **[1 mark]**.

e) (i) structure **[1 mark]**

(ii) Only one optical isomer/enantiomer will bind with the target receptor OR increased chance of side effects (because the other isomer bonds to another receptor elsewhere) **[1 mark]**.

Section Seven — Analysis and Synthesis

Page 107 — Identifying Functional Groups

1 a) Substance A is a neutral substance, so it can't be a carboxylic acid **[1 mark]**. It doesn't react with Fehling's solution, so it's not an aldehyde **[1 mark]**. But it does react with 2,4-DNPH, indicating it's a carbonyl compound **[1 mark]**, so it must be a ketone **[1 mark]**. The reaction with iodine tells you that it contains a methyl carbonyl group **[1 mark]**.

b) The red precipitate with Fehling's solution shows that it's an aldehyde **[1 mark]**, and the lack of crystals with iodine shows that it doesn't contain a methyl carbonyl group **[1 mark]**.

2 a) Add orange bromine solution **[1 mark]** to each liquid and shake. The alkene will decolourise the bromine **[1 mark]**, but the other liquid will not react **[1 mark]**.

You've got to give the results for both substances in this type of question.

b) The alcohol is a tertiary alcohol **[1 mark]**.

Answers

Page 109 — UV–Visible Spectra

1 a) This molecule has a conjugated system of double bonds [1 mark], forming delocalised areas of electrons [1 mark], and lone pairs of electrons on its oxygen atoms [1 mark], which absorb light in the UV–visible region [1 mark]. The molecule will absorb visible light rather than the higher energy UV light due to the long conjugated system [1 mark].

b) The wavelengths of light absorbed are those which are not purple [1 mark].

c) The chromophore [1 mark].

Page 111 — Atomic Emission Spectroscopy

1 The energy given is for 1 mole so it must be converted into the energy for 1 atom.

$519 \text{ kJ mol}^{-1} = \frac{519000}{6.02 \times 10^{23}} = 8.62 \times 10^{-19} \text{ J atom}^{-1}$ [1 mark]

Then you can use the formula E = hf to find the frequency.
E = hf [1 mark]
$8.62 \times 10^{-19} = 6.63 \times 10^{-34} \times f$

$f = \frac{8.62 \times 10^{-19}}{6.63 \times 10^{-34}} = 1.3 \times 10^{15} \text{ Hz}$ [1 mark]

6.02×10^{23} is an utterly humungous number — it's the number of particles in a mole (Avogadro's number).

2 $c = f\lambda$ [1 mark]

For the UV light: $3 \times 10^{8} = f \times 300 \times 10^{-9}$

$f = \frac{3 \times 10^{8}}{300 \times 10^{-9}} = 1 \times 10^{15} \text{ Hz}$ [1 mark]

$E = hf = 6.63 \times 10^{-34} \times 1 \times 10^{15}$
$= 6.63 \times 10^{-19} \text{ J}$ [1 mark]

For the infrared light: $3 \times 10^{8} = f \times 800 \times 10^{-9}$

$f = \frac{3 \times 10^{8}}{800 \times 10^{-9}} = 3.75 \times 10^{14} \text{ Hz}$ [1 mark]

$E = hf = 6.63 \times 10^{-34} \times 3.75 \times 10^{14} = 2.49 \times 10^{-19} \text{ J}$ [1 mark]
So UV light contains more energy (per photon) than IR light [1 mark].

Page 113 — Chromatography and Electrophoresis

1 a) R_f = spot distance ÷ solvent distance [1 mark]
= 7 ÷ 8 = 0.875 [1 mark].
There's no units as it's a ratio.

b) Substance A has moved further up the plate because it's less strongly adsorbed [1 mark] onto the stationary phase [1 mark] than substance B [1 mark].
Make sure you mention substance B — it's easy to forget.

Page 115 — Mass Spectrometry

1 a) 78 [1 mark].

b) A has a mass of 15, so it's a methyl group, CH_3 [1 mark].
B has a mass of 29, so it's most likely to be an ethyl group, C_2H_5 [1 mark].
C has a mass of 43, which is most likely to be a propyl group, C_3H_7 [1 mark].
Fragment B also has the same mass as a CHO group — but you know it's an alkyl halide, and there are no CHO groups in alkyl halides.

c) The M peak and the M+2 peak look to be in the ratio 3:1 [1 mark]. This suggests the halogen is chlorine [1 mark].

d) You know it's an alkyl halide and that the halogen is chlorine. So if we take 35 (the A_r of the main isotope of chlorine) away from 78 (the M_r of the alkyl halide containing the main chlorine isotope), the alkyl bit must have a mass of 43 [1 mark]. All alkyl bits follow the general formula C_nH_{2n+1}. n must be 3 to make this add up to 43. So the molecular formula is C_3H_7Cl [1 mark]. To give fragment C [1 mark], the chlorine atom must be on an end carbon [1 mark].

[1 mark]

Page 117 — Infrared Spectrometry

1 a) A's due to an O–H group in a carboxylic acid [1 mark].
B's due to a C=O as in an aldehyde, ketone, acid or ester [1 mark].
C's due to a C–O as in an alcohol, ester or acid [1 mark].
D's also due to a C–O as in an alcohol, ester or acid [1 mark].

b) The spectrum suggests it's a carboxylic acid — it's got a COOH group [1 mark]. This group has a mass of 45, so the rest of the molecule has a mass of 29 (74 – 45), which is likely to be C_2H_5 [1 mark]. So the molecule could be C_2H_5COOH — propanoic acid [1 mark].

Page 119 — NMR Spectroscopy — The Basics

1 a) E.g. deuterated water/CCl_4 [1 mark]. This solvent has no single protons [1 mark], so doesn't produce any peaks on the spectrum [1 mark].

b) 1.0 p.p.m., 3.5 p.p.m. and 5.0 p.p.m. [1 mark for all three].

c) The single hydrogen at δ = 5 p.p.m. is the H atom in the OH group of the alcohol [1 mark]. The three hydrogens at a shift of 1 p.p.m. are likely to be in a CH_3 group [1 mark]. The two hydrogens at a shift of 3.5 p.p.m. are likely to be from a CH_2OH group [1 mark]. The structure's likely to be CH_3CH_2OH [1 mark].
The CH_2OH group has three hydrogens, but only the two attached to the carbon cause this chemical shift. The one attached to the oxygen has a different chemical shift.

Page 121 — More NMR Spectroscopy

1 a) A CH_2 group adjacent to a halogen [1 mark].
You gotta read the question carefully — it tells you it's an alkyl halide. So the group at 3.6 p.p.m. can't have oxygen in it. It can't be halogen-CH_3 either, as this has 3 hydrogens in it.

b) A CH_3 group [1 mark].

c) CH_2 added to CH_3 gives a mass of 29, so the halogen must be chlorine with a mass of 35 [1 mark]. So a likely structure is CH_3CH_2Cl [1 mark].

d) The quartet at 3.6 p.p.m. is caused by 3 protons on the adjacent carbon [1 mark]. The n +1 rule tells you that 3 protons give 3 + 1 = 4 peaks [1 mark].
Similarly the triplet at 1.3 p.p.m. is due to 2 adjacent protons [1 mark] giving 2 + 1 = 3 peaks [1 mark].

Answers

Page 123 — Combined Spectral Analysis

1) a)

	Carbon	Hydrogen	Oxygen
Percentage	54.55	9.09	36.36
Divide by A_r	$54.55 \div 12 = 4.546$	$9.09 \div 1 = 9.09$	$36.36 \div 16 = 2.273$
Divide by smallest	2	4	1

[1 mark]
So the empirical formula's C_2H_4O [1 mark]. This has a mass of 44. The molecular mass is 88, so the molecular formula is $C_4H_8O_2$ [1 mark].

b) *The absorption at 1740 cm⁻¹ corresponds to a C=O bond in an aldehyde, ketone, ester or carboxylic acid [1 mark], while the absorption at 1240 cm⁻¹ indicates a C–O bond [1 mark]. These absorptions point to an ester [1 mark].*

c)
$$H_3C-C \overset{O}{\underset{O-C_2H_5}{\big<}} \qquad H_5C_2-C \overset{O}{\underset{O-CH_3}{\big<}}$$ *[1 mark for each]*

d) *The boiling point of a compound is unique and would identify which of these structures is correct [1 mark].*
OR the NMR spectrum would show up the correct structure as the protons in each are in different environments [1 mark].

Page 125 — Organic Synthesis

1 *Step 1: The methanol is refluxed [1 mark] with $K_2Cr_2O_7$ [1 mark] and sulphuric acid [1 mark] to form methanoic acid [1 mark].*
Step 2: The methanoic acid is reacted under reflux [1 mark] with ethanol [1 mark] using an acid catalyst [1 mark].

2 *Step 1: React propane with bromine [1 mark] in the presence of UV light [1 mark]. Bromine is toxic and corrosive [1 mark] so great care should be taken. Bromopropane is formed [1 mark].*
Step 2: Bromopropane is then refluxed [1 mark] with sodium hydroxide solution [1 mark], again a corrosive substance so take care [1 mark], to form propanol [1 mark].

Page 127 — Practical Techniques

1 a) *The purer sample will have the lower boiling point [1 mark], so the sample that boils at 64 °C is purer [1 mark].*

b) *To purify the sample you could dissolve it in propanone [1 mark] and allow it to partially crystallise [1 mark]. You'd then filter the crystals [1 mark], then wash them with propanone [1 mark] and dry them [1 mark].*

c) *The purity could be checked by measuring the boiling point [1 mark], OR by spectroscopic means [1 mark].*

Page 129 — The Chemical Industry

1 a) *Whether there's a distribution network available (road, rail, sea links, etc.).*
Where they can release waste effluent and what local regulations are.
How easy it is to get the raw materials to the site.
Whether a workforce is available.
The cost of the site. [1 mark each, maximum 2 marks]

b) *Fixed cost — e.g. cost of construction/equipment, possibly labour costs.*
Variable cost — e.g. cost of raw materials, waste effluent treatment, distribution of product.
[1 mark for a fixed cost, 1 mark for a variable cost]

c) *The plant is operating under-capacity / is not making its maximum quantity [1 mark]. The fixed costs / construction and labour costs remain the same [1 mark], but less of the product is made and sold [1 mark].*

d) *Non-stop process; can operate 24 hours per day; lower labour costs; automated process; larger quantities can be made; better control over variation in quality [1 mark each, maximum 2 marks].*

Questions about the chemical industry are about the real world — you might have to dust off your common sense. Don't forget, industries just want to make money.

Section Eight — Biochemistry

Page 131 — Carbohydrates

1 a) *$C_{12}H_{22}O_{11}$ [1 mark]*
Glucose has the molecular formula $C_6H_{12}O_6$, but when two of them join together, a water molecule's lost. So all you have to do is double $C_6H_{12}O_6$ and take away H_2O.

b) *α-1,4-glycosidic link [1 mark for α-1,4 and 1 mark for glycosidic]*

c) *Hydrolysis [1 mark] using enzymes or hot aqueous acid [1 mark].*
Water is lost when maltose forms from α–D–glucose. So water has to be put back by hydrolysis to reform α–D–glucose.

2 a) *Amylose [1 mark] and amylopectin [1 mark]. Amylose is unbranched [1 mark] and amylopectin is highly branched by α–1,6–glycosidic links [1 mark].*

b) (i) *To store energy [1 mark]*
(ii) *Structural material [1 mark]*
(iii) *To store energy [1 mark]*

Page 133 — Lipids and Membrane Structures

1 a) *E.g. Both have a non-polar hydrocarbon part/ester bonds [1 mark]. Phosphoglycerides have a phosphate group/polar, hydrophilic head, and triglycerides don't [1 mark].*

b)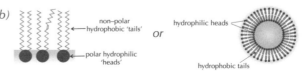

Polar heads are attracted to water [1 mark]. Van der Waals forces hold non-polar tails together/non-polar tails are protected from water by polar heads [1 mark].
[1 mark for diagram of bilayer. 1 mark for correctly labelled hydrophilic or polar head and 1 mark for hydrophobic or non-polar tail.]

Answers

2 a) (i) Enzyme/lipase/hot aqueous acid or alkali *[1 mark]*
(ii) Fatty acids *[1 mark]* and glycerol *[1 mark]*.
Glycerol (propane-1,2,3-triol) and fatty acids are always produced
when triglycerides are fully hydrolysed.
b) $M_r(C_{57}H_{110}O_6) = (12 \times 57) + (1 \times 110) + (16 \times 6) = 890$
[1 mark]
% O in $C_{57}H_{110}O_6 = (96/890) \times 100 = 10.8\%$ *[1 mark]*
c) Triglycerides have a greater proportion of carbon by
mass *[1 mark]*. Carbohydrates are partially oxidised
already *[1 mark]*. So carbohydrates release <u>less</u> energy per
gram *[1 mark]*.

Page 135 — Proteins
1 a) Hydrogen bonding *[1 mark]*

b)
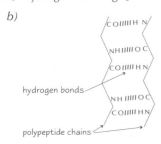

*[1 mark for general shape of layers and 1 mark for showing
hydrogen bonding.]*

2 a) The arrangement of two or more polypeptide chains *[1 mark]*
and non-protein components which are bound to the
polypeptide chains *[1 mark]*. Haemoglobin has four
polypeptide chains *[1 mark]* and four non-protein haem
groups *[1 mark]*.
b) O_2 from lungs combines reversibly with Fe^{2+} *[1 mark]*.
The lone pair of electrons on O_2 is used to form a weak,
dative bond to Fe^{2+} *[1 mark]*. O_2 is acting as a ligand/each
haemoglobin molecule can carry four O_2 molecules
[1 mark].

Page 137 — Enzymes
1 a) Competitive *[1 mark]*
b) Malonate competes *[1 mark]* for the active site, preventing
succinate getting to it *[1 mark]*.
If the inhibitor was a completely different shape from the substrate
molecule, it'd be a non-competitive inhibitor and would bond with the
substrate molecule away from the active site.
c) It would increase proportionally until it reached a maximum
rate *[1 mark]*.

2 a) Eventually, all the active sites on the enzyme would be
working at maximum rate *[1 mark]*.
b) It will decrease *[1 mark]*. Enzymes work best at an optimum
pH/changes in pH can cause groups such –COOH or $-NH_2$
in the active site to accept or lose protons *[1 mark]*. This
causes a change in the active site's shape/induces changes in
the interactions between substrate and active site/affects
hydrogen bonding *[1 mark]*.
pH changes are bad news for enzymes — it denatures them.
Changes in temperature don't do them much good either.

Page 139 — DNA and RNA Structure
1 a) Ribose *[1 mark]*
b) (i) A nucleotide is a monomer/consists of a single sugar-
phosphate base unit whilst a nucleic acid is a polymer/
consists of many nucleotides chemically bonded *[1 mark]*.

(ii)

*[1 mark for a phosphate and base joined to a pentose
sugar. 1 mark for joining the phosphate and base onto
the correct points, relative to one another.]*

2 a) DNA *[1 mark]* because thymine was found/there was no
uracil, U *[1 mark]*.
b) C pairs with G, so 26% is C *[1 mark]*. This gives a total of
52% of C and G, leaving 48% for complementary base pairs
A and T. 24% A and 24% T *[1 mark]*.
c) –TCGG–
[2 marks for all correct, deduct 1 mark for each mistake.]

Page 141 — DNA Replication and Protein Synthesis
1 a) DNA does not contain uracil/only RNA contains uracil
[1 mark].
b) Four amino acids (or for three amino acids, plus partially for
two others) *[1 mark]*. Each amino is coded for by a triplet of
bases *[1 mark]*.
c) –TTCCACGTAGCT– *[2 marks for all correct, 1 mark deducted
for each mistake.]*
Remember — thymine in DNA is replaced by uracil in RNA and that uracil
pairs with adenine.

2 a) Translation *[1 mark]*
b) A tRNA has an anticodon at one end and glycine attached at
the other *[1 mark]*. The anticodon on the tRNA involved is
–CCA– (it's complementary to the mRNA codon) *[1 mark]*.
The tRNA anticodon binds to the codon on the mRNA *[1
mark]*. The same thing happens with the next codon and a
peptide link is formed with the adjacent amino acid *[1 mark]*.
c) There are $3 \times 73 = 219$ bases coding for the 73 amino acids/
there's a triplet of bases for each amino acid *[1 mark]*. The
other 3 bases code for the termination/stop triplet *[1 mark]*.

Index

A

alpha helix 97, 134
absorbance 66, 67
absorption of energy
    66, 95, 108, 118
absorption spectrum 110
acids 28, 29, 32, 34, 41, 60
acid anhydrides 82
acid catalysts 78
acid dissociation constant 30
acid hydrolysis 75, 80
acidic buffers 36, 37
acidic solutions 42
acidified potassium dichromate
    106
acidity reactions 38
activating group 89
activation energy
    10, 12, 13, 15, 41, 68
active site 13, 136
activity 136
acylation 92
acyl chlorides 79, 82, 86, 88
addition polymers 100-101
addition reaction 87
adenine 98, 138, 140
adsorption 12, 112
agonist drugs 104
alcohol in blood or urine 112
alcohols 77, 106, 116
aldehydes
    74, 76, 78, 106, 107, 116
alignment of protons 118
aliphatic amines 91
alkalis 32, 40
alkaline buffers 36
alkaline hydrolysis 106
alkanes 77
alkenes 70, 100
alkoxide ions 88
alkylbenzene 86
alkyl groups 72, 106
alloys 59, 69
alternative reaction pathway 12
aluminium chloride 42
aluminium halides 86
aluminium oxide 41
amides 93
amines 90-93
amino acids
    70, 96-98, 113, 134,
    139, 141
ammonia 12, 25, 90, 92
ammonia solution 39, 57, 64
amphoteric 39, 41, 43, 96
amylopectin 131
amylose 131
anode 52
antagonist drugs 104
anti-cancer drugs 63
anticlockwise rule 50, 52
anticodons 139
antiseptic 88, 89
aqua complex 65
aromatic amines 90, 91
aromatic compounds 84, 125

Arrhenius equation 15
aspirin 83, 117
atactic polymer 101
atomic number 54
atomisation enthalpy 4
autocatalysis 68
azo dyes 94

B

β-pleated sheet 134
Balmer series 110
barrier methods 53
bases
    28, 29, 34, 41, 69, 138,
    139
base hydrolysis 80
base peak 114
base triplets 98, 139
basic metal oxides 69
batch production 128
batteries 52
beer 13
Benedict's solution 76
benzene 84, 87, 88, 108,
    125
benzenediazonium chloride
    125
benzoic acid 87
beta-pleated sheet 134
bidentate ligands 62-64
bilayers 133
bimolecular layers 133
biodegradable 103
biological washing powders 13
blast furnace 69
blood 37, 69
boiling point 122, 127
bond dissociation enthalpy 2
bond enthalpies 5
bonding 43
Born-Haber cycles 2, 4
Brady's reagent 76, 107
brass 59, 61
breaking bonds 5, 10
breathalyser 117
bromine 106, 115
bromination 87
bromine water 89, 106
Brønsted-Lowry acids/bases 28
bronze 59
Buchner flask and funnel 126
buffer solutions 36, 37
burette 32, 60, 61
butene 70
by-products 129

C

C=C double bonds 70
calibration graphs 66
calomel electrode 49
carbohydrates 130-132
carbocations 72
carbon 43, 69
carbon dioxide 25, 37
carbon monoxide 12, 69
carbon tetrachloride 43

carbon-13 115
carbonates 3
carbonyl compounds 75-77
carbonyl group 74, 107
carboxyhaemoglobin 69
carboxylate ions 88
carboxylates 79
carboxyl group 78
carboxylic acids
    74, 77-79, 107, 116,
    132
carotene 108
carpets 102
catalysts
    12-14, 18, 24, 25, 55,
    68, 128
catalytic converters 12
cathode 52, 62
cationic surfactants 90
cell membranes 133
cell potential 48-51
cellobiose 130
cellulose 131
charge density 3, 38
cheese 13
chemical analysis 11
chemical industry 128
chemical shift 118, 121
chiral molecules 70, 71, 96
chlorides 42
chlorine 40, 42, 46, 115
chloroalkanes 106
chlorobenzene 125
chloromethane 72
chromate(VI) ions 58
chromatography 112
chrome plating 53
chromium 54, 58, 69
chromophore 95, 108
cis isomers 70
cis-trans isomerism 63, 70
closed system 20
clothes 102
Co²⁺ 51
Co³⁺ 51
cobalt 57
codons 139, 141
coinage bronze 59
collision theory 10
colorimeter 11, 66, 67, 108
colour 56, 64, 66
colour change 11, 94
coloured ions 55
combustion analysis 123
competitive inhibitors 13, 137
complementary base pairing
    98, 140
complex ions 55, 62-65
concentration
    10, 14, 17, 24, 26, 35,
    60, 61, 66, 67
concentration-time graphs
    10, 16, 17
condensation polymers
    96, 102-103, 131

condensation reactions
    130, 132, 134, 138
conductivity 43
conductors 43
confectionery industry 137
conjugate acid/base 28
conjugate pairs 28
conjugated electron systems
    108, 109
construction costs 128
contact process 12, 68
continuous production 128
convergence limit 111
coordinate bonds
    38, 42, 62, 64, 90
coordination numbers
    62, 64, 66
co-polymerisation 101
copper 54, 58, 59, 61
copper iodide 59
copper(I) 52, 61, 59
copper(II) 59, 66
co-products 129
covalent bonds 4, 38, 62
covalent substances
    5, 7, 40-43
crystallisation 65
cyclohexane 87
cyclohexene 87
cytosine 98, 138

D

d orbitals 68
d-block 54
dative covalent bonds 62, 90
delocalisation 88
delocalised structure of
    benzene 84
delocalised electrons 88, 108
denaturation 13, 135, 136
deoxyhaemoglobin 134
deoxyribose sugar 138
designing medicines 124
detector 66
deuterated solvents 119
deuterium 119
deuterium oxide 120
diamines 102
diamond 43
diazonium salts 94
dicarboxylic acids 102
dichromate(VI) ions 58, 60
dichloromethane 72
diglycerides 132
dimers 42
2,4-dinitrophenylhydrazine (2,4-
    DNPH) 76, 107
diols 102
dipeptides 96
dipolar ions 96
dipole-dipole interactions
diprotic acids 33
disaccharides 130
discovery of hydrogen 111
disinfectant 88
disorder 8

Index

displacement reactions 59
disposal of polymers 101
disproportion 46, 52, 59
dissolving 6
disulphide bridge 97, 134
DNA
    98, 99, 113, 138-140
DNA polymerase 140
double bonds 70, 88, 108
double helix 98, 138
doublet 120
drug design 104, 105
drug testing 105
drunk drivers 117
ductility 43
dynamic equilibria 20

E
eclipse 111
EDTA 62, 64, 66
electrical conductivity 11
electrochemical cells 48, 49
electrochemical series 50, 51
electrodes 52
electrode potential charts 50
electrolysis 62
electrons 46, 108, 110
electron
    affinity 4
    density 38, 88
    shielding 118
electronegative 46, 73, 88
electronic configurations 54,
    66
electronic transitions 108
electrophile 72
electrophilic addition 72
electrophilic substitution 85,
    86
electrophoresis 113
electrostatic attraction 2, 42
electrostatic forces 7
elemental percentage
    composition 123
emf 48
emission spectra 110, 111
empirical formula 123
en 63
enantiomers 70, 71
end point 32, 60, 61
endothermic reactions
    2, 5, 6, 9, 24, 64
energy 2, 3, 5, 8, 9, 19,
    108, 110
energy gap 66
energy levels 55, 66,
    108-110, 118
enthalpy 2-7, 9
enthalpy change 2, 4-6
enthalpy change
    of atomisation 2, 4
    of formation 2
    of hydration 2, 6, 7, 40
    of neutralisation 29
    of solution 2
    of solvation 6

enthalpy cycles 6
enthalpy of ionisation 2
enthalpy profiles 12, 68
entropy 8, 9
entropy change 6, 51, 64
enzymes 12, 13, 136
equilibra 20, 21, 24,
    26-30, 34, 36, 37, 39,
    51, 57, 65
equilibrium concentration
    20, 21
equilibrium constants
    20-24, 26, 27, 29,
    34
equivalence point 32-35
essential oils 81
ester 78, 80, 88, 116, 132
esterification 132
esters in paints 112
ethanal 75
ethane-1,2-diamine 62-64
ethanedioic acid 33
ethanol 117
ethanoic anhydride 83
ethanoyl chloride 88
ethyl ethanoate 116
eutrophication 44
excitation of electrons 108
exothermic reactions
    2, 3, 5, 6, 24, 29
exponential relationship 15,
    16
external circuit 48
extrapolation 67

F
fabric conditioner 90
Fajan's rules 4
fat 81, 132
fatty acids 81, 132
Fe^{2+} 51, 54, 68, 69
Fe^{3+} 51, 54, 68
feasible reactions 9, 51
feedstock 128, 129
Fehling's solution 76, 107
fermenter 99
filtration 126
fingerprint region 116
first electron affinity 2
first ionisation enthalpy 2
first order reaction 14, 16, 19
fishy smell 90
fixed costs 129
flavour 80
food chain 27
formulas of complexes 67
fractional distillation
    92, 126, 127
fractionating column 126
fragmentation patterns 114
free energy change 9
free radical substitution 18, 72
free-floating nucleotides 140
frequency 66, 110, 111

Friedel-Crafts
    acylation 86
    alkylation 86
fructose 130, 137
fruit juice 13

G
galvanising 53
gas equilibria 22, 23
gas syringe 11
gas/liquid chromatography 112
gaseous non-metal oxides 69
gene therapy 99
genetic code 98
genetic engineering 99
genetic fingerprinting 113
geometric isomerism 63
germanium 43
giant ionic lattices 2, 40-42
giant structures 40-42
glucose 130, 137
glucose isomerase 137
glue 80
glycerol 132
glycogen 131
gradient 15
gradient of concentration/time
    graph 10, 16, 17
graphite 43
Grignard reagents 77
ground state 110
Group 2 7
    elements 4
    hydroxides 7
    sulphates 7
    carbonates 3
Group 4 43
guanine 98, 138

H
Haber process
    12, 25, 55, 68, 128
haem 69
haemoglobin 62, 69, 134
hair products 90
half-cells 48, 49
half-equations
    47, 50, 52, 56, 57
half-equivalence point 34
half-life 16, 17
halides 106
haloalkanes 72, 77, 91
halogen carriers 86
heat exchangers 129
heavy metal ions 13, 135,
    137
hens' teeth 52
Hess's Law 4, 6
heterogeneous catalysts
    12, 15, 68
heterogeneous equilibria 24
heterolytic fission 72
1,2,3,4,5,6-hexachlorocyclo-
    hexane 87
hexadentate 62, 64
Hoffman degradation 93

homogeneous catalysts 12, 68
homogeneous equilibria 24
homolytic fission 72
homolytic free radical addition
    100
horoscope 114
humps 68
hydrated iron(III) oxide 52
hydrides 46
hydrochloric acid 64
hydrogen bonds
    7, 74, 78, 97, 98,
    134-136, 138
hydrogen cyanide 75
hydrogen nuclei 118
hydrogen peroxide 57, 58
hydrogencarbonate 107
hydrolysis 38, 39, 42, 78, 93,
    130, 132, 134
hydroxide precipitates 64
hydroxides 7, 41
hydroxonium ions 28, 29
hydroxynitriles 75
hydroxyl groups 75, 106

I
immobilised enzymes 137
impurities 69
indicators 32, 34, 35
inert atmosphere 58
inert pair effect 43
infrared radiation 116
infrared spectroscopy 116, 117
inhibition 13, 137
initial rate of reaction 10, 17
initial rates method 17
initiation reaction 72
integration trace 120
intermediate ion 68
iodide ions 61, 68
iodine 68, 76, 107
ionic
    bonding 2, 4, 135
    charge 3
    compounds 2, 4, 7
    half-equations 47
    interactions 134
    lattice 2, 3, 6, 41
    product of water 29, 30
    radius 3, 40, 55
ionisation energy/enthalpy
    2, 4, 111
ionisation enthalpy 4
ionisation of acids 30
iron
    12, 52, 53, 55, 68, 69, 86
iron halides 86
iron ore 53
iron(II) ions 56
iron(II) hydroxide 52
iron(III) chloride solution 88
Iron(III) hydroxide 52
iron(III) oxide 52
isoelectric point 96, 113

Index

isomerism
    cis-trans 63, 70
    geometric 63, 70
    optical 63, 70, 71
    stereo 63, 70
    structural 70
isoprenaline 105
isotactic polymers 101
isotopes 115

K

K_a 31, 34, 37
K_c 20, 21, 24, 51
Kekulé structure 84
ketones 74, 76, 106, 107, 116
kevlar 102
K_{In} 34
kinetic energy 10
K_p 22-24
K_{stab} 65

L

labile protons 120
lactic acid 75
lattice enthalpy 2-4, 6, 7
Le Chatelier's principle
    24, 28, 36, 58
lead 43
Lewis acid/base 38
Lewis acid-base reactions 38
Liebig condenser 126
ligands 57, 58, 62, 63, 66,
    67, 69
ligand exchange
    64, 65, 67, 94
light 66
lime 69
limestone ($CaCO_3$) 25
linear shape 62
lipases 132
litmus 106
'lock and key' model 136
log 29, 30, 34
logarithmic form of Arrhenius
    equation 15
lone pairs 62, 63, 69
Lyman series 110, 111

M

M peak 114, 115
M+1 peak 115
M+2 peak 115
M+4 peak 115
magnesium
    chloride 42
    hydroxide 40
    oxide 40
    sulphide 69
malleability 43
maltose 130
manganate(VII) ions 47, 60
margarine 68, 81
mass spectrometer 114
mass spectra 114
mass/charge ratio 114

maximum absorbance 67
Maxwell-Boltzmann distribution
    10, 12
mean bond enthalpies 5
mechanisms 18, 19
medicines 104-105
melting point 41, 42, 122
messenger RNA 139, 140
metal
    carbonates 39
    chlorides 42
    hydroxide 39
    ions 38
    oxides 41, 69
metal-aqua ions 38, 39, 64
metallic properties 43
metalloids 43
metals 43
methyl carbonyl group 76,
    107
methyl orange 32, 34, 95
methyl red 34
mild steel 128
minimising costs 25
Mn^{2+} 68
mobile phase 112
molarities 35
mole fractions 22
molecular ions 114
molecular structure 114
moles 35, 60, 65, 67
Monastral blue 66
monodentate ligands 62, 64
monoglycerides 132
monomers 100
mononucleotides 138
monoprotic acids 30
monosaccharides 130, 131
mordant dyes 94
M_r 114
mRNA 139, 140
multidentate ligands 62, 66, 69
multiplets 120

N

n + 1 rule 120
neutral solutions 29, 42
neutralisation 29, 32, 34, 35
nickel 12, 68
ninhydrin solution 96
nitriles 78, 91, 93
nitrobenzene 85, 125
nitronium ions 85
NMR spectroscopy 118,
    120, 121
non-competitive inhibitors
    13, 137
non-metal oxides 41, 69
non-metals 43
non-polar solvents 7
noradrenaline 105
nuclear charge 40
nuclear magnetic resonance
    118
nucleic acids 138
nucleophiles 73, 75, 77, 92

nucleophilic addition 74, 75
nucleophilic
    addition-elimination 83
nucleophilic substitution 73, 92
nucleotides 98, 138
nylon 102

O

oceans 25
octahedral shape 62, 64
oil paint 81
oils 81
optical isomerism 63, 70, 71
optimisation 128
orbitals 66, 108, 110
order of reaction 14, 16-19
organic synthesis 124, 125
overall charge 62
oxidation
    46 -50, 52, 53, 57, 68,
    74, 78, 106
oxidation number/state
    40, 43, 46, 51, 56, 58,
    59, 62, 68, 94
oxide layer 69
oxides 40, 41, 43, 69
oxidising agents
    43, 46, 47, 57, 60,
    61, 68, 106
oxoanions 58
oxygen 46, 69
oxyhaemoglobin 69, 134

P

p-orbitals 43, 88
paints 66, 81
paper chromatography 96, 112
partial pressure 22, 23
particles 10
partition 27, 112
partition coefficient 27
Paschen series 110
pectinase 13
pentose sugar 98
peptide bond/link
    96, 97, 134, 141
Period 3 40-42
Period 4 54
periodic table 54
peroxides 46
peroxodisulphate 68
pest control 27
pesticides 27
PET 102
pH 13, 30, 31, 34-37, 58,
    135
pharmacophores 104, 105
pH calculations 30, 31
pH curves 32-35
pH meter 35
phenol 88
phenol red 34
phenolphthalein 32, 34
phenoxide ions 88
phenylamine 125
phenylketone 86

phosphoglycerides 133
phospholipids 133
phosphorus(V) chloride 82, 106
phosphorus(V) oxide 93
phosphorus-32 113
photography 62
physical state 8
pigments 66, 111
pipette 32, 61
pK_a 31, 34
pK_w 29
Planck's constant 66, 111
planar molecule 84
plane-polarised light 71, 96
plasmid 99
plasticiser 80
plasticisation 101
platin 63
platinum 12
platinum electrode 48
poisoned catalyst 12
polar solvents 7, 80
polarisation 3, 4, 40-42, 77
polyamides 93, 102
polychloroethene/PVC 100
polydentate 62
polyesters 102
polyethene 100
polymerisation 100
polymorphism 94
polynucleotide 78, 139
polypeptides 97, 103, 141
poly(phenylethene)/polysterene
    100
polysaccharides 131
polytetrafluoroethene (PTFE)
    100
position of equilibrium 24
potassium dichromate 60, 61,
    74, 106
potassium hydrogencarbonate
    107
potassium iodide 59, 61, 107
potassium manganate(VII) 60
precipitates 25, 26, 39, 94
predicting a structure 122
preparing a complex compound
    65
pressure 24, 25
primary alcohols 74, 78, 106
primary aliphatic amines 90
primary haloalkanes 73
primary structure of proteins
    134
printing ink 80
production processes 128
propogation reaction 72
proteases 13
proteins
    96, 97, 134, 136, 139,
    141
proton acceptors 28
protonation 95
proton donors 28
proton-free solvents 119
proton transfer 28

Index

PTFE 100
purification of solids 126
purines 98
pyranose rings 130
pyrimidines 98

Q

quantum number 110
quartet 120
quaternary ammonium salts 90, 92
quaternary structure 134

R

racemic mixture 71
radicals 12
Raney nickel catalyst 87
rate constant 14, 15
rate equation 14-19
rate of reaction 10-15, 18, 19, 25, 108
rate-concentration graphs 16
rate-determining step 18
raw materials 128
reaction mechanisms 18, 19
reactive dyes 94
reactivity 40
receptor sites 104
recrystallisation 126
recycling 53
redox reactions 46-48, 58, 60, 68, 69
redox titrations 60
reducing agents 43, 46, 56, 60, 74, 79
reducing costs 129
reduction 46 - 48, 50, 52, 58, 61, 74, 106
reflectance spectroscopy 109
reflected light 66
refluxing 74, 77, 80
relative isotopic abundance 114
relative peak area 121
rennin 13
resonance 118
restoration of paintings 109, 111
retention times 112
reversible reactions 20, 24
rhodium 12
ribose 130, 138, 140
ribosomal RNA 139
ribosomes 139, 141
RNA 138, 140
RNA polymerase 140
rRNA 139, 141
rust 52

S

s-orbital 43
sacrificial methods 53
safety precautions 124
salbutamol 105

salicylic acid 117
salts 36, 41
saponification 132
saturated fatty acid 81
saturated solution 26
scandium 54
second electron affinity 2
second ionisation enthalpy 2
second order reaction 14, 17, 19
secondary alcohols 106
secondary haloalkanes 73
secondary standard electrode 49
secondary structure 134
self-replication 138, 140
semi-conservative replication 140
semiconductors 43
separating components of a mixture 127
separating funnel 27
separating liquids 126
separating solids from liquids 126
shampoo 37
side-reactions 129
silicon 43
silicon tetrachloride 43
silver bromide 62
silver complexes 62
silver nitrate solution 106
silver plating 62
simple molecular structure 40, 42
slag 69
soap 81, 132
sodium
    carbonate 33, 39
    chloride 42
    hydrogencarbonate 107
    hydroxide 40, 64, 106
    phenoxide 89
    thiosulphate solution 61
solubility product 26
solubility trends 7
solutes 27
solvents 27
solvent extraction 27
smell 80
specific base pairing 138
specificity 136
spectral analysis 122, 123
spectroscopy 108, 109, 122
spin 118
spin-spin coupling 120
splitting patterns 121
spontaneous endothermic reactions 9
square planar shape 62, 63
stability constants 65
standard concentration 51
standard conditions 2, 49
standard electrode potentials 48-50, 52, 56
standard hydrogen electrode

49
standard temperature 51
starch 61, 131
state 12, 68
stationary phase 112
steel 53, 69
stereoisomerism 63
strong acids 28, 29, 32
strong bases 28, 29
structure of proteins 97
subshells 54, 59, 66
substrates 13, 136
sugar-phosphate backbone 98
sulphates 7
sulphur 40
sulphur dichloride oxide 82
sulphur dioxide 56
sulphur trioxide 85
sulphuric acid 68
syndiotactic polymers 101
synthesis routes 124

T

tangent 10, 16, 17
temperature 10, 13-15, 20, 24 - 26, 135
termination reactions 72
tertiary alcohols 106
tertiary haloalkanes 73
tertiary structure 134
terylene/PET 102
testing of drugs 105
tetrachlorides 43
tetrachloromethane 72
tetrahedral shape 62, 64
tetramethylsilane (TMS) 118
thermal decomposition 3
thin-layer chromatography 112
thiosulphate ions 61, 62
thymine 98, 138-140
tin 43, 53, 59
titanium oxide 66
titrations 11, 32-35, 60, 61
TMS 118
Tollens' reagent 62, 76, 107
total oxidation state 62
trans isomers 70
transcription 140
transfer RNA 139
transition elements 54-69
transition metal compounds 38
transition metals 54-69
transition state 19
translation 141
transmission spectroscopy 109
trichloromethane 72
triglycerides 132
triglyceryl esters 132
triple bond 108
triplet 120
tRNA 139, 141
two-way chromatography 112

U

ultraviolet light 108
unidentate ligands 62
unsaturated fatty acid 81
unsaturation 106
uracil 138-140
UV light 108

V

vaccines, production of 99
van der Waals forces 41, 42, 74, 78, 97, 103, 134-136
vanadium 54, 56
vanadium(V) oxide/pentoxide 12, 68
variable costs 129
variable oxidation states 55
visible light 66, 108
visible spectrum 66
volatile organic substances 126
voltage 48, 49
voltmeter 48

W

washing powders 37
waste effluent 129
waste heat 129
waste products 129
wavelength 110, 116
wavenumber 116
weak acid 28-30, 32, 34, 36, 37
weak alkali 32
weak base 28, 29
white paint 66
willow bark 117

Y

yeast 13
yellow/orange azo compound 125
yield 25, 128

Z

zero order reactions 10, 14, 17
zinc 54, 56, 58, 59
zwitterions 96, 113

The Periodic Table

1.0
H
Hydrogen
1

Relative
Atomic Mass →

Atomic number →

Periods	Group I	Group II
1		
2	6.9 **Li** Lithium 3	9.0 **Be** Beryllium 4
3	23.0 **Na** Sodium 11	24.3 **Mg** Magnesium 12
4	39.1 **K** Potassium 19	40.1 **Ca** Calcium 20
5	85.5 **Rb** Rubidium 37	87.6 **Sr** Strontium 38
6	132.9 **Cs** Caesium 55	137.3 **Ba** Barium 56
7	223 **Fr** Francium 87	226.0 **Ra** Radium 88

45.0 **Sc** Scandium 21	47.9 **Ti** Titanium 22	50.9 **V** Vanadium 23	52.0 **Cr** Chromium 24	54.9 **Mn** Manganese 25	55.8 **Fe** Iron 26	58.9 **Co** Cobalt 27	58.7 **Ni** Nickel 28	63.5 **Cu** Copper 29	65.4 **Zn** Zinc 30
88.9 **Y** Yttrium 39	91.2 **Zr** Zirconium 40	92.9 **Nb** Niobium 41	95.9 **Mo** Molybdenum 42	98 **Tc** Technetium 43	101.1 **Ru** Ruthenium 44	102.9 **Rh** Rhodium 45	106.4 **Pd** Palladium 46	107.9 **Ag** Silver 47	112.4 **Cd** Cadmium 48
138.9 **La** Lanthanum 57	178.5 **Hf** Hafnium 72	181.0 **Ta** Tantalum 73	183.9 **W** Tungsten 74	186.2 **Re** Rhenium 75	190.2 **Os** Osmium 76	192.2 **Ir** Iridium 77	195.1 **Pt** Platinum 78	197.0 **Au** Gold 79	200.6 **Hg** Mercury 80
227.0 **Ac** Actinium 89									

Group III	Group IV	Group V	Group VI	Group VII	Group 0
					4.0 **He** Helium 2
10.8 **B** Boron 5	12.0 **C** Carbon 6	14.0 **N** Nitrogen 7	16.0 **O** Oxygen 8	19.0 **F** Fluorine 9	20.2 **Ne** Neon 10
27.0 **Al** Aluminium 13	28.1 **Si** Silicon 14	31.0 **P** Phosphorus 15	32.1 **S** Sulphur 16	35.5 **Cl** Chlorine 17	39.9 **Ar** Argon 18
69.7 **Ga** Gallium 31	72.6 **Ge** Germanium 32	74.9 **As** Arsenic 33	79.0 **Se** Selenium 34	79.9 **Br** Bromine 35	83.8 **Kr** Krypton 36
114.8 **In** Indium 49	118.7 **Sn** Tin 50	121.8 **Sb** Antimony 51	127.6 **Te** Tellurium 52	126.9 **I** Iodine 53	131.3 **Xe** Xenon 54
204.4 **Tl** Thallium 81	207.2 **Pb** Lead 82	209.0 **Bi** Bismuth 83	209 **Po** Polonium 84	210 **At** Astatine 85	222 **Rn** Radon 86

The Lanthanides

The Actinides

140.1 **Ce** Cerium 58	140.9 **Pr** Praseodymium 59	144.2 **Nd** Neodymium 60	145 **Pm** Promethium 61	150.4 **Sm** Samarium 62	152.0 **Eu** Europium 63	157.3 **Gd** Gadolinium 64	158.9 **Tb** Terbium 65	162.5 **Dy** Dysprosium 66	164.9 **Ho** Holmium 67	167.3 **Er** Erbium 68	168.9 **Tm** Thulium 69	173.0 **Yb** Ytterbium 70	175.0 **Lu** Lutetium 71
232.0 **Th** Thorium 90	231.0 **Pa** Protactinium 91	238.0 **U** Uranium 92	237.0 **Np** Neptunium 93	244 **Pu** Plutonium 94	243 **Am** Americium 95	247 **Cm** Curium 96	247 **Bk** Berkelium 97	251 **Cf** Californium 98	254 **Es** Einsteinium 99	257 **Fm** Fermium 100	256 **Md** Mendelevium 101	254 **No** Nobelium 102	260 **Lr** Lawrencium 103